Applications of Soil Physics

Applications of Soil Physics

DANIEL HILLEL

DEPARTMENT OF PLANT AND SOIL SCIENCES
UNIVERSITY OF MASSACHUSETTS
AMHERST, MASSACHUSETTS

 1980

ACADEMIC PRESS

A Subsidiary of Harcourt Brace Jovanovich, Publishers

New York   London   Toronto   Sydney   San Francisco

ACADEMIC PRESS, INC.
111 Fifth Avenue, New York, New York 10003

United Kingdom Edition published by
ACADEMIC PRESS, INC. (LONDON) LTD.
24/28 Oval Road, London NW1 7DX

Library of Congress Cataloging in Publication Data

Hillel, Daniel.
    Applications of soil physics.

    Bibliography: p.
    Includes index.
    1. Soil physics. 2. Soil moisture. 3. Soil
management. I. Title.
S592.3.H53   631.4'3   80−535
ISBN 0−12−348580−0

PRINTED IN THE UNITED STATES OF AMERICA

80 81 82 83   9 8 7 6 5 4 3 2 1

Dedicated to my beloved children
Adi, Ron, Sari, Ori, and Shira
who have filled my cup with joy
and made all my trials worthwhile

Contents

14. Solute Transport during Infiltration into Homogeneous Soil

by D. E. Elrick

Bibliography

Preface

This volume is a close sequel to, and should best be considered in conjunction with, a companion text entitled "Fundamentals of Soil Physics." The two volumes share more than author, publisher, and date of publication. They both derive from and supersede an earlier text ("Soil and Water: Physical Principles and Processes") published about a decade ago, and thus have a common philosophy, terminology, and format. Both are directed at the same constituency of upper-level undergraduate and graduate students of the environmental, engineering, and agronomic sciences. However, although this second treatise is based implicitly upon the fundamental principles enunciated in the first one, it differs sufficiently in aim and scope to justify separate publication. Whereas its forerunner primarily describes general principles, the thrust of the present work is to extend and direct those principles toward the understanding of phenomena that are likely to be encountered in the field, as a further step toward the definition and eventual solution of problems having practical importance.

The first and larger part of this book provides a systematic description of the field-water cycle and its management. It includes chapters on infiltration and runoff, redistribution and drainage, evaporation and transpiration, as well as irrigation and tillage. The second part of the book presents advanced expositions of transpiration from plant canopies, freezing phenomena, scaling and similitude analysis, spatial variability of soil physical properties, and movement of solutes during infiltration. As principal author, I consider myself fortunate to have been able to enlist the participation of several of my most highly regarded colleagues, whose contributions constitute the last five chapters. These authors are, in alphabetical order, David E. Elrick, Edward E. Miller, Robert D. Miller, John L. Monteith, Donald R. Nielsen, and Arthur W. Warrick. Although no single book can ever be considered complete unto itself, it is my hope that the array of topics included herein provides a fairly comprehensive introductory survey of the applications of soil physics in the light of contemporary knowledge. In writing this, I am fully aware of the ephemeral nature of what I called "contemporary knowledge." As research progresses, even the most advanced exposition soon recedes into fading obsolescence. This may be a sad fate for authors to contemplate, but a happy one for science. So be it, then.

There is a special fascination in the topic of soil and water which, in any case, transcends the state of the science at any one moment. Perhaps it even antecedes every child's early interest in mud pies. Those of us who read the Bible cannot but note the powerfully symbolic account of creation in the first chapter of Genesis, which describes how the waters were divided and separated from dry land, and how man himself was created out of, and prefated to return to, "affar," which is, literally, the material of the soil. The primeval association of man with soil is manifested most strongly in the name *Adam*, derived directly from *adama*, a Hebrew word with the composite connotation of earth, land, and soil. Other ancient traditions evoke equally strong associations.

Our civilization seems to have drifted away from that intimacy with the soil which was the mark of our forebears in every land and culture. For that, we may be paying a heavier price than we realize. Seeing this, a philosopher and poet named Nietzsche felt driven to proclaim (in "Thus Spake Zarathustra"): "Man and man's earth are unexhausted and undiscovered. Wake and listen! Verily, the earth shall yet be a source of recovery. Remain faithful to the earth, with the power of your virtue. Let your gift-giving love and your knowledge serve the meaning of the earth." Perhaps our most precious possession and resource, both physical and spiritual, is contained in that most common substance that we sometimes call "dirt," but that is in fact the mother lode of life and the purifying medium wherein waste is recycled and productivity regenerated.

Would Montaigne forgive me? He was the man who wrote: "No one is exempt from saying silly things; the mischief is to say them deliberately."

Peruse me, O Reader, if you find delight in my work. . . . And come, men, to see the wonders which may be discovered in nature by such studies.

Leonardo da Vinci
Madrid Codex I

1 Soil Physics Explores the Hidden Turmoil in the Field

A poet gazing through his window might view the field lying outside as a place of pastoral serenity and restfulness. Not so the soil physicist. What he sees in the field is not rest but unceasing turmoil, a dynamic system in which matter and energy are in a constant state of flux. The soil physicist sees the radiant energy reaching the field and observes how this energy is transformed and what processes it powers. To be useful, the knowledge required by the soil physicist of this dynamic system must be quantitative. He strives to measure and to relate the variable rates of those simultaneous processes and to predict how these rates might change under possible control measures. Of particular importance are the processes of water and solute movement in the soil, and their combined effect on the growth of plants.

By some strange turn of historical and geographical fate, a sizable fraction of the world's population happens to be living in the globe's arid and semi-arid zones, where, by a cruel quirk of nature, the requirements of living things for water are greatest even while the supplies of water by natural precipitation are least. This discrepancy strongly affects agricultural plants, which, by their physiology and the nature of their interaction with the field environment, must transpire constantly and in fact must draw from the soil and transmit to the unquenchably thirsty atmosphere hundreds of times more water than they need strictly for their own growth. Thus in the arid zone the scales are weighted heavily against agriculture, and the imbalance must be rectified by intensive irrigation and water conservation. Nor is the problem confined to the arid zone alone. Such are the vagaries of climate that even so-called humid regions suffer periodic dry spells, or droughts.

It is the ability of the soil to serve as a reservoir for water and the nutrients dissolved in it which must bridge the gap between plant requirements, which are practically incessant, and the supply of water, which is intermittent and may be infrequent. But the soil is a leaky reservoir which loses water downward by seepage and upward by evaporation. To manage the system so as to maximize water use efficiency, we must monitor the balance of incoming versus outgoing water and the consequent change of moisture as well as nutrient storage in the root zone. This requires not merely a qualitative understanding of how the system operates, but also a quantitative knowledge of its mechanisms and the rates of its governing processes.

Being a vital link in the larger chain of interconnected media and processes comprising the biosphere, the soil interacts both with the atmosphere and with underlying strata. The soil also interacts in numerous ways with surface and underground bodies of water. Especially important is the interrelation between the soil and the microclimate. Radiant energy reaching the field is partly reflected and partly absorbed, depending on surface conditions. The energy absorbed is transformed into soil heat, "sensible" heat of the air, and latent heat of evapotranspiration. Only a minute fraction goes to photosynthesis, which, however, is the vital process of all agriculture, indeed of the entire biological cycle.

Modern society generates waste, and the problem of how to dispose of various waste materials has become increasingly acute in recent years. There is growing interest in the possibility of applying such materials to the land, in an attempt to utilize the soil's ability to filter, retain, buffer, immobilize, decompose, or otherwise mitigate the hazards of polluting agents. However attractive the notion of the soil as "living filter," the sad fact is that our knowledge of the processes involved is still woefully inadequate and serious misconceptions abound. Soils have been credited with an amazing, even mystical, capacity to purify contamination, albeit on the strength of very little conclusive long-term evidence. As often as not, the soil is only a way station in a continuous cycle, and its limited capacity to dispose of harmful pollutants can easily be overtaxed or abused. We mention this problem even though it does not lie within the recognized domain of soil physics but rather in the interdisciplinary realm which includes such related fields as soil chemistry and microbiology. Physical transport phenomena, however, are almost invariably involved and seldom very well defined in practice.

Part I:

THE FIELD WATER CYCLE
AND ITS MANAGEMENT

The important role of the soil in the hydrologic cycle can hardly be overemphasized. Particularly crucial to this role is the soil surface zone, where the interaction of atmospheric water takes place with the lithosphere. It is here that the complex partitioning between rainfall (or irrigation), infiltration, runoff, evapotranspiration, and deep seepage is initiated and sustained. This zone is also a primary site for the management and control by man of that all-important resource, water.

The movement of water in the field can be characterized as a continuous, cyclic, repetitive sequence of processes, without beginning or end. However, we can conceive of the cycle as if it begins with the entry of water into the soil by the process of infiltration, continues with the temporary storage of water in the soil, and ends with its removal from the soil by drainage, evaporation, or plant uptake. Several fairly distinct stages of the cycle can be recognized, and, although these stages are interdependent and may at times be simultaneous, we shall attempt, for the sake of clarity, to describe them separately in the following several chapters.

All the rivers run into the sea,
yet the sea is not full;
Unto the place whence the rivers come,
thither they return again.

Ecclesiastes 1:7

2 *Infiltration and Surface Runoff*

A. Introduction

When water is supplied to the soil surface, whether by precipitation or irrigation, some of the arriving water penetrates the surface and is absorbed into the soil, while some may fail to penetrate but instead accrue at the surface or flow over it. The water which does penetrate is itself later partitioned between that amount which returns to the atmosphere by evapotranspiration and that which seeps downward, with some of the latter reemerging as streamflow while the remainder recharges the groundwater reservoir.

Infiltration is the term applied to the process of water entry into the soil, generally by downward flow through all or part of the soil surface.[1] The rate of this process, relative to the rate of water supply, determines how much water will enter the root zone, and how much, if any, will run off. Hence the rate of infiltration affects not only the water economy of plant communities, but also the amount of surface runoff and its attendant danger of soil erosion. Where the rate of infiltration is restricted, plants may be denied sufficient moisture while the amount of erosion increases. Knowledge of the infiltration process as it is affected by the soil's properties and transient conditions, and by the mode of water supply, is therefore a prerequisite for efficient soil and water management.

---

[1] Water may enter the soil through the entire surface uniformly, as under ponding or rain, or it may enter the soil through furrows or crevices. It may also move up into the soil from a source below (e.g., a high water table).

Comprehensive reviews of the principles governing the infiltration process have been published by Philip (1969a) and by Swartzendruber and Hillel (1973).

B. "Infiltration Capacity" or Infiltrability

If we sprinkle water over the soil surface at a steadily increasing rate, sooner or later the rising supply rate will exceed the soil's limited rate of absorption, and the excess will accrue over the soil surface or run off it (Fig. 2.1). The *infiltration rate* is defined as the volume flux of water flowing into the profile per unit of soil surface area. This flux, with units of velocity, has also been referred to as "infiltration velocity." For the special condition wherein the rainfall rate exceeds the ability of the soil to absorb water, infiltration proceeds at a maximal rate, which Horton (1940) called the soil's "infiltration capacity." This term was not an apt choice, as pointed out by Richards (1952), since it implies an *extensive* aspect (e.g., one speaks of the *capacity* of a reservoir, when referring to its *total volume*) rather than an *intensive* aspect (e.g., a flow rate in terms of volume per units of area and time), as is more appropriate for a flux. Richards then proposed "infiltration rate" instead of "infiltration capacity," with "infiltration velocity" instead of "infiltration rate," but this suggestion has not been widely adopted.

More recently, Hillel (1971) has coined the term *infiltrability* to designate the infiltration flux resulting when water at *atmospheric pressure* is made *freely available* at the soil surface. This single-word replacement avoids the extensity–intensity contradiction in the term infiltration capacity and allows the use of the term infiltration rate in the ordinary literal sense to represent the surface flux under any set of circumstances, whatever the rate or pressure at which the water is supplied to the soil. For example, the infiltration rate can be expected to exceed infiltrability whenever water is ponded over the soil to a depth sufficient to cause the pressure at the surface to be significantly

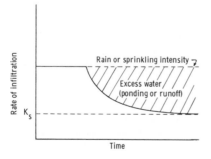

Fig. 2.1. Dependence of the infiltration rate upon time, under an irrigation of constant intensity lower than the initial value, but higher than the final value, of soil infiltrability.

greater than atmospheric pressure. On the other hand, if water is applied slowly or at a subatmospheric pressure, the infiltration rate may well be smaller than the infiltrability. In other words, as long as the rate of water delivery to the surface is smaller than the soil's infiltrability, water infiltrates as fast as it arrives and the supply rate determines the infiltration rate, i.e., the process is *supply controlled* (or *flux controlled*). However, once the delivery rate exceeds the soil's infiltrability, it is the latter which determines the actual infiltration rate, and thus the process becomes *surface controlled* or *profile controlled*. Horton, considered by many to be the father of modern physical hydrology, hypothesized that the soil surface zone determines what he called the infiltration-capacity, more or less apart from conditions within the soil profile. On the other hand, Childs (1969) regarded the infiltration process as a consequence of both the hydraulic conductivity and the hydraulic gradient prevailing in the soil's surface zone, allowing for the possibility that the gradient might be affected by the conditions existing throughout the profile.

If a shallow layer of water is instantaneously applied, and thereafter maintained, over the surface of an initially unsaturated soil, the full measure of soil infiltrability comes into play from the start. Many trials of infiltration under shallow ponding have shown infiltrability to vary, and generally to decrease, in time. Thus, the cumulative infiltration, being the time integral of the infiltration rate, has a curvilinear time dependence, with a gradually decreasing slope (Fig. 2.2).

Soil infiltrability and its variation with time are known to depend upon the initial wetness and suction, as well as on the texture, structure, and uniformity (or layering sequence) of the profile. In general, soil infiltrability is high in the early stages of infiltration, particularly where the soil is initially quite dry, but tends to decrease monotonically and eventually to approach asymptotically a constant rate, which is often termed the *final infiltration capacity*[2] but which we prefer to call the *steady-state infiltrability*.

The decrease of infiltrability from an initially high rate can in some cases result from gradual deterioration of the soil structure and the consequent partial sealing of the profile by the formation of a dense surface crust, from the detachment and migration of pore-blocking particles, from swelling of clay, or from entrapment of air bubbles or the bulk compression of soil air if it is prevented from escaping as it is displaced by incoming water. Primarily, however, the decrease in infiltrability results from the inevitable decrease in the *matric suction gradient* (constituting one of the forces drawing water into the soil) which occurs as infiltration proceeds.

---

[2] The adjective "final" in this context does not signify the end of the process (since infiltration can persist practically indefinitely if profile conditions permit), but it does indicate that soil infiltrability has finally attained a constant value from which it appears to decrease no more.

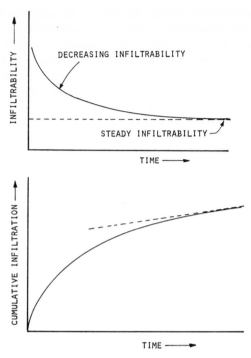

Fig. 2.2. Time dependence of infiltrability and of cumulative infiltration under shallow ponding.

If the surface of an initially dry soil is suddenly saturated, as, for instance, if the surface is ponded, the matric suction gradient acting in the surface layer is at first very steep. As the wetted zone deepens, however, this gradient is reduced, and, as the wetted part of the profile becomes thicker and thicker, the suction gradient tends eventually to become vanishingly small. In a horizontal column, the infiltration rate eventually tends to zero. However, in downward flow into a vertical column under continuous ponding, the infiltration rate (equal to the infiltrability) can be expected to settle down to a steady gravity-induced, rate which, as we shall later show, is practically equal to the saturated hydraulic conductivity if the profile is homogeneous and structurally stable (Figure 2.2). In summary, soil infiltrability depends on the following factors:

(1) Time from the onset of rain or irrigation. The infiltration rate is apt to be relatively high at first, then to decrease, and eventually to approach a constant rate that is characteristic for the soil profile.

(2) Initial water content. The wetter the soil is initially, the lower will

be the initial infiltrability (owing to smaller suction gradients) and the quicker will be the attainment of the final (constant) rate, which itself is generally independent of the initial water content.

(3) Hydraulic conductivity. The higher the saturated hydraulic conductivity of the soil is, the higher its infiltrability tends to be.

(4) Soil surface conditions. When the soil surface is highly porous and of "open" structure the initial infiltrability is greater than that of a uniform soil, but the final infiltrability remains unchanged, as it is limited by the lower conductivity of the transmission zone beneath. On the other hand, when the soil surface is compacted and the profile covered by a surface crust of lower conductivity the infiltration rate is lower than that of the uncrusted (uniform) soil. The surface crust acts as an hydraulic barrier, or bottleneck, impeding infiltration. This effect, which becomes more pronounced with a thicker and denser crust, reduces both the initial infiltrability and the eventually attained steady infiltrability. A soil of unstable structure tends to form such a crust during infiltration, especially as the result of the slaking action of beating raindrops. In such a soil, a plant cover or a surface mulch of plant residues can serve to intercept and break the impact of the raindrops and thus help to prevent surface sealing.

(5) The presence of impeding layers inside the profile. Layers which differ in texture or structure from the overlying soil may retard water movement during infiltration. Perhaps surprisingly, clay layers and sand layers can have a similar effect, although for opposite reasons. The clay layer impedes flow owing to its lower *saturated* conductivity, while a sand layer retards the wetting front (where unsaturated conditions prevail) owing to the lower *unsaturated* conductivity of the sand at equal matric suction. Flow into a dry sand layer can take place only after the pressure head has built up sufficiently for water to move into and fill the large pores of the sand.

Methods to measure soil infiltrability were reviewed by Bertrand (1965). These methods include flooding, applying artificial rainfall, and analysis of watershed hydrographs.

C. Profile Moisture Distribution during Infiltration

If we examine a homogeneous profile at any moment during infiltration under ponding, we shall find that the surface of the soil is saturated, perhaps to a depth of several millimeters or centimeters, and that beneath this zone of complete saturation is a less than saturated, lengthening zone of apparently uniform wetness, known as the *transmission zone*. Beyond this zone there is a *wetting zone*, in which soil wetness decreases with depth at a steepening

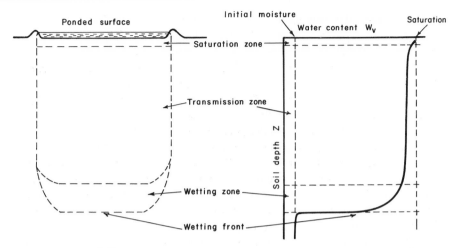

Fig. 2.3. The infiltration moisture profile. At left, a schematic section of the profile; at right, the water content versus depth curve. The common occurrence of a saturation zone as distinct from the transmission zone may result from the structural instability of the surface zone soil.

gradient down to a *wetting front*, where the moisture gradient is so steep[3] that there appears to be a sharp boundary between the moistened soil above and the relatively dry soil beneath.

The typical moisture profile during infiltration, first described by Bodman and Coleman (1944) by and Coleman and Bodman (1945), is shown schematically in Figure 2.3. Later investigations have cast some doubt as to whether a saturation zone distinct from the transmission zone necessarily exists, or whether it is merely an experimental artifact or anomaly resulting from the looseness, structural instability, slaking, or swelling of the soil at the surface. The saturation zone might also result from the air-entry value, or from air entrapment. The surface soil, being unconfined and subject to the disruptive and slaking action of raindrops and turbulent water, often experiences aggregate breakdown and colloidal dispersion, resulting in the formation of an impeding crust, which, in turn, affects the moisture profile below.

However, if we continue periodically to examine the moisture profile of a stable soil during infiltration, we shall find that the nearly saturated transmission zone lengthens (deepens) continuously, and that the wetting zone and wetting front move downward continuously (though at a diminishing

---

[3] The reason for the steepening gradient is that as the water content decreases the hydraulic conductivity generally decreases exponentially. Since the flux is the product of the gradient and the conductivity, it follows that to get a certain flux moving in the soil the gradient must increase as the conductivity decreases.

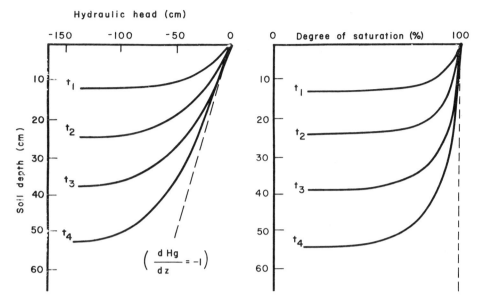

Fig. 2.4. Water-content profiles (at right) and hydraulic-head profiles (at left) at different times (t_1, t_2, t_3, t_4) during infiltration into a uniform soil ponded at the surface. dH_g/dZ is the gravitational head gradient. In this figure the possible existence of a saturation zone distinct from the transmission zone is disregarded.

rate), with the latter becoming less steep as it moves deeper into the profile. Typical families of successive moisture and hydraulic-head profiles are shown in Figure 2.4.

With the foregoing qualitative description of infiltration as a background, we can now proceed to consider some of the quantitative aspects of the process as it occurs under various conditions.

D. Infiltrability Equations

Numerous formulations, some entirely empirical and others theoretically based, have been proposed over the years in repeated attempts to express infiltrability as a function of time or of the total quantity of water infiltrated into the soil. Following the comprehensive critique of these equations given by Swartzendruber and Hillel (1973), we shall now present some of the more widely applied equations in their historical order of appearance. We use the symbol I to represent the cumulative volume of water infiltrated in time t per unit area of soil surface and the symbol i for the infiltrability as a volume flux (i.e., the volume of water entering a unit soil surface area per unit time). Thus $i = dI/dt$ and $I = \int_0^t i \, dt$.

The earliest equation was introduced by Green and Ampt (1911),

$$i = i_c + b/I \tag{2.1}$$

Here i_c and b are the characterizing constants, with i_c being the asymptotic steady infiltration flux reached when t (and hence I) become large. Not that at $t = 0$ I is also zero, so that Eq. (2.1) predicts i to be infinite initially and then to decrease gradually to its eventual value i_c. A more detailed elucidation of the Green and Ampt approach will be given in the next section.

The next equation is that of Kostiakov (1932),

$$i = Bt^{-n} \tag{2.2}$$

where B and n are constants. This strictly empirical formulation also provides an infinite initial i but implies that i approaches zero as t increases, rather than a constant nonzero i_c. This could have relevance for purely horizontal water absorption (in the absence of a gravity gradient) but is clearly deficient for downward infiltration.

The third equation is due to Horton (1940),

$$i = i_c + (i_0 - i_c)e^{-kt} \tag{2.3}$$

where i_c, i_0, and k are the characterizing constants. At $t = 0$ the infiltrability is not infinite but takes on the finite value i_0. The constant k determines how quickly i will decrease from i_0 to i_c. This form is also integrable and provides I as an explicit function of t. However, this equation is cumbersome in practice, since it contains three constants which must be evaluated experimentally.

The fourth equation is that of Philip (1957c),

$$i = i_c + s/2t^{1/2} \tag{2.4}$$

where i_c and s are the characterizing constants. This equation is a truncated form of the series Eq. (2.34) to be presented in Section F. Once again, the infiltrability is represented as infinite at zero time, and, as in Eq. (2.1), only two constants are required. Equation (2.4) will integrate to provide either I as an explicit function of t or t as an explicit function of I.

The fifth equation was proposed by Holtan (1961):

$$i = i_c + a(M - I)^n \tag{2.5}$$

where i_c, a, M, and n are constants. Holtan further specified M as the water-storage capacity of the soil above the first impeding stratum (total porosity minus the antecedent soil water, expressed in units of equivalent depth), but the meaning of M for a soil without an impeding stratum was not made clear.

Furthermore, what is usually not stated explicitly about Eq. (2.5) is that it can only be construed as holding for the range $0 \leq I \leq M$, since $i = i_c$ can only occur at the single point $I = M$. When I exceeds M, then the quantity $(M - I)^n$ becomes either positive and increasing, negative and decreasing, or imaginary, depending upon whether n is even, odd, or fractional, respectively. Thus, in addition to needing the condition $0 \leq I \leq M$ on Eq. (2.5), to be complete one must also state

$$i = i_c \qquad \text{for} \quad I > M \tag{2.6}$$

since there is no reason whatever to suppose that infiltration should cease once $I = M$. Hence, instead of a single equation good for all $I > 0$, the Holtan expression must be recognized as the two-form mathematical specification represented by Eq. (2.5) and (2.6). Equation (2.5) is integrable, but only provides t as an explicit function of I and not vice versa.

The Green–Ampt and Philip equations both arise out of mathematical solutions to well-defined physically based theories of infiltration. The Horton and Holtan equations, on the other hand, are essentially empirical expressions selected to have the correct qualitative shape. For this reason, therefore, they are not quite so inherently restrictive as to the mode of water application, since they do not imply surface ponding from time zero on, as do the Green–Ampt and Philip equations. In fact, for a flux-controlled type of beginning infiltration, the finite initial infiltration rate of the Horton and the Holtan equations is obviously the more realistic. The larger number of characterizing constants in these two equations, however, can at times hinder their usefulness, while at other times it might help to provide a better description of the phenomenon. Both of these aspects have been recognized and demonstrated (Skaggs *et al.*, 1969).

E. The Green and Ampt Approach

A simple, approximate, yet still useful approach to the infiltration process was suggested as early as 1911 by Green and Ampt in their classic paper on flow of air and water through soils. Their approach has been found to apply quite satisfactorily for certain cases of infiltration into initially dry soils, particularly of the coarse-textured type, which exhibit a sharp wetting front (e.g., Hillel and Gardner, 1970). The solution gives no information about details of the moisture profile during infiltration, but does offer estimates of the infiltration rate and the cumulative infiltration functions of time, i.e., $i(t)$ and $I(t)$.

The principal assumptions of the Green and Ampt approach are that there exists a distinct and precisely definable wetting front and that the matric

suction at this wetting front remains effectively constant, regardless of time
and position. Furthermore, this approach assumes that, behind the wetting
front, the soil is uniformly wet and of constant conductivity. The wetting
front is thus viewed as a plane separating a uniformly wetted infiltrated zone
from an as yet totally uninfiltrated zone. In effect, this supposes the K–ψ
relation (hydraulic conductivity versus suction head) to be discontinuous,
i.e., to change abruptly, at the suction value prevailing at the wetting front,
from a high value at lower suctions to a very much lower value at higher
suctions.

These assumptions simplify the flow equation, making it amenable to
analytical solution. For horizontal infiltration, a Darcy-type equation can
be applied directly:

$$i = \frac{dI}{dt} = K \frac{H_0 - H_f}{L_f} \tag{2.7}$$

where i is the flux into the soil and through the transmission zone, I the
cumulative infiltration, K the hydraulic conductivity of the transmission
zone, H_0 the pressure head at the entry surface, H_f the effective pressure head
at the wetting front, and L_f the distance from the surface to the wetting front
(the length of the wetted zone). If the ponding depth is negligible and the
surface is thus maintained at a pressure head of zero, we obtain simply

$$\frac{dI}{dt} = -K \frac{H_f}{L_f} = K \frac{\Delta H_p}{L_f} \tag{2.8}$$

where ΔH_p is the pressure-head change from the surface to the wetting front.
This suggests that the infiltration rate varies linearly with the reciprocal of
the distance to the wetting front.

Since a uniformly wetted zone is assumed to extend all the way to the
wetting front, it follows that the cumulative infiltration I should be equal to
the product of the wetting front depth L_f and the wetness increment $\Delta \theta =
\theta_t - \theta_i$ (where θ_t is the transmission-zone wetness during infiltration and θ_i
is the initial profile wetness which prevails beyond the wetting front):

$$I = L_f \Delta \theta \tag{2.9}$$

(In the special case where θ_t is saturation and θ_i is zero, $I = fL_t$, where f
is the porosity.) Therefore,

$$\frac{dI}{dt} = \Delta \theta \frac{dL_f}{dt} = K \frac{\Delta H_p}{L_f} = K \frac{\Delta \theta \, \Delta H_p}{I} \tag{2.10}$$

where dL_f/dt is the rate of advance of the wetting front. The infiltration rate is thus seen to be inversely related to the cumulative infiltration. Rearranging Eq. (2.10), we obtain

$$L_f \, dL_f = K \frac{\Delta H_p}{\Delta \theta} \, dt = \tilde{D} \, dt \qquad (2.11)$$

where the composite term $(K \, \Delta H_p / \Delta \theta)$ can be regarded as an effective diffusivity $\tilde{D}$ for the infiltrating profile. Integration gives

$$\frac{L_f^2}{2} = K \frac{\Delta H_p}{\Delta \theta} t = \tilde{D} t \qquad (2.12)$$

$$L_f = \sqrt{2Kt \, \Delta H_p / \Delta \theta} = \sqrt{2\tilde{D}t} \qquad (2.13)$$

or

$$I = \Delta \theta \sqrt{2\tilde{D}t}, \quad i = \Delta \theta \sqrt{\tilde{D}/2t} \qquad (2.14)$$

Thus the depth of the wetting front is proportional to $\sqrt{t}$, and the infiltration rate is proportional to $1/\sqrt{t}$.

With gravity taken into account, the Green and Ampt approach gives

$$\frac{dI}{dt} = \Delta \theta \frac{dL_f}{dt} = K \frac{H_0 - H_f + L_f}{L_f} \qquad (2.15)$$

which integrates to

$$Kt/\Delta \theta = L_f - (H_0 - H_f) \ln[1 + L_f/(H_0 - H_f)] \qquad (2.16)$$

As t increases, the second term on the right-hand side of Eq. (2.16) increases more and more slowly in relation to the increase in L_f, so that, at very large times, we can approximate the relationship by

$$L_f \cong Kt/\Delta \theta + \delta \qquad (2.17)$$

or

$$I \cong Kt + \delta$$

where δ can eventually be regarded as a constant.

The Green and Ampt relationships are essentially empirical, since the value of the effective "wetting-front suction" must be found by experiment. For infiltration into initially dry soil, it may be of the order of 50–100 cm H_2O, or 0.05–0.1 bar (Green and Ampt, 1911; Hillel and Gardner, 1970). However, in actual field conditions, particularly where the initial moisture is not uniform, H_f may be undefinable. In some situations, the wetting front may be so diffuse that its exact location at any particular time cannot be approximated satisfactorily.

The Green and Ampt model was designated the *delta-function approximation* by Philip (1966a), because it corresponds exactly to the nonlinear diffusion description of infiltration in the special case where the diffusivity is wholly concentrated at the wet end of the moisture range (a so-called Dirac delta function). According to Philip, this model represents a reasonable approximation of reality in situations where the wetting front is sharp and the moisture profile differs little from a step function: e.g., one-dimensional systems with water supplied under positive hydrostatic pressure to soils which are coarse-textured, initially dry, and free from the complications of air entrapment.

F. Modern Approaches to Infiltration Theory

1. HORIZONTAL INFILTRATION

We begin by combining the Darcy equation with the continuity equation to obtain the general flow equation for water in soil (see Hillel, 1980, Chapters 8 and 9). For one-dimensional flow the appropriate form of the flow equation is

$$\frac{\partial \theta}{\partial t} = \frac{\partial}{\partial x}\left(K \frac{\partial H}{\partial x}\right) \tag{2.18}$$

where θ is the volumetric wetness, t is time, x is distance along the direction of flow, and H is the hydraulic head. The simplest application of this equation is in the description of the infiltration (termed absorption, Philip, 1969a) of water in the horizontal direction. In this case, the gravity force is zero, and water is drawn into the soil by matric suction gradients only. If the soil is homogeneous, the diffusivity equation can be applied directly [subject to the limitations pointed out in Chapter 9 of Hillel (1980)]:

$$\frac{\partial \theta}{\partial t} = \frac{\partial}{\partial x}\left[D(\theta) \frac{\partial \theta}{\partial x}\right] \tag{2.19}$$

This equation can be solved readily for the special conditions of flow into an infinitely long column of uniform initial wetness θ_i, where the plane of entry ($x = 0$) is instantaneously brought to, and thereafter maintained at, a higher wetness θ_0. These conditions are written formally in the following way:

$$\theta = \theta_i, \qquad x \geq 0, \qquad t = 0$$
$$\theta = \theta_0, \qquad x = 0, \qquad t > 0 \tag{2.20}$$

We can now simplify Eq. (2.19) by introducing the composite variable $\lambda(\theta)$ defined as:

$$\lambda(\theta) = xt^{-1/2} \tag{2.21}$$

Note that λ is a function of θ only. When substituted into Eq. (2.19) it transforms that equation from a partial to an ordinary differential equation (the only variables being θ and λ):

$$-\frac{\lambda}{2}\frac{d\theta}{d\lambda} = \frac{d}{d\lambda}\left[D(\theta)\frac{d\theta}{d\lambda}\right] \tag{2.22}$$

This is known as the *Boltzmann transformation*, as it was first introduced by Ludwig Boltzmann in 1894.

Equation (2.21) can be rewritten

$$x = \lambda(\theta)t^{1/2} \tag{2.23}$$

During infiltration into an initially dry and stable soil, the advancing wetting front is usually visible and indicates a moving zone of practically constant wetness. In this situation, $\lambda(\theta)$ can be assumed to remain constant, and therefore x (taken to be the distance to the wetting front) is proportional to $t^{1/2}$. A plot of the distance to the wetting front against $t^{1/2}$ should therefore give a straight line with a slope equal to λ.

The cumulative volume of water which has entered through a unit area of the entry plane $(x = 0)$ in time t is given by

$$I = \int_{\theta_i}^{\theta_o} x \, d\theta \tag{2.24}$$

where I is the cumulative infiltration, θ_i the initial volumetric wetness of the unwetted soil, and θ_o the saturation wetness. Substitution of the Boltzmann variable [Eq. (2.21)] gives:

$$I = t^{1/2} \int_{\theta_i}^{\theta_o} \lambda(\theta) \, d\theta = st^{1/2} \tag{2.25}$$

wherein

$$s = \int_{\theta_i}^{\theta_o} \lambda(\theta) \, d\theta = I/t^{1/2} \tag{2.26}$$

The coefficient s is a constant (a definite integral) and is called the sorptivity (Philip, 1969). Note that s depends on both θ_i and θ_o.

Therefore, it is defined only in relation to a fixed initial state θ_i and an imposed boundary condition θ_o. The dimensions of sorptivity are length per square-root of time $(LT^{-1/2})$. From Eq. (2.25) we note that a plot of cumulative infiltration I versus $t^{1/2}$ should give a straight line with a slope equal to s.

Introducing the Boltzmann transformation into Eq. (2.19) also entails transformation of the boundary conditions. Accordingly, Eq. (2.20) becomes

$$\theta = \theta_i, \qquad \lambda \to \infty$$

$$\theta = \theta_0, \qquad \lambda = 0 \tag{2.27}$$

Philip (1955) described an iterative procedure for the numerical solution of Eq. (2.22) subject to (2.27). A numerical solution is required because the $D(\theta)$ relationship makes Eq. (2.22) nonlinear. A quasi-analytical technique was introduced by Parlange (1971) and computer-based simulation procedures were developed by deWit and van Keulen (1972) and Elrick (1979).

Bruce and Klute (1956) published a technique based on the method of Matano (1933) for determining $D(\theta)$ from experimentally obtained moisture distributions [i.e., $\theta(\lambda)$] for horizontal infiltration. If we multiply Eq. (2.22) by $d\lambda$ and integrate from $\lambda = \infty$ to $\lambda(\theta)$ (i.e., from θ_i to θ) we obtain:

$$-\frac{1}{2}\int_{\theta_i}^{\theta} \lambda \, d\theta = D(\theta) \, d\theta/d\lambda \Big|_{\theta_i}^{\theta} = D(\theta) \, d\theta/d\lambda \tag{2.28}$$

wherein $D(\theta) \, d\theta/d\lambda$ at $\theta = \theta_i$ is zero.

Solving for $D(\theta)$ gives

$$D(\theta) = -\frac{1}{2}\frac{d\lambda}{d\theta}\int_{\theta_i}^{\theta} \lambda \, d\theta \tag{2.29}$$

In practice, $D(\theta)$ can be obtained from experimental $\theta(\lambda)$ curves by determining the appropriate slopes and areas, as illustrated in Fig. 2.5.

For certain known functions of $D(\theta)$, a solution is obtainable in closed form for the ordinary differential equation resulting from the Boltzmann transformation, $y = \frac{1}{2}x\sqrt{D_i t}$ (Gardner and Mayhugh, 1958). The solution indicates that the rate of advance of any wetness value, and the infiltration rate itself, are reciprocally proportional to the square root of time, while the distance of advance of any wetness value, as well as the cumulative infiltration, are directly proportional to the square root of time. For the entry of water into soils, exponential D functions of the form

$$D = D_i \exp(\theta - \theta_i) \tag{2.30}$$

(where D_i is the diffusivity corresponding to the initial wetness θ_i) were found to give good agreement between theory and experiment for a number of soils. Other functions exist which give solutions but few actually apply to soils.

The time dependences of the cumulative infiltration I and of the infiltration

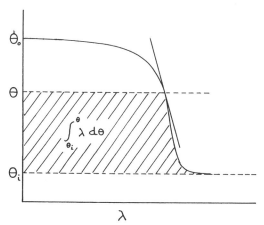

Fig. 2.5. Moisture distribution $\theta(\lambda)$ during horizontal infiltration into a stable homogenous soil of uniform initial wetness θ_i. At $t = 0$, water is introduced at the entry plane $x = 0$, so that soil wetness there is instantaneously changed to and thereafter maintained at θ_0. Note that the tangent line gives the value of $d\lambda/d\theta$ and the shaded area gives the value of the integral $\int_{\theta_i}^{\theta} \lambda \, d\theta$. $D(\theta)$ can thus be calculated according to Eq. (2.29).

rate i can be expressed in terms of the *weighted mean diffusivity*[4] D (Crank, 1956), which is a constant value of the diffusivity giving the same flux as the variable diffusivity which in fact operates during the actual flow process. Assuming the diffusivity constant allows Eq. (2.19) to be solved analytically. The following equations are obtained:

$$i = \tfrac{1}{2}(\theta_0 - \theta_i)\sqrt{\bar{D}/\pi t} \tag{2.31}$$

and

$$I = \int_0^t i \, dt = (\theta_0 - \theta_i)\sqrt{\bar{D}t/\pi} \tag{2.32}$$

which again indicate square root of time behavior. Time t can be eliminated from these equations to obtain the relation of infiltration rate to cumulative infiltration:

$$i = (\theta_0 - \theta_i)^2 \bar{D}/\pi I \tag{2.33}$$

As mentioned earlier, when water moves into a relatively dry soil, a distinct wetting front is often observed, this front being in effect a moving boundary between the already wetted and the as yet unwetted parts of the

---

[4] According to Crank (1956), the weighted mean diffusivity for sorption processes is given by

$$\bar{D} = \tfrac{5}{3}[1/(\theta_0 - \theta_i)]^{5/3} \int_{\theta_i}^{\theta_0} D(\theta)(\theta - \theta_i)^{2/3} \, d\theta$$

soil. From the considerations given, it can be inferred that the steepness, or sharpness, of the wetting front is related to the difference between the diffusivity of the wetted soil (near the entry surface) and that of the relatively dry soil ahead of the wetting front. Thus a coarse-textured soil, which characteristically shows steeper decrease of D with decreasing θ, typically exhibits a sharper and more distinct wetting front than fine-textured soils. Similarly, the wetting front is sharper during infiltration into dry than into moist soils.

2. VERTICAL INFILTRATION

Downward infiltration into an initially unsaturated soil generally occurs under the combined influence of suction and gravity gradients. As the water penetrates deeper and the wetted part of the profile lengthens, the average suction gradient decreases, since the overall difference in pressure head (between the saturated soil surface and the unwetted soil inside the profile) divides itself along an ever-increasing distance. This trend continues until eventually the suction gradient in the upper part of the profile becomes negligible, leaving the constant gravitational gradient in effect as the only remaining force moving water downward. Since the gravitational head gradient has the value of unity (the gravitational head decreasing at the rate of 1 cm with each centimeter of vertical depth below the surface), it follows that the flux tends to approach the hydraulic conductivity as a limiting value. In a uniform soil (without crust) under prolonged ponding, the water content of the wetted zone approaches saturation.[5]

Darcy's equation for vertical flow is

$$q = -K\frac{dH}{dz} = -K\frac{d}{dz}(H_p - z) \qquad (2.34)$$

where q is the flux, H the total hydraulic head,[6] H_p the pressure head, z the vertical distance from the soil surface downward (i.e., the depth), and K the hydraulic conductivity. At the soil surface, $q = i$, the infiltration rate. In an unsaturated soil, H_p is negative and can be expressed as a suction head ψ. Hence,

$$q = K\frac{d\psi}{dz} + K \qquad (2.35)$$

[5] In practice, because of air entrapment, the soil-water content may not attain total saturation but some maximal value lower than saturation which has been called "satiation." Total saturation is assured only when a soil sample is wetted under vacuum.

[6] Recalling Chapters 7 and 8 in Hillel (1980), we note that the hydraulic head H is the sum of a pressure head H_p and a gravity head H_g. If the datum plane is at the soil surface ($z = 0$) and z is depth $H = H_p - z$.

Combining these formulations of Darcy's equation, (2.34) and (2.35) with the continuity equation $\partial\theta/\partial t = -\partial q/\partial z$ gives the general flow equation

$$\frac{\partial\theta}{\partial t} = \frac{\partial}{\partial z}\left(K\frac{\partial H}{\partial z}\right) = -\frac{\partial}{\partial z}\left(K\frac{\partial\psi}{\partial z}\right) - \frac{\partial K}{\partial z} \qquad (2.36)$$

If soil wetness θ and suction head ψ are uniquely related, then the left-hand side of Eq. (2.36) can be written $\partial\theta/\partial t = (\partial\theta/\partial\psi)(\partial\psi/\partial t)$, which transforms Eq. (2.36) into

$$C\frac{\partial\psi}{\partial t} = \frac{\partial}{\partial z}\left(K\frac{\partial\psi}{\partial z}\right) + \frac{\partial K}{\partial z} \qquad (2.37)$$

where $C (= -\partial\theta/\partial\psi)$ is defined as the *specific* (or differential) *water capacity* (i.e., the change in water content in a unit volume of soil per unit change in matric potential).

Alternatively, we can transform the right-hand side of Eq. (2.36) once again using the chain rule to render $-\partial\psi/\partial z = -(\partial\psi/\partial\theta)(\partial\theta/\partial z) = (1/C)(\partial\theta/\partial z)$. *We thus obtain*

$$\frac{\partial\theta}{\partial t} = \frac{\partial}{\partial z}\left(\frac{K}{C}\frac{\partial\theta}{\partial z}\right) - \frac{\partial K}{\partial z} \qquad \text{or} \qquad \frac{\partial\theta}{\partial t} = \frac{\partial}{\partial z}\left(D\frac{\partial\theta}{\partial z}\right) - \frac{\partial K}{\partial z} \qquad (2.38)$$

where D, once again, is the soil-moisture diffusivity, which we propose calling the *hydraulic diffusivity* (see Hillel, 1980, Chapter 9). Equations (2.36), (2.37), and (2.38) can all be considered as forms of the *Richards equation* (Swartzendruber, 1969).

Note that the above three equations contain two terms on their right-hand sides, the first term expressing the contribution of the suction, (or wetness) gradient, and the second term expressing the contribution of gravity. Whether the one or the other term predominates depends on the initial and boundary conditions and on the stage of the process considered. For instance, when infiltration takes place into an initially dry soil, the suction gradients at first can be much greater than the gravitational gradient, and the initial infiltration rate into a horizontal column tends to approximate the infiltration rate into a vertical column. Water from a furrow will therfore tend at first to infiltrate laterally almost to the same extent as vertically (Fig. 2.6). On the other hand, when infiltration takes place into an initially wet soil, the suction gradients are small from the start and become negligible much sooner (Fig. 2.7).

The first mathematically rigorous solution of the flow equation applied to vertical infiltration was given by Philip (1957a). A more recent comprehensive review of infiltration theory was provided by the same author (Philip, 1969a). His original solution pertained to the case of an infinitely

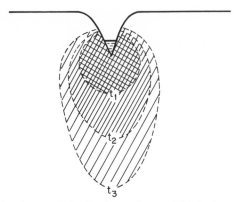

Fig. 2.6. Infiltration from an irrigation furrow into an initially dry soil. The wetting front is shown after different periods of time ($t_1 < t_2 < t_3$). At first the strong suction gradients cause infiltration to be nearly uniform in all directions; eventually the suction gradients decrease and the gravitational gradient predominates.

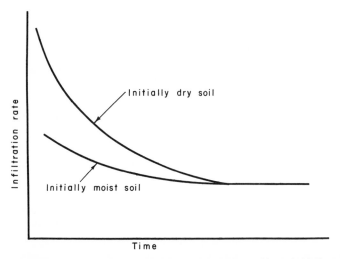

Fig. 2.7. Infiltrability as a function of time in an initially dry and in an initially moist soil.

deep uniform soil of constant initial wetness θ_i, assumed at time zero to become submerged under a thin layer of water that instantaneously increases soil wetness at the surface from its initial value to a new value θ_0 (near saturation) that is thereafter maintained constant. Mathematically, these conditions are stated as

$$t = 0, \quad z > 0, \quad \theta = \theta_i$$
$$t \geq 0, \quad z = 0, \quad \theta = \theta_0 \tag{2.39}$$

His solution took the form of a power series:

$$z(\theta,t) = \sum_{n=1}^{\infty} f_n(\theta)t^{n/2}$$

$$= f_1(\theta)t^{1/2} + f_2(\theta)t + f_3(\theta)t^{3/2} + f_4(\theta)t^2 + \cdots \quad (2.40)$$

where z is the depth to any particular value of wetness θ and the coefficients $f_n(\theta)$ are calculated successively from the diffusivity and conductivity functions. This solution indicates that at small times the advance of any θ value proceeds as $\sqrt{t}$ (just as in horizontal infiltration), while at larger times the downward advance of soil wetness approaches a constant rate $(K_0 - K_i)/(\theta_0 - \theta_i)$, where K_0 and K_i are the conductivities at the wetness values of θ_0 (wetted surface) and θ_i (initial soil wetness), respectively.

Philip's solution also describes the time dependence of cumulative infiltration I in terms of a power series:

$$I(t) = \sum_{n=1}^{\infty} j_n(\theta)t^{n/2}$$

$$= st^{1/2} + (A_2 + K_0)t + A_3 t^{3/2} + A_4 t^2 + \cdots + A_n t^{n/2} \quad (2.41)$$

in which the coefficients $j_n(\theta)$ are, again, calculated from $K(\theta)$ and $D(\theta)$, and the coefficient s is called the *sorptivity*. Differentiating Eq. (2.32) with respect to t we obtain the series for the infiltration rate $i(t)$:

$$i(t) = \tfrac{1}{2}st^{-1/2} + (A_2 + K_0)$$
$$+ \tfrac{3}{2}A_3 t^{1/2} + 2A_4 t + \cdots + \tfrac{n}{2}A_n t^{n/2-1} \quad (2.42)$$

In practice, it is generally sufficient for an approximate description of infiltration to replace Eq. (2.41) and (2.42) by two-parameter equations of the type

$$I(t) = st^{1/2} + At, \qquad i(t) = \tfrac{1}{2}st^{-1/2} + A \quad (2.43)$$

where t is not too large. In the limit as t approaches infinity, the infiltration rate decreases monotonically to its final asymptotic value $i(\infty)$. Philip (1969a) pointed out that this does not imply that $A = K_0$, particularly not at small or intermediate times. However at *large times* (for which the infinite series does not converge) it is possible to represent Eq. (2.43) as

$$I = st^{1/2} + Kt, \qquad i = \tfrac{1}{2}st^{-1/2} + K \quad (2.44)$$

where K is the hydraulic conductivity of the soil's upper layer (the transmission zone), which in a uniform soil under ponding is nearly equal to the saturated conductivity K_s. For very large values of time Philip's analysis yields

$$i = K \quad (2.45)$$

This theoretical analysis was used to interpret the earlier experimental findings of Bodman and Coleman (1944).

Recall that the sorptivity has already been defined (according to Philip, 1969a) in terms of the horizontal infiltration equation:

$$s = I/t^{1/2} \tag{2.46}$$

As such, it embodies in a single parameter the influence of the matric suction and conductivity on the transient flow process that follows a step-function change in surface wetness or suction. Strictly speaking, one should write $s(\theta_0, \theta_i)$ or $s(\psi_0, \psi_i)$, since s has meaning only in relation to an initial state of the medium and an imposed boundary condition. Philip also defined an "intrinsic sorptivity"—a parameter which takes into account the viscosity and surface tension of the fluid.

It should be obvious from the foregoing that the effects of ponding depth and initial wetness (Fig. 2.7) can be significant during early stages of infiltration, but decrease in time and eventually tend to vanish in a very deeply wetted profile. Typical values of the "final" (steady) infiltration rate are shown in Table 2.1. These values merely give an order of magnitude, while in actual cases the infiltration rate can be considerably higher (particularly during the initial stages of the process and in well-aggregated or cracked soils) or lower (as in the presence of a surface crust).

Table 2.1

Soil type	Steady infiltration rate (mm/hr)
Sands	> 20
Sandy and silty soils	10–20
Loams	5–10
Clayey soils	1–5
Sodic clayey soils	< 1

G. Infiltration into Layered Profiles

Notwithstanding the instructive value of the solutions described thus far to the flow equation, pertaining to infiltration into uniform profiles, it is obvious that in nature infiltration seldom, if ever, takes place under such ideal conditions. More typically, infiltration processes occur in soils which are uniform neither in texture nor in initial wetness. The solution of the flow equation for such conditions is apparently impossible by analytical means, and hence progress in the development of more realistic theories of infiltration was frustrated for a long time. Experimental studies, however, have shown that the process of infiltration can be greatly affected by soil profile heterogeneity. For example, Miller and Gardner (1962), who conducted experiments on the effects of thin layers of different texture sandwiched into otherwise uniform profiles, showed that although in any conducting soil the matric suction and hydraulic head must be continuous throughout the profile regardless of layering sequence, the wetness and conductivity may exhibit abrupt discontinuities at interlayer boundaries. The steady-state downflow of water through a two-layer profile had earlier been analyzed by Takagi (1960), who showed that where the upper layer is less pervious than the lower negative pressures (suctions) develop in the lower layer, and these can remain constant throughout a considerable depth range.

The advent of high-speed computers and of computer-based numerical methods for solving nonlinear differential equations subject to complex boundary conditions made possible a new approach to infiltration theory. No longer is the theory shackled to the restrictive assumptions which had been necessary in order to formulate the process mathematically in closed form, as required for the attainment of analytical solutions. One by one these restrictive conditions have been relaxed as increasingly sophisticated numerical methods were developed so that gradually more complex, time-variable, space-variable, and multidimensional processes could be handled.

An early demonstration that the computer can be used effectively to obtain a mathematical description of infiltration in heterogeneous soil profiles with strata of differing properties and initial wetness values was given by Hanks and Bowers (1962). In their model the initial wetness θ_i was allowed to vary with depth z, and different $K(\psi)$ and $\theta(\psi)$ functions were assigned to soil layers of different thicknesses. Later Wang and Lakshminarayana (1968) modified this formulation to make more explicit the expression of heterogeneity, i.e., the dependence of K and ψ on z and θ in the form $K = K(\theta, z)$ and $\psi = \psi(\theta, z)$. Both models were still based on the surface-ponding boundary condition. More recent investigations have provided even more comprehensive descriptions of infiltration into variously constituted soil profiles (e.g., Hillel and Talpaz, 1977; Hillel, 1977). Space limitations preclude a detailed presentation of the numerous models now available. Hence the following discussion will provide only a qualitative description of some of the features characterizing infiltration into layered profiles.

One typical situation is that of a coarse layer of higher saturated hydraulic conductivity overlying a less conductive finer-textured layer. In such a case the infiltration rate is at first controlled by the coarse layer, but when the wetting front reaches and penetrates into the finer-textured layer, the infiltration rate can be expected to drop and tend to that of the finer soil alone. Thus, in the long run, it is the layer of lesser conductivity which controls the process. If infiltration continues for long, then positive pressure heads (a "perched water table") can develop in the coarse soil, just above its boundary with the impeding finer layer.

In the opposite case of infiltration into a profile with a fine-textured layer over a coarse-textured one, the initial infiltration rate is again determined by the upper layer. As water reaches the interface with the coarse lower layer, however, the infiltration rate may decrease. Water at the wetting front is normally under suction, and this suction may be too high to permit entry into the relatively large pores of the coarse layer. This explains the observation (Miller and Gardner, 1962) that the wetting-front advance stops for a time (though infiltration at the surface does not stop) until the pressure head at the interface builds up sufficiently for water to penetrate into the coarse material. Thus, under unsaturated flow conditions a layer of sand or gravel in a medium-or fine-textured soil, instead of enhancing water movement in the profile, may actually impede it. The lower layer, in any case, cannot become saturated, since the restricted rate of flow through the less permeable upper layer cannot sustain flow at the saturated hydraulic conductivity of the coarse lower layer (except when the externally applied pressure, i.e., the ponding depth, is large or where both the layers are under the water table).

H. Infiltration into Crust-Topped Soils

A very important special case of a layered soil is that of a profile which develops a thin crust, or seal, at the surface. Such a seal can develop under the beating action of raindrops (Ekern, 1950; McIntyre, 1958; Tackett and Pearson, 1965), or as a result of the spontaneous slaking and breakdown of soil aggregates during wetting (Hillel, 1960).

The action of a raindrop striking at an exposed soil surface is believed to be related to its kinetic energy E_k:

$$E_k = \tfrac{1}{2}mv^2 \tag{2.47a}$$

Here m is raindrop mass and v velocity. The total action of a rainstorm might be expected therefore to be a function of the sum of the kinetic energies of all the drops:

$$E_{k,tot} = \tfrac{1}{2}\sum m_i v_i^2 \tag{2.47b}$$

wherein m_i, v_i are the masses and velocities of raindrops of successive size groups. As a raindrop of any given size falls earthward through the atmosphere, it is accelerated by gravity but encounters the viscous resistance of the atmosphere, which increases with velocity. When the resistance equals the downward force of gravity (*mg*), acceleration stops and the drop continues to fall at a constant, "terminal" velocity which depends on its mass. The spectrum of drop sizes, and hence also of terminal velocities, varies from storm to storm and form place to place. The impact of raindrops on the soil can be mitigated by the presence of a protective cover of vegetation or mulch (which can intercept the drops before they strike ground) and is also influenced by surface roughness, slope, incident angle, and standing water on the surface.

Another important factor affecting the formation of a surface seal (crust) is the chemical composition of the infiltrating solution. A high sodium adsorption ratio (SAR), coupled with a low overall concentration of salts, can induce dispersion and swelling of the clay present in the soil's surface layer, which in turn can have a strong effect on soil infiltrability. The dispersed aggregates collapse and close the interaggregate cavities, and migrating clay particles tend to lodge in soil pores beneath the surface. Evidence of these phenomena was provided by Chen and Banin (1975), by Frenkel *et al.* (1978), and by Oster and Schroer (1979). The effect of sodium in the added water is particularly pronounced at the soil surface because of the mechanical action resulting from the kinetic energy of the applied water and because of the absence of a confining matrix. Extreme clay dispersion and clogging can occur at SAR values as low as 10 if solution concentration is lower than 2 mole/m^3 (Oster and Schroer, 1979). Particularly destructive to surface soil

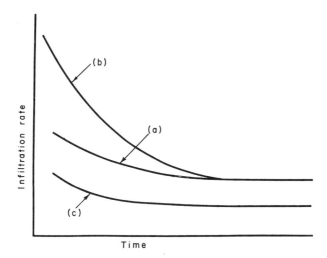

Fig. 2.8. Infiltrability as a function of time (a) in a uniform soil, (b) in a soil with a more permeable upper layer, and (c) in a soil covered by a surface crust.

structure is the alternation of sodic irrigation water with rainfall. Where such conditions occur drastic reduction of soil infiltrability can result.

Surface crusts are characterized by greater density, finer pores, and lower saturated conductivity than the underlying soil. Once formed a surface crust can greatly impede water intake by the soil (Fig. 2.8), even if the crust is quite thin (say, not more than several millimeters in thickness) and the soil is otherwise highly permeable. Failure to account for the formation of a crust can result in gross overestimation of infiltration.

An analysis of the effect of a developing surface crust upon infiltration was carried out by Edwards and Larson (1970), who adapted the Hanks and Bowers (1962) numerical solution to this problem. Hillel (1964), and Hillel and Gardner (1969, 1970) used a quasianalytical approach to calculate fluxes during steady and transient infiltration into crust-capped profiles from knowledge of the basic hydraulic properties of the crust and of the underlying soil.

The problem is relatively simple in the case of steady infiltration. Steady-state conditions require that the flux through the crust q_c be equal to the flux q_u through the subcrust transmission zone:

$$q_c = q_u$$

or

$$K_c(dH/dz)_c = K_u(dH/dz)_u \tag{2.48}$$

where K_c, $(dH/dz)_c$, K_u, and $(dH/dz)_u$ refer to the hydraulic conductivity and hydraulic-head gradient of the crust and underlying transmission zone, respectively. The gradient through the transmission zone tends to unity when steady infiltration is approached, as the suction gradient decreases with the increase in wetting depth, eventually leaving the gravitational gradient as the only effective driving force. In the absence of a suction-head gradient in the zone below the crust, we obtain (with the soil surface as our reference level)

$$q = K_u(\psi_u) = K_c(H_0 + \psi_u + z_c)/z_c \tag{2.49}$$

where $K_u(\psi_u)$ is the unsaturated hydraulic conductivity of the subcrust zone, a function of the suction head ψ_u which develops in this zone, beginning just under the hydraulically impeding crust; H_0 is the positive hydraulic head imposed on the surface by the ponded water and z_c is the vertical thickness of the crust.

Where the ponding depth H_0 is negligible and the crust itself is very thin and of low conductivity (e.g., where z_c is very small in relation to the suction ψ_u which forms at the subcrust interface), we can assume the approximation

$$q_u = q_c = K_c\psi_u/z_c \tag{2.50}$$

The condition that the crust remain saturated even while its lower part will be under suction is that its critical air entry suction ψ_a not be exceeded (i.e., $\psi_u < \psi_a$).

This, together with the condition that the subcrust hydraulic-head gradient approximate unity, leads to the approximation

$$K_u/\psi_u = K_c/z_c = 1/R_c \qquad (2.51)$$

i.e., the ratio of the hydraulic conductivity of the underlying soil transmission zone to its suction is approximately equal to the ratio of the crust's (saturated) hydraulic conductivity to its thickness. The latter ratio is the reciprocal of the hydraulic resistance per unit area of the crust R_c.[7] Also, by Eq. (2.48),

$$q = K_u(\psi_u) = \psi_u/R_c \qquad (2.52)$$

Employing a K–ψ relationship of the type $K = a\psi^{-n}$ (where a and n are characteristic constants of the soil), Hillel and Gardner (1969) obtained

$$q = a^{1/(n+1)}/R_c^{n/(n+1)} = B/R_c^{n/(n+1)}$$
$$\psi_u = (aR_c)^{1/(n+1)} = BR_c^{1/(n+1)} \qquad (2.53)$$

where $B = a^{1/(n+1)}$ is a property of the subcrust soil. The theoretical consequences of this theory are illustrated in Figure 2.9, which shows how the steady infiltrability can be expected to decrease with increasing hydraulic resistance of the crust. Gardner (1956) has shown that the values of a and of n generally increase with increasing coarseness, textural as well as structural, of the soil. Typically, sandy soils have n values of 4 or more, whereas clayey soils may have n values of about 2.

Both the crust and the underlying soil are seen to affect the infiltration rate and suction profile, and the crust-capped soil is thus viewed as a self-adjusting system in which the physical properties of the crust and underlying soil interact in time to form a steady infiltration rate and moisture profile. In this steadily infiltrating profile, the subcrust suction which develops is such as to create a gradient through the crust and a conductivity in the subcrust zone which will result in an equal flux through both layers.

The problem is rather more complicated in the prevalent case of transient infiltration into an initially unsaturated profile, during which the flux, the wetting depth, the subcrust suction, and the conductivity might all be changing with time.

Assuming the Green and Ampt conditions (Section E) and H_0 to be negligible, Hillel and Gardner (1970) recognized three stages during transient infiltration into crusted profiles: an initial stage, in which the rate is finite and

---

[7] A distinction is made between the hydraulic resistance per unit area, defined as above, and the hydraulic resistivity, which is equal to the reciprocal of the conductivity.

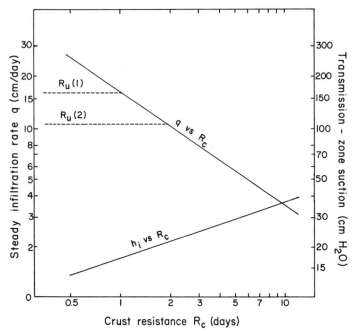

Fig. 2.9. Theoretical effect of crust resistance upon flux and subcrust suction during steady infiltration into crust-capped columns of a uniform soil with $n = 2$, $a = 4.9 \times 10^3$. The broken lines (1) and (2) indicate the hypothetical effect of subcrust hydraulic resistance $R_u(1) < R_u(2)$. The decreasing q–R_c curve applies only where the hydraulic conductance of the subcrust layers is not limiting. (After Hillel and Gardner, 1969.)

dependent on crust resistance R_c and on an effective subsoil suction; an intermediate stage, in which cumulative infiltration I increases approximately as the square root of time; and a later stage, in which I can be expressed as the sum of a steady and a transient term, the latter becoming negligible at long times. Cumulative infiltration I was shown to decrease with increasing R_c, particularly in coarse-textured and coarse-structured soils.

Where the gravity effect is negligible (e.g., in horizontal flow or during the early stages of vertical infiltration into an initially dry medium of high matric suction), the infiltration versus time relationship was given by

$$I = \sqrt{K_u^2 R_c^2 (\Delta\theta)^2 + 2K_u H_f (\Delta\theta)t} - K_u R_c \qquad (2.54)$$

Where the gravity effect is significant, the expression given is

$$L_f = K_u t / \Delta\theta + (H_f - K_u R_c) \ln\{[H_f + (K_u t / \Delta\theta) + \delta(t)]/H_f\} \quad (2.55)$$

where the correction term $\delta(t)$ becomes negligibly small as t increases. Thus,

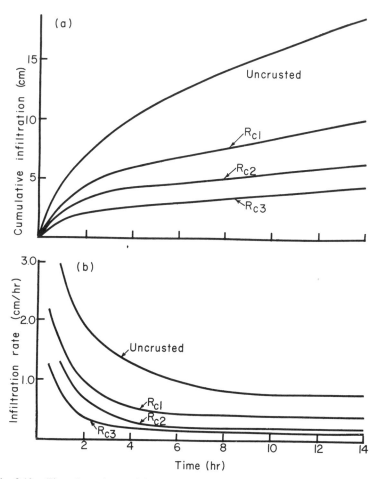

Fig. 2.10. Time dependence of (a) cumulative infiltration and of (b) infiltration rate for uncrusted and crusted columns of Negev loess. Crust resistance values R_{c1}, R_{c2}, R_{c3} are 3.2, 9.1, and 17 days, respectively. (After Hillel and Gardner, 1969.)

L_f can be expressed as the sum of a steady and a transient term. Some experimental results are shown in Figure 2.10.

The foregoing analysis takes no account of the possibility of unstable flow conditions. Where such conditions occur, the description given may not be valid even approximately. Unfortunately, there is as yet insufficient knowledge, both theoretical and phenomenological, to predict just when unstable flow begins during infiltration into crust-covered soils and how it might relate to antecedent moisture conditions and hydraulic properties of both the crust and the subcrust zones.

I. Instability of Wetting Fronts during Infiltration

An interesting and still incompletely understood phenomenon is that of *unstable flow*, particularly *wetting-front instability*, which occurs most notably in the transition of infiltrating water from a fine-textured to a coarse-textured layer. Rather than advance as a smooth front, the percolating water may concentrate at certain locations to break through in the form of fingerlike or tonguelike protrusions into the coarse-textured layer. Flow through the coarse-textured layer thereafter occurs through "chimneys" or "pipes" within the layer rather than through the entire layer uniformly (Fig. 2.11). This phenomenon has been recognized and studied by Hill and Parlange (1972), Raats (1973), and Philip (1975).

Wetting front instability can invalidate our best-laid theories of one-dimensional infiltration. That by itself may not seem to be much of a disaster to nontheoreticians. However, the phenomenon is not merely of academic interest, and can indeed have important practical consequences. At the very least, the loss of our predictive power means that we cannot manage our system with the necessary degree of certainty. For example, unstable flow can allow small "outlaw" streams of water to flow directly, and carry raw pollutants, into the groundwater while bypassing the soil's filtration and purification mechanisms.

The existence of instability at the moving interface between two immiscible fluids of different densities or viscosities (such as oil and water) has long been recognized (Taylor, 1950; Saffman and Taylor, 1958; Wooding, 1969). Interfacial instability can also occur between two miscible fluids, as in the penetration of a saline solution into a porous medium filled with fresh water (Bachmat and Elrick, 1970). However, the occurrence of instability during

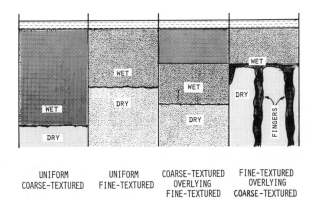

UNIFORM	UNIFORM	COARSE-TEXTURED	FINE-TEXTURED
COARSE-TEXTURED	FINE-TEXTURED	OVERLYING	OVERLYING
		FINE-TEXTURED	COARSE-TEXTURED

Fig. 2.11. Appearance of "fingers" during infiltration into a profile with a fine-textured layer overlying a coarse-textured layer (schematic).

the flow of water in unsaturated soil has only lately come to the attention of soil physicists. A very convincing pictorial demonstration of this effect at the transition from a layer of fine-textured material into an underlying layer of coarser material (of larger pores) was published by Hill and Parlange (1972). They postulated that the formation of fingers in this case is due to an instability at an air–water interface which is gravity driven and is more pronounced as the pore size of the coarser sand increases. They further postulated that the phenomenon is governed by a characteristic length scale L of the coarse-textured medium, which they based on the ratio of the diffusivity and conductivity values, and that fingers can appear only if the width and breadth dimensions of the system are sufficiently larger than L. Thus they attempted to explain why earlier studies of infiltration through crusts, carried out in narrow columns, failed to detect the phenomenon. In the experiments of Hill and Parlange, the fingers moved with a constant velocity, corresponding closely to the ratio of the saturated hydraulic conductivity of the coarse sublayer to its antecedent wetness (K/θ). Moreover, the cross-sectional areas of the fingers were nearly uniform (about 6 cm^2) for the particular material. In these fingers, the infiltrating water moved in saturated columns surrounded by an unsaturated cylinder of partially wetted soil. The overall flow rate through the system was governed by the number of such fingers per unit total area of interface. The intriguing question is thus why any perturbations which might appear in the wetting front are in some cases flattened out by lateral diffusion so that the front remains stable, while in other cases they become self-generating fingers which destabilize the wetting front.

A general explanation was attempted by Raats (1973), who pointed out that profile layering is not a prerequisite for wetting front instability, which may also be due to increased air pressure ahead of the wetting front (e.g., Peck, 1965) or to water repellency of soil layers. Raats based his analysis on the Green and Ampt model (Section E) and offered the hypothesis that stability depends on the change of velocity u of the wetting front with distance L: a small perturbation in an initially planar wetting front will tend to grow if u increases with L and will tend to disappear if u decreases with L. No fundamental reason is offered for this hypothesis, but its consequences are explored for various sets of conditions likely to cause instability, including infiltration of water under ponding with compression of air ahead of the wetting front and infiltration into a soil in which antecedent wetness increases with depth. In the case of a surface crust, Raats' hypothesis leads to the following criterion for the onset of instability:

$$\Omega > (L - h_c)/K_L \qquad (2.56)$$

in which Ω is the hydraulic resistance of the crust (its thickness divided by its hydraulic conductivity), L is the depth to the wetting front, h_c is the pressure head at the wetting front, and K_L is the hydraulic conductivity of the subcrust soil. Accordingly, instability is favored by a high crust resistance, a small value of $-h_c$ (i.e., a low suction, or high antecedent wetness, of the subcrust soil) and a large value of K_L.

An important qualification is in order: The Green and Ampt model used as the basis for Raats' theory is most appropriate for soils of coarse texture with uniform, wide pores. In finer-textured soils with a wide distribution of pore sizes, wetness varies gradually with suction so that wetting fronts are not so sharp and perturbations in the wetting front may be dampened by lateral movement due to suction gradients.

The problem was subsequently reassessed by Philip (1975), who applied the methods of hydrodynamic stability analysis (Lin, 1955). He used the same Green and Ampt (delta-function) model as did Raats, but proposed a different criterion of stability, stating that "the physical attribute of the system which is fundamental to the question of stability is the water pressure gradient behind the wetting front. When this gradient assists the flow, the flow is stable; when this gradient opposes it, the flow is unstable." Perturbations of the wetting front are characterized in terms of a wavelength number, and a second criterion of instability is that this characteristic of a perturbation be greater than a critical value ($= 2\pi\sqrt{2\gamma/\rho g G}$, where γ is the air–water surface tension, ρ the water density, g the gravitational acceleration, and G the pressure gradient at the wetting front). Instability may thus be inhibited in laboratory columns narrower than this value. In examining the implications of his criteria for particular cases, Philip suggested that flow can become unstable (a) if the wetting front reaches a stratum which is so water-repellent that the pressure head needed to penetrate it is positive, (b) if air entrapment and compression ahead of the wetting front cause a sufficient rise in air pressure there, (c) if the air-entry pressure is more negative than the wetting front moisture potential during redistribution, and (d) if hydraulic conductivity increases in depth at a sufficient rate. In all the various instances of instability which Philip identified, it is essential that the flow be gravity assisted. Where gravity plays no part, as in horizontal absorption, or in gravity-opposed flow such as capillary rise into soils, there is no mechanism whereby the pressure gradient can oppose the direction of flow and hence instability cannot arise. Finally, Philip admits to the limitations of an analysis based on the delta-function model, which may not apply to real soils with gradual (rather than sharp) wetting fronts.

J. Rain Infiltration

When rain or sprinkling intensity exceeds soil infiltrability, in principle the infiltration process is similar to the case of shallow ponding. If rain intensity is less than the initial infiltrability value of the soil but greater than the final value, then at first the soil will absorb at less than its potential rate and the flow of water in the soil will occur under unsaturated conditions; however, if the rain is continued at the same intensity, and as soil infiltrability decreases, the soil surface will eventually become saturated and henceforth the process will continue as in the case of ponding infiltration. Finally, if rain intensity is at all times lower than soil infiltrability (i.e., lower than the effective saturated hydraulic conductivity), the soil will continue to absorb the water as fast as it is applied without ever reaching saturation. After a long time, as the suction gradients become negligible, the wetted profile will attain a wetness for which the conductivity is equal to the water supply rate, and the lower this rate, the lower the degree of saturation of the infiltrating profile. This effect is illustrated in Fig. 2.12.

The process of infiltration under rain or sprinkler irrigation was studied by Youngs (1964) and by Rubin and Steinhardt (1963, 1964), Rubin *et al.* (1964), and Rubin (1966). The latter author, who used a numerical solution of the flow equation for conditions pertinent to this problem, recognized three modes of infiltration due to rainfall: (1) *nonponding infiltration*, involving rain not intense enough to produce ponding, (2) *preponding infiltration*, due to rain that can produce ponding but that has not yet done so, and (3) *rainpond infiltration*, characterized by the presence of ponded water. Rainpond infiltration is usually preceded by preponding infiltration, the transition

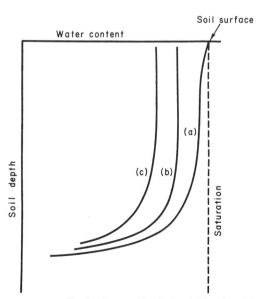

Fig. 2.12. The water-content distribution profile during infiltration (a) under ponding, (b) under sprinkling at relatively high intensity, and (c) under sprinkling at a very low intensity.

between the two being called *incipient ponding.* Thus, nonponding and preponding infiltration rates are dictated by rain intensity, and are therefore *supply controlled* (or *flux controlled*), whereas rainpond infiltration rate is determined by the pressure (or depth) of water above the soil surface as well as by the suction conditions and conductivity relations of the soil. Where the pressure at the surface is small, rainpond infiltration, like ponding infiltration in general, is *profile controlled.*

In the analysis of rainpond or ponding infiltration, the surface boundary condition generally assumed is that of a constant pressure at the surface, whereas in the analysis of nonponding and preponding infiltration, the water flux through the surface is considered to be equal either to the rainfall rate or to the soil's infiltrability, whichever is the lesser. In actual field conditions, rain intensity might increase and decrease alternately, at times exceeding the soil's saturated conductivity (and its infiltrability) and at other times dropping below it. However, since periods of decreasing rain intensity involve complicated hysteresis phenomena (Hillel, 1980), the analysis of variable-intensity rainstorms is rather difficult.

Rubin's analysis was based on the assumption of no hysteresis. The falling raindrops were taken to be so small and numerous that rain could be treated as a continuous body of "thin" water reaching the soil surface at a specifiable rate. Soil air was regarded as a continuous phase, at atmospheric pressure. The soil was assumed to be uniform and stable (i.e., no fabric changes such as swelling or surface crusting).

We shall briefly review the consequences of this analysis in qualitative terms. If a constant pressure head is maintained at the soil surface (as in rainpond infiltration), then the flux of water into this surface must be constantly decreasing with time. If a constant flux is maintained at the soil surface, then the pressure head at this surface must be constantly increasing with time. Infiltration of constant-intensity rain can result in ponding only if the *relative rain intensity* (i.e., the ratio of rain intensity to the saturated hydraulic conductivity of the soil) exceeds unity. During nonponding infiltration under a constant rain intensity q_r, the surface suction will tend to a limiting value ψ_{lim} such that $K(\psi_{\text{lim}}) = q_r$.

Under rainpond infiltration, the wetted profile consists of two parts: an upper, water-saturated part; and a lower, unsaturated part. The depth of the saturated zone continuously increases with time. Simultaneously, the steepness of the moisture gradient at the lower boundary of the saturated zone (i.e., at the wetting zone and the wetting front) is continuously decreasing (these phenomena accord with those of infiltration processes under ponding, as described in the previous sections of this chapter). The higher the rain intensity is, the shallower is the saturated layer at incipient ponding and the steeper is the moisture gradient in the wetting zone.

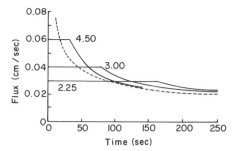

Fig. 2.13. Relation between surface flux and time during infiltration into Rehovot sand due to rainfall (solid lines) and flooding (dashed line). The numbers labeling the curves indicate the magnitude of the relative rain intensity. (After Rubin, 1966.)

Figure 2.13 describes infiltration rates into a sandy soil during preponding and rainpond infiltration under three rain intensities. The horizontal parts of the curves correspond to preponding infiltration, and the descending parts to rainpond infiltration periods. As pointed out by Rubin (1966), the rainpond infiltration curves are of the same general shape and approach the same limiting infiltration rate, but they do not constitute horizontally displaced parts of a single curve and do not coincide with the infiltration rate under flooding, which is shown as a broken line in the same graph.

A rainstorm of any considerable duration typically consists of spurts of high-intensity rain punctuated by periods of low-intensity rain. During such respite periods, surface soil moisture tends to decrease because of internal drainage, thus reestablishing a somewhat higher infiltrability. The next spurt of rainfall is therefore absorbed more readily at first, but soil infiltrability quickly falls back to, or even below, the value it had at the end of the last spurt of rain. The recovery of infiltrability under an intermittent (as opposed to a continuous) supply of water was demonstrated by Busscher (1979). A complete description would, of course, necessitate taking account of the hysteresis phenomenon in the alternately wetting-and-draining surface zone.

Additional theoretical studies of rain infiltration have been carried out by Braester (1973) and by Bruce and Whisler (1972). The latter developed a computer-based model to account for the infiltration of variable-intensity rainstorms into heterogeneous soil profiles. The same authors also carried out extensive experiments on rain infiltration into laboratory columns and in the field. While the results appeared to corroborate theoretical predictions reasonably well for conditions of an undisturbed sod cover, problems were encountered with bare soils subject to crusting from raindrop impact. Further studies have been carried out by Mein and Larson (1973). A combined theoretical and experimental study of rain infiltration was reported by Swartzendruber and Hillel (1975), who used a variable-intensity sprinkling infiltrometer (Rawitz *et al.*, 1972) to test a set of equations designed to predict the time of appearance and the quantity of *surface water excess* (i.e., the excess of rainfall over infiltration) for different rain intensities. Still more comprehensive theoretical studies of rain infiltration and subsequent soil moisture storage for different soil textures and variously composed soil profiles were carried out by Hillel and van Bavel (1976) and by Hillel and Talpaz (1977).

Although considerable progress has been achieved in recent years, the process of rain infiltration has not yet been studied in sufficient detail in the field to establish the applicability of existing theories. Complications due to the discreteness of raindrops (which causes alternate saturation and redistribution at the surface), as well as to the highly variable nature of rainstorm intensities and raindrop energies and the unstable nature of many (perhaps even most) soils, can cause anomalies which idealized theories are unable to take into account. Additional complications can arise in cases of air occlusion and when the soil exhibits profile or areal heterogeneity.

K. Some Topics of Current Research on Infiltration

Thus far in this chapter, we have endeavored to develop the basis for a systematic understanding of infiltration processes and theories. We began with the simplest case of horizontal (gravity-free) infiltration into homogeneous columns and continued with vertical infiltration into both homogeneous and heterogeneous profiles. In taking this approach we have followed the general trend in infiltration research, which is to direct attention to more and more complex sets of conditions likely to be encountered in the field and to attempt to formulate appropriate theories which might be more realistic than the simplistic and highly idealized formulations of yesteryear.

One topic of current interest is that of infiltration against a compressing body of air (as in the presence of a high water table) and the resulting counterflow of air. An early study of this problem was made by Free and Palmer (1940). A theoretical approach was developed by LeFur (1962) and more recently by Peck (1965), Adrian and Franzini (1966), Brusthern and Morel-Seytoux (1970) and Morel-Seytoux and Noblanc (1972), who formulated a set of equations to account for the concurrent motions of water and air, which they described as the *two-phase flow of immiscible fluids* in porous media. In these equations, the fluxes, pressures, densities, and permeabilities pertaining to the two fluids are considered simultaneously. The development of theory has been accompanied by experimental studies, first in the laboratory (Peck, 1965) and more lately in the field as well (Dixon and Linden, 1972).

Quite another topic of increasing contemporary interest is that of infiltration into *swelling soils* (also called *expansive soils*). Following the experimental work by Smiles and Rosenthal (1968), Philip (1969a, 1972) developed a comprehensive theory of hydrostatics and hydrodynamics in swelling soils. Some aspects of his theory were disputed by Youngs and Towner (1970) and later reexamined critically and reformulated by Miller (1975). Philip postulated that the total potential of water in a swelling soil (excluding solute effects) includes a third component (in addition to the two components of hydraulic potential in a rigid soil), namely, the *overburden potential*. It arises because of the forces resisting increase in bulk volume following increase in the soil's *moisture ratio* (i.e., the volume of water per unit volume of solid matter). Once its pores are filled with water, a layer of rigid soil can absorb no more but merely transmits any additional water supplied to it, whereas a layer of swelling soil continues to imbibe water for some time after initial saturation, as it expands against whatever constraining pressure it encounters until the swelling and constraining pressures are in balance.

To apply Darcy's law to the motion of water relative to the moving soil matrix (rather than to fixed space), the equation had to be recast in terms of a *material coordinate system* anchored to the matrix. An interesting prediction of the theory, later verified experimentally (Smiles, 1974), is that during infiltration into a swelling soil the relative importance of gravity is reduced, and the absorption phase is prolonged, in comparison with infiltration into a rigid soil. Thus, cumulative infiltration into a swelling soil can remain proportional to the square root of time for many hours, even days. During "ordinary" episodes of rainfall or irrigation, therefore, infiltration may be regarded in effect as an absorption process. Experimental results also suggest that hysteresis is not nearly so significant in swelling soils as it is in coarse granular soils (Smiles, 1976).

A very challenging problem related to infiltration into swelling soils is how to account for the role of cracks. As pointed out by Miller (1975), when a field of swelling soil is allowed to dry on top, in due time a system of regularly spaced surface cracks tends to appear. As evaporation continues, these cracks gradually deepen so that the regions of soil between the cracks really consist of separate, free-standing vertical columns. Secondary cracks may eventually appear on the tops and sides of these columns. The depth of crack penetration can vary from several centimeters to perhaps a meter (and in extreme cases even more), depending on the nature of the clay and the climate (i.e., the duration of the evaporation period). Infiltration into such cracked soils, in its initial stages at least, is obviously very different from the orderly one-dimensional process described by most existing theories. Until the soil reswells and the columns coalesce to close the cracks, much of the applied water (especially if it is ponded over the surface, as, for example, in flood irrigation) bypasses the surface zone as it runs directly into the cracks, whence it is absorbed three dimensionally into the deeper strata. This geometrically complex process has thus far defied theoretical analysis. If not easily amenable to an analytical approach, however, this problem too, like so many other formerly intractable ones, may yield eventually to the brute power of the computer.

There yet remains the very real problem of the areal (i.e., horizontal) heterogeneity of both rainfall and soil in the field. A simple application of one-dimensional (vertical) infiltration theory cannot by itself provide a complete understanding of phenomena on a field scale. The occurrence and extent of spatial variability in soil-water properties in ordinary fields is now well documented (e.g., Nielsen *et al.*, 1973) and its possible consequences can no longer be ignored. For example, a region of low infiltrability would yield surface-water excess and runoff much sooner than a region of high infiltrability. If the latter region were so located as to receive the runoff from the former region, infiltration of water might still be complete for the two regions considered together as a unit. If, on the other hand, the more infiltrable region were located upslope from the less infiltrable region, the field as a unit would be producing surface-water excess and runoff just as soon as this occurred for the less infiltrable region. A systematic approach toward such composite fields was recently attempted by Hillel and Hornberger (1979).

The infiltration of solute-laden water is another important topic on which research has begun to focus only recently (Elrick *et al.*, 1979).

L. Surface Runoff

Whenever the rate of water supply to the soil surface (whether it be by rain, irrigation, or snow melt) exceeds the rate of infiltration, free water, called *surface water excess*, tends to accumulate over the soil surface. This water collects in depressions, often called *pockets*, thus forming puddles, the total volume of which, per unit area, is called the *surface storage capacity*. It depends upon the geometric irregularities of the surface as well as upon the overall slope of the land (Fig. 2.14). Only when surface storage is filled and the puddles begin to overflow can actual runoff begin. The term *surface runoff* thus represents the portion of the water supply to the surface which neither is absorbed by the soil nor accumulates on the surface, but which runs downslope and eventually collects in channels variously called *rills* or *gullies*. These channels generally form a treelike pattern of connecting or converging branches with numerous confluences leading to larger and larger streams.

Fig. 2.14. Effect of surface roughness and slope on pocket storage of rainfall excess. The water thus accumulated eventually infiltrates after cessation of the rain.

As pointed out by Eagleson (1970), there exists a whole spectrum of channel geometries and flow types. On the one extreme is the thin sheetlike runoff called *overland flow*. It is likely to be the primary type in surface runoff from small natural areas or fields having little topographic relief. The next distinctive type is found in the smallest stream channels, which gather the overland flow in a continuous fashion along their length to form the lowest order of *stream flow*. As these smallest streams merge with one another, they form streams of "higher order" which have concentrated tributaries as well as continuous lateral inflows.

Overland flow is considered as two-dimensional laminar flow in the initial condition, becoming turbulent when the depth and velocity of the flow increase to a critical value, at which turbulence entering the laminar flow domain will not dampen out. Vennard (1961) gives a Reynolds number of 500 as a criterion for the onset of this condition. (Reynolds number is a dimensionless index expressing the product of the mean velocity of the fluid and its effective hydraulic radius divided by its kinematic viscosity).

To obtain a simplified formulation of overland flow, we begin with the basic equation governing the rate of flow (discharge) in a channel:

$$Q = VA \qquad (2.57)$$

where Q is the flow rate, V the average velocity, and A the cross-sectional area of flow.

Conservation of matter (continuity) requires that, in the absence of rainfall and infiltration, the change in water level h with time be equal to the negative of the change of flow rate with distance, i.e.,

$$\partial h/\partial t = -\partial Q/\partial x \qquad (2.58)$$

If rain (r) and infiltration (i) rates are taken into account, we have

$$\partial h/\partial t + \partial Q/\partial x = r - i \qquad (2.59)$$

Mean velocity and cross-sectional area of flow depend on shape and size of channel. Since flow is inhibited by shear stresses resulting from the immobility of the water in immediate contact with the sides and bottom of the channel, the hydraulic resistance of a channel is generally proportional to

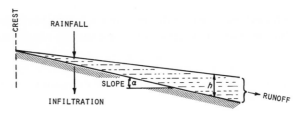

Fig. 2.15. Surface-water excess and "sheet" overland flow (schematic).

the area of the perimeter of contact per unit volume of flowing liquid (or to the length of the perimeter of contact per units cross-sectional area of flow). This contact area can be characterized in terms of the *hydraulic radius*, or *hydraulic mean depth*, of a channel, defined as the cross-sectional area divided by the wetted perimeter.

Now consider an infinitely wide channel without sides. Such a "wide-open channel" is represented by a laterally uniform sloping surface with a layer of water flowing over it, as illustrated in Figure 2.15. Here the effective hydraulic radius is h. As an approximation, let us assume that the two forces acting on the flowing water are (1) the component of the water's weight acting in the direction of the bed slope, and (2) the shear stresses developed at the solid-to-water boundary. If the velocity is more or less uniform along the slope, then the two forces must be approximately balanced, and we get

$$\tau_0 = \rho g h \alpha \qquad (2.60)$$

where ρ is the liquid's density, g the acceleration of gravity, h the water depth, α the slope, and τ_0 the bottom shear stress.[8]

Shear stress τ_0 is known to be related to the average velocity squared:

$$\tau_0 = a\rho V^2 \qquad (2.61)$$

where a is a proportionality factor. Combining the last two equations gives

$$V = (g/a)^{1/2}h^{1/2}\alpha^{1/2} = Ch^{1/2}\alpha^{1/2} \qquad (2.62)$$

where C is the *Chezy coefficient*, named after the French engineer who first discovered this relationship some 200 years ago. A large amount of empirical

---

[8] A more rigorous approach is to start from the momentum equation:

$$\frac{\partial V}{\partial t} + \frac{\partial V}{\partial x} + g\frac{\partial h}{\partial x} = (r - f)\frac{V}{h} - \frac{\tau_0}{\rho h} + g\alpha$$

Here the underlying assumptions are that the bottom slope is small, the velocity distribution uniform, and the overpressure from rainfall negligible. If we also assume that rainfall and infiltration have little effect on flow dynamics and that the slope of the free surface ($g\,\partial h/\partial x$) and the inertial terms ($\partial V/\partial t + V\,\partial V/\partial x$) are small compared to friction ($\tau_0/\rho h$) and bottom slope ($g\alpha$), we arrive at $\tau_0 = \rho g h \alpha$.

work conducted since has indicated that Chezy's coefficient C is not a constant but depends on water depth h and surface roughness n as follows:

$$C = h^{1/6}/n \qquad (2.63)$$

Combining this with Eq. (2.60) gives the useful equation generally attributed to Robert Manning (1891):

$$V = h^{2/3}\alpha^{1/2}/n \qquad (2.64)$$

Multiplying by h generally converts the left side of this equation to discharge per unit width:

$$Q = h^{5/3}\alpha^{1/2}/n \approx h^2\alpha^{1/2}/n \qquad (2.65)$$

In practice, a power of 2 for h rather than $\frac{5}{3}$ has been found to give satisfactory results for overland flow. Measured valued of the roughness coefficient have been tabulated (e.g., Sellin, 1969) and range between 0.01 for extremely smooth channels to 0.1 for flood plains with a growth of heavy timber. A value of about 0.03 seems to be reasonable for soil surfaces (either bare or with a cover of short grass).

The last version of Manning's equation can be combined with Eq. (2.59) to yield the *kinematic wave equation*:

$$\partial h/\partial t + \beta h^{m-1} \partial h/\partial x = r - i \qquad (2.66)$$

where $\beta = \alpha^{1/2}/n$, and $m = 2$ for Eq. (2.65).

The foregoing set of equations was used by Hillel and Hornberger (1979) to model overland flow from texturally heterogeneous fields.

In agricultural fields, runoff is generally undesirable, since it results in loss of water and often causes erosion, the amount of which increases with increasing rate and velocity of runoff. The way to prevent erosion is to protect the soil surface against raindrop splash (e.g., by mulching), to increase soil infiltrability and surface storage, and to obstruct overland flow so as to prevent it from gathering velocity. Maintenance and stabilization of soil aggregation will minimize slaking and detachment of soil particles by raindrops and running water. A crusted or compacted soil generally has a low infiltration rate and therefore will produce a high rate of runoff. Proper tillage, especially on the contour, can increase infiltration and surface storage capacity, thus reducing runoff (Burwell and Larson, 1969).

M. Runoff Inducement

Uncontrolled runoff is never desirable, and since its control requires additional investment in most agricultural situations it is indeed preferable that no runoff should occur from the land. In areas where natural rainfall is on the borderline of being insufficient for dryland farming, it is particularly important that as much of the precipitation as possible infiltrate into the soil. However, in many arid regions large tracts of land remain unused owing to insufficient or unstable rainfall, unarable soils (too shallow, stony, or saline), or unsuitable topography. Rain falling on such land is almost invariably lost. Often rainfall amounts per storm are insufficient to either recharge the groundwater or to support an economically valuable crop. This water generally infiltrates to a shallow depth only and is quickly lost either by direct evaporation or by transpiration of ephemeral vegetation, usually weeds. Some of this water may also appear as runoff, causing destructive flash floods typical of the desert, although natural runoff seldom exceeds about 10% of annual precipitation (Myers, 1963). If total rainfall amount is appreciable but mostly occurring in such small amounts as to make control and collection of the runoff water an uneconomic proposition, both the water that infiltrates and that appearing as runoff is lost. However, when it is feasible, the possibility of controlling and increasing the amount of surface runoff obtainable from such lands can be of great importance, particularly where water is scarce and the runoff thus obtained can augment the water supply of the region.

The collection and utilization of runoff have been practiced by desert peoples since antiquity. In large areas of the Middle East it is still possible today to find functional ancient systems of terraced stream beds which were used for farming, whose water supply was often augmented by runoff water collected on the adjoining hillsides and brought to the fields by means of sophisticated conveyance systems (Evenari and Koller, 1956; Evenari et al., 1958; Lowdermilk, 1954). There is evidence that the builders of these systems understood the inadequacy of natural runoff and therefore perfected techniques for inducing soil crusting and increased runoff. Noticing apparently that the soil had a natural tendency to crust, which was mitigated, however, by a protective cover of stones (*desert pavement*), the ancient inhabitants of the Negev region in Israel collected these stones from the soil surface into geometrically arranged mounds, thereby exposing the intervening soil surface to the impact of raindrops, which caused the formation of a dense crust (Tadmor et al., 1957). This method of decreasing soil infiltrability was both laborious and of temporary effectiveness since erosion of the unstable soil crust eventually exposed additional stones. As a consequence, the ancient runoff farmers required a water-contributing area 20 times or more larger than the area to which water was brought for utilization. Similar techniques were used to collect runoff water in cisterns for subsequent use by man and domestic animals.

Modern technology holds the promise of more efficient runoff inducement than was possible in ancient times. Runoff can now be increased severalfold by means of mechanical treatment (stone clearing, smoothing, and compaction) as well as by a variety of chemical treatments to seal and stabilize the surface. The soil surface can be made water repellent or impermeable either by the application of mechanical barriers to water movement, such as plastic, aluminum foil, concrete, or sheet metal, or by the artificial formation of an impervious soil crust with the aid of various materials such as sodic salts and petroleum distillation byproducts (Myers, 1961, 1963; Hillel, 1967).

Where rainfall is insufficient but the soil is otherwise arable, it is possible to utilize the land in a modern system of runoff farming, where part of the land surface is shaped and treated appropriately for runoff inducement while another portion of the land receives the runoff water so produced (Fig. 2.16). The irrigation may be carried out either immediately with runoff water flowing directly off the contributing plots to the receiving plots or water can be stored for future use in reservoirs. Although the former method is less expensive, it is also less flexible since the irrigation is applied immediately following precipitation, regardless of whether the root zone storage has been depleted sufficiently to effectively store all the water applied.

Several systems have been tried in respect to size and arrangement of the contributing area in relation to the water-receiving area. A small watershed can be treated in its entirety so as to provide the maximal amount of water at the outflow of the basin for conveyance to irrigated fields. On a still smaller scale, strips of land can be treated on a slope so as to contribute their share of rainfall as runoff to adjacent lower lying "runon strips" in which crops can be grown (Hillel, 1967). Microwatersheds are the third possible approach, wherein each single contributing area serves a single tree or row of plants (Shanan *et al.*, 1970; Aase and Kemper, 1968; Fairbourn and Kemper, 1971).

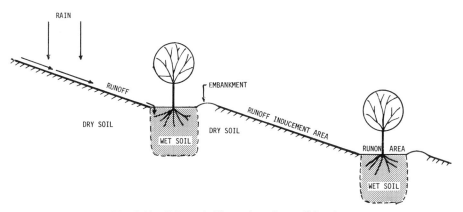

Fig. 2.16. Schematic illustration of run-off farming.

Experience with these modern runoff farming systems is still too limited to allow assessment of their relative effectiveness in various conditions of topography, climate, soil, and crop.

Sample Problems

1. The infiltration rate under shallow ponding was monitored as a function of cumulative rainfall and found to be 20 mm/hr when a total of 100 mm had infiltrated. If the eventual steady rate of infiltration is 5 mm/hr, estimate the infiltration rate at a cumulative infiltration of 200 and 400 cm. Use the Green and Ampt theory.

We refer to Eq. (2.1):

$$i = i_c + b/I$$

where i is the infiltration rate, i_c the eventual steady rate, I the cumulative infiltration, and b a constant. Substituting the values given above, and solving for b,

$$20 \text{ mm/hr} = 5 \text{ mm/hr} + (b \text{ mm}^2/\text{hr})/100 \text{ mm},$$
$$b = 100 \,(20 - 5) = 1500 \text{ mm}^2/\text{hr}$$

We can now use the calculated value of b to estimate the infiltration rate at any value of cumulative infiltration.

At I = 200 mm

$$i = 5 + 1500/200 = 12.5 \text{ mm/hr}$$

At I = 400 mm

$$i = 5 + 1500/400 = 8.75 \text{ mm/hr}$$

2. Given the soil conditions described in the preceding problem, how much water can be delivered to the root zone of a crop without exceeding the soil's infiltrability if the sprinkling irrigation rate is 15 or 25 mm/hr? What is the highest steady sprinkling rate we can use if we wish to provide an irrigation of 250 mm in the shortest possible time?

At a sprinkling rate of 15 mm/hr,

$$15 \text{ min/hr} = 5 \text{ mm/hr} + (1500 \text{ mm}^2/\text{hr})/I \text{ mm},$$
$$I = 1500/(15 - 5) = 150 \text{ mm}$$

At a sprinkling rate of 25 mm/hr,

$$I = 1500/(25 - 5) = 75 \text{ mm}$$

Note that the higher the sprinkling rate, the smaller the total amount of water which can infiltrate without exceeding the soil's infiltrability (and thus causing flooding or runoff).

To infiltrate 250 mm by steady sprinkling without exceeding infiltrability, we can apply

$$i = 5 + \frac{1500}{250} = 11 \text{ mm/hr}$$

3. A horizontal infiltration trial was conducted with a soil-filled tube having a cross-sectional area of 50 cm². After 15 min, cumulative infiltration totaled 150 cm³. What is the expectable cumulative infiltration and infiltration rate at the end of 1, 4, and 16 hr?

We use Philip's horizontal infiltration ("absorption") theory to calculate the sorptivity s according to Eq. (2.26):

$$s = I/t^{1/2}$$

where t is time and I is cumulative infiltration.

$$s = \frac{150 \text{ cm}^3/50 \text{ cm}^2}{(15 \text{ min} \times 60 \text{ sec/min})^{1/2}} = \frac{3.0 \text{ cm}}{30 \text{ sec}^{1/2}} = 0.1 \text{ cm/sec}^{1/2}$$

We can now apply this value of s to estimate the cumulative infiltration at various times, again using Eq. (2.26):

$$I = st^{1/2}$$

At the end of 1 hr

$$I = 0.1 \text{ cm/sec}^{1/2} \times (3600 \text{ sec})^{1/2} = 6.0 \text{ cm}$$

At the end of 4 hr

$$I = 0.1 \text{ cm/sec}^{1/2} \times (14400 \text{ sec})^{1/2} = 12.0 \text{ cm}$$

At the end of 16 hr:

$$I = 0.1 \text{ cm/sec}^{1/2} \times (57600 \text{ sec})^{1/2} = 24.0 \text{ cm}$$

The corresponding infiltration rates are obtainable by differentiating the above equation with respect to time t:

$$i = dI/dt = s/2t^{1/2}$$

At 1 hr

$$i = (0.1 \text{ cm/sec}^{1/2})/2(3600 \text{ sec})^{1/2} = 8.33 \times 10^{-4} \text{ cm/sec} \cong 30 \text{ mm/hr}$$

At 4 hr

$$i = (0.1 \text{ cm/sec}^{1/2})/2(14400 \text{ sec})^{1\,2} = 4.166 \times 10^{-4} \text{ cm/sec} \cong 15 \text{ mm/hr}$$

At 16 hr

$$i = (0.1 \text{ cm/sec}^{1/2})/2(57600 \text{ sec})^{1/2} = 2.0833 \times 10^{-4} \text{ cm/sec} \cong 7.5 \text{ mm/hr}$$

4. If the effective saturated hydraulic conductivity of the soil in the preceding problem is 2×10^{-4} cm/sec, estimate the cumulative infiltration and infiltration rate values for a vertical column at 1, 4, and 16 hr. If the initial wetness is (0.05 cm³ H₂O/(cm³ soil) and the saturation wetness is 0.45 cm³/cm³, estimate the depth of the wetting front at the same times.

To calculate cumulative infiltration into a vertical column, we use Eq. (2.44):

$$I = st^{1/2} + Kt$$

At 1 hr (3600 sec)

$$\begin{aligned} I &= 0.1 \text{ cm/sec}^{1/2} \times 60 \text{ sec}^{1/2} + 2 \times 10^{-4} \text{ cm/sec} \times 3600 \text{ sec} \\ &= 6.0 \text{ cm} + 0.72 \text{ cm} = 6.72 \text{ cm} \end{aligned}$$

At 4 hr (14400 sec)

$$\begin{aligned} I &= 0.1 \text{ cm/sec}^{1/2} \times 120 \text{ sec}^{1/2} + 2 \times 10^{-4} \text{ cm/sec} \times 14400 \text{ sec} \\ &= 12 \text{ cm} + 2.88 \text{ cm} = 14.88 \text{ cm} \end{aligned}$$

At 16 hr (57600 sec)

$$\begin{aligned} I &= 0.1 \text{ cm/sec}^{1/2} \times 240 \text{ sec}^{1/2} + 2 \times 10^{-4} \text{ cm/sec} \times 57600 \text{ sec} \\ &= 24 \text{ cm} + 11.52 \text{ cm} = 35.52 \text{ cm} \end{aligned}$$

To calculate the infiltration rate into a vertical column, we differentiate Eq. (2.44) with respect to time:

$$i = dI/dt = s/2t^{1/2} + K$$

At 1 hr

$$\begin{aligned} i &= (0.1 \text{ cm/sec}^{1/2})/(2 \times 60 \text{ sec}^{1/2}) + 2 \times 10^{-4} \text{ cm/sec} \\ &= 8.33 \times 10^{-4} \text{ cm/sec} + 2 \times 10^{-4} \text{ cm/sec} = 10.33 \times 10^{-4} \text{ cm/sec} \\ &\cong 37.19 \text{ mm/hr} \end{aligned}$$

At 4 hr

$$\begin{aligned} i &= (0.1 \text{ cm/sec}^{1/2}(/(2 \times 120 \text{ sec}^{1/2}) + 2 \times 10^{-4} \text{ cm/sec} \\ &= 4.166 \times 10^{-4} \text{ cm/sec} + 2 \times 10^{-4} \text{ cm/sec} = 6.166 \times 10^{-4} \text{ cm/sec} \\ &\cong 22.22 \text{ mm/hr} \end{aligned}$$

At 16 hr

$$\begin{aligned} i &= (0.1 \text{ cm/sec}^{1/2})/(2 \times 240 \text{ sec}^{1/2}) + 2 \times 10^{-4} \text{ cm/sec} \\ &= 2.083 \times 10^{-4} \text{ cm/sec} + 2 \times 10^{-4} \text{ cm/sec} = 4.083 \times 10^{-4} \text{ cm/sec} \\ &\cong 14.7 \text{ mm/hr} \end{aligned}$$

To calculate the depth of the wetting front L_f we use a variant of Eq. (2.9):

$$L_f = I/\Delta\theta$$

At 1 hr

$$L_f = (6.72 \text{ cm})/(0.45 - 0.05) = 16.8 \text{ cm}$$

At 4 hr

$$L_f = (14.88 \text{ cm}/0.4 = 37.2 \text{ cm}$$

At 16 hr

$$L_f = (35.52 \text{ cm})/0.4 = 88.8 \text{ cm}$$

Note: The student is invited, for his own edification, to plot the calculated values of I, i, and L_f against time and against the square root of time. The relationships involved will become much clearer in the process. A comparison between the corresponding data of Problems 3 and 4 illustrates the difference between horizontal and vertical infiltration.

Kepler may have got the idea of a universal gravity from the thought that all things in nature must attract one another because they are infused with a share of God's universal love.

Jacob Bronowski[1]

3   Internal Drainage and Redistribution Following Infiltration

A. Introduction

When rain or irrigation ceases and surface storage is depleted by evaporation or infiltration, the infiltration process comes to an end, since no more water enters into the soil. Downward water movement within the soil, however, does not cease immediately and may in fact persist for a long time as soil water redistributes within the profile. The soil layer wetted to near saturation during infiltration does not retain its full water content, as part of its water moves down into the lower layers under the influence of gravity and possibly also of suction gradients. In the presence of a high water table, or if the profile considered is initially saturated throughout, this postinfiltration movement is herein termed *internal drainage*. In the absence of groundwater, or where the water table is too deep to affect the relevant depth zone, and if the profile is not initially wetted to saturation throughout its depth, this movement is called *redistribution*, since its effect is to redistribute soil water by increasing the wetness of successively deeper layers at the expense of the infiltration-wetted layers of the soil profile.

In some cases, the rate of redistribution decreases rapidly, becoming imperceptible after several days so that thereafter the initially wetted part of the soil appears to retain its moisture, unless this moisture is evaporated or is taken up by plants. In other cases, redistribution may continue at an appreciable, though diminishing, rate for many days or weeks.

[1] In Bronowski (1977).

The importance of the redistribution process should be self-evident, as it determines the amount of water retained at various times by the different depth zones in the soil profile and hence can affect the water economy of plants. The rate and duration of downward flow during redistribution determine the effective *soil water storage*, a property that is vitally important, particularly in relatively dry regions, where water supply is infrequent and plants must rely for long periods on the unreplenished reservoir of water within the rooting zone. Even in relatively humid regions, where the water supply by precipitation would seem to be sufficient, inadequate soil moisture storage can deprive the crops to be grown of a major portion of the water supply and can cause crop failure. Finally, the redistribution process is important because it often determines how much water will flow through the root zone (rather than be detained, or temporarily stored, within it) and hence how much leaching of solutes will take place.

As we shall explain more fully in the sections to follow, soil-water storage is generally not a fixed quantity or a static property but a dynamic phenomenon, determined by the time-variable rates of soil-water inflow to, and outflow from, the relevant soil volume.

B. Internal Drainage in Thoroughly Wetted Profiles

We have already made a distinction between the postinfiltration movement of soil water in cases where a groundwater table is present fairly close to the soil surface (i.e., at a depth not exceeding a few meters) and in cases where groundwater is either nonexistent or too deep to affect the state and movement of soil moisture in the root zone.

At the groundwater table, also called the *phreatic surface*, soil water is at atmospheric pressure. Beneath this level the hydrostatic pressure exceeds atmospheric pressure, while above this level soil water is under suction. Internal drainage in the presence of a water table tends to a state of equilibrium, in which the suction at each point corresponds to its height above the free water level. The downward drainage flux decreases as the hydraulic gradient decreases, both approaching zero in time (provided, of course, there is no further addition of water by another episode of infiltration, or abstraction of water by evapotranspiration—processes which would prevent the attainment of static equilibrium). At equilibrium, if attainable, soil wetness would increase in depth to a value of saturation just above the water table, a depth distribution which would mirror the soil moisture characteristic curve. Phenomena involving falling water tables and groundwater drainage will be treated in our next chapter.

In the present section, we wish to concentrate upon the internal drainage of profiles initially wetted to near saturation throughout their depth. We shall treat this process as if it were unaffected by any water table, which, if it exists, is assumed to lie too far below the zone of interest to be of any direct consequence. Internal drainage beyond the root zone has also been referred to as *deep percolation*.

In the hypothetical case of a very deeply wetted profile, one might justifiably assume that internal drainage, in the absence of any suction gradients, occurs under the influence of gravity alone. If so, the downward flux through any arbitrary plane (say, the bottom of the root zone) should be equal to the hydraulic conductivity and should diminish in time as the conductivity diminishes due to reduction of water content in the infiltration-wetted but since-unreplenished soil above the plane considered. This can be stated formally as follows

Recall the one-dimensional (vertical) form of Darcy's equation:

$$q = -K(\theta)\frac{\partial}{\partial z}(-\psi - z) \tag{3.1}$$

where q is flux, $K(\theta)$ hydraulic conductivity (a function of wetness θ), z depth, and ψ matric suction. With $\partial\psi/\partial z$ assumed to be zero, we have simply

$$q = K(\theta) \tag{3.2}$$

If we happen to know the functional dependence of K on θ, we can formulate q as an explicit function of θ. For instance, if the function is exponential, say,

$$K(\theta) = ae^{b\theta} \tag{3.3}$$

where a and b are constants, we get

$$q = ae^{b\theta} \quad \text{or} \quad \ln q = A + b\theta \tag{3.4}$$

where $A = \ln a$. Thus, even a small decrease in θ can result in a steep (logarithmic) decrease of the flux q.

A simple approach to the internal drainage process is possible if the soil profile can be assumed to drain uniformly, i.e., if the soil is equally wet and its wetness diminishes at an equal rate throughout the draining profile. Observations have shown this to be a reasonable approximation in many cases. In the absence of any flow through the upper soil surface, the flux q_b through any plane at depth z_b must equal the rate of decrease of total water content W, where $W = \theta z_b$ (the product of the wetness θ and depth z_b):

$$q_b = K(\theta) = -dW/dt = -z_b\, d\theta/dt \tag{3.5}$$

As shown in Chapter 9 of Hillel (1980), this equation can be used to measure the hydraulic conductivity as a function of wetness in a covered plot in the field. Note that the downward flux increases in proportion to depth. Note also that it diminishes in time, as does the rate of decrease of soil wetness, in accordance with the functional decrease of conductivity with the remaining soil wetness. Once the characteristic $K(\theta)$ function is established for a given soil, it becomes possible to predict the time dependence of θ and of drainage rate q for any depth.

Another approximation is to assume that the downward drainage flux is proportional to the total amount of water remaining in the draining profile. Thus

$$-dW/dt = \lambda W \qquad (3.6)$$

where, as before, $-dW/dt$ is the time rate of decrease of profile water content (storage) W and λ is a proportionality constant. Equation (3.6) can be integrated by separation of variables:

$$\int \frac{dW}{W} = -\lambda \int dt \qquad (3.7)$$

to give

$$\ln W = -\lambda t + c \qquad (3.8)$$

wherein c is the integration constant. The last equation can be rewritten as

$$W = e^c/e^{\lambda t} \qquad (3.9)$$

At $t = 0$ (the beginning of the process), $W = e^c$, a constant which we can designate W_i (initial water content of the profile). Thus

$$W = W_i e^{-\lambda t} \qquad \text{or} \qquad W/W_i = e^{-\lambda t} \qquad (3.10)$$

which is the well-known equation used to describe decay processes, such as radioactive decay. The proportionality constant λ is also called the *decay constant*, and it can be interpreted as the fraction of the remaining soil water content which drains per unit time. To characterize the stability of soil moisture storage, use can be made of the concept of *half life* $t_{1/2}$, which is the time required for the cumulative drainage of half the initial amount of water present in the soil (i.e. the value of t in the last equation at which $W = W_i/2$). Thus, at $t = t_{1/2}$, $e^{\lambda t} = 2$. Hence,

$$t_{1/2} = \ln 2/\lambda = 0.693/\lambda \qquad (3.11)$$

If Eq. (3.6) is not by itself a good description of the internal drainage process, as W tends rather unrealistically to approach zero in time, the equation can perhaps be made more realistic by providing for soil water content to approach asymptotically some finite residual value W_r rather than

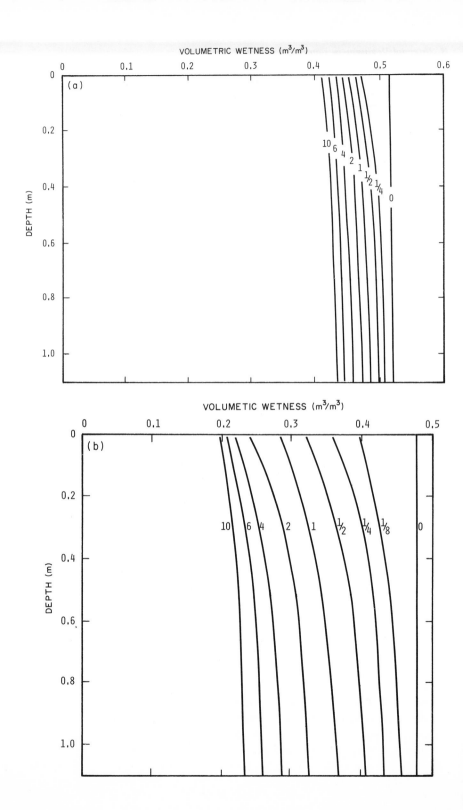

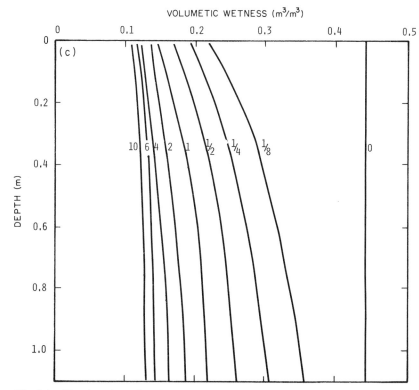

Fig. 3.1. The changing soil moisture distribution during drainage from initially-saturated uniform profiles of (a) clay, (b) loam, and (c) sand. The numbers indicate duration of the process (days). (After Hillel and van Bavel, 1976.)

zero. As an example, let us assume that W_r corresponds to the product of the soil depth considered, z_b, and the equilibrium value of soil wetness at a matric suction value of, say, $\frac{1}{3}$ bar. The value of W_r obviously depends on soil texture, as do the values of W_i, λ, and indeed $t_{1/2}$.

For a more fundamental approach to the problem of predicting the dependence of flux and wetness on time, we must go to the general flow equation, which can be stated as

$$\frac{\partial \theta}{\partial t} = -\frac{\partial}{\partial z}\left[K(\theta)\,\frac{\partial \psi}{\partial z} \right] - \frac{\partial K(\theta)}{\partial z} \tag{3.12}$$

This equation can be solved subject to the following conditions:

$$
\begin{array}{llll}
t = 0, & z \geq 0, & \theta = \theta_s & \text{(initially saturated profile)} \\
t > 0, & z = 0, & q = 0 & \text{(no further infiltration or} \\
& & & \text{evaporation)} \\
t > 0, & z = z_b, & q = K(\theta_b) & \text{(flux equal to conductivity at} \\
& & & \text{bottom of profile, } z_b) \tag{3.13}
\end{array}
$$

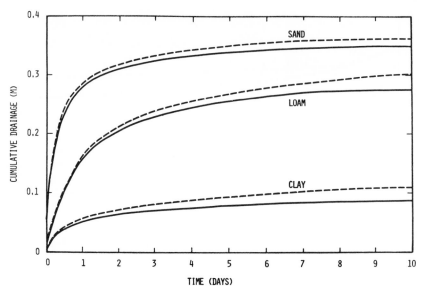

Fig. 3.2. Cumulative gravity drainage from initially-saturated uniform profiles (1.16 m deep) of sand, loam, and clay. Dashed lines—drainage without evaporation; solid lines—simultaneous drainage and evaporation. (After Hillel and van Bavel, 1976.)

Results obtained by numerical solution of the equation for the conditions shown (Hillel and van Bavel, 1976) are illustrated for soils of different textures in Fig. 3.1–3.3, which show the pattern of decrease of soil moisture storage during internal drainage. The sandy soil, although it contains less water at saturation, is seen to drain much more rapidly at first. Thus, it loses half again as much water as the loam and nearly five times as much water as the clay during the first two days. Thereafter these differences decrease and are eventually reversed as further drainage from the sand slows down to a very low rate while drainage from the loam and the clay persists at appreciable rates for many more days.

Figure 3.3 is of particular interest, as it shows the local volumetric wetness at the arbitrary depth of 0.4 m (chosen to represent the zone of principal root activity) as a function of time during the internal drainage of deeply wetted, uniform profiles of sandy, loamy, and clayey soils. After two days of drainage from saturation without evaporation, the wetness values were about 45, 29, and 16% for the clayey, loamy, and sandy soils, respectively. Simultaneous evaporation from the soil surface reduced these values by about 2% in all three cases. These wetness values represent suction heads of 1.3–1.1 m for the clay and loam and about 0.8 m for the sand. Thereafter continuing drainage and evaporation tended to extract more water from this zone of the clay and loam profiles than from the sand, for a subsequent loss

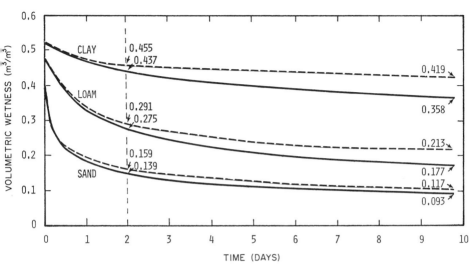

Fig. 3.3. Volumetric wetness at depth of 41 cm as function of time in initially-saturated uniform profiles of sand, loam, and clay. Dashed lines—drainage without evaporation; solid lines—simultaneous drainage and evaporation. (After Hillel and van Bavel, 1976.)

of an additional 8% or so on the tenth day from the clay and loam, as against about 4% from the sand. The suction head values prevailing at this indicative depth after 10 days of internal drainage without evaporation were about 1.8 m in the clay and loam and about 1.3 m in the sand. With evaporation, the corresponding values were about 4, 2.5, and 1.8 m for the clay, loam, and sand, respectively.

Two approaches are possible to the problem of how to characterize the internal drainage process for a given soil profile. One is to formulate a characteristic curve or function to describe the process in dynamic terms (i.e., the dependence of total soil moisture storage W, or of average wetness $\bar{\theta}$, or of flux q, upon time, starting at zero time from a state of saturation). Alternatively, one may choose to characterize the process in static terms by specifying some limiting value of soil wetness at an arbitrarily fixed time, or when the flux has diminished to some small value considered to be negligible. The first approach provides more information about and insight into the process of internal drainage but is more complicated and difficult in practice. The second approach is much more frequently taken by field workers who desire to have a single value to use as a basis for estimating soil moisture storage. The field capacity concept is an outgrowth of that desire. This concept will be discussed in detail in Section F of this chapter.

Implicit in the foregoing discussion was the supposition that the soil profile is texturally or structurally uniform in depth. Quite a different situation can prevail when the profile is layered, as demonstrated by Hillel and Talpaz (1977). An impeding layer inside or below the root zone can greatly hamper internal drainage. In many cases such a layer can be harmful to crops, as it might cause retention of excessive moisture, and perhaps of salts, in the root zone. In some cases, however, as in rapidly draining sands, it might be desirable to form an artificial barrier at some depth to promote greater retention of moisture within reach of plant roots. Methods for doing this were described by Erickson (1972).

C. Redistribution of Soil Moisture in Partially Wetted Profiles

Perhaps more typical than the condition described in the preceding section (namely, that of a uniformly and deeply wetted profile) are cases in which the end of the infiltration process consists of a wetted zone in the upper part of the profile and an unwetted zone beneath. The postinfiltration movement of water in such a profile can truly be called *redistribution*, as the relatively dry deeper layers (those beyond the infiltration wetting front) draw water from the upper ones. The time-variable rate of redistribution depends not only on the hydraulic properties of the conducting soil (as does the rate of internal drainage, previously described) but also on the initial wetting depth, as well as on the relative dryness of the bottom layers. When the initial wetting

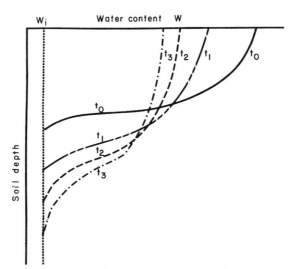

Fig. 3.4. The changing moisture profile in a medium-textured soil during redistribution following an irrigation. The moisture profiles shown are for 0, 1, 4, and 14 days after the irrigation. w_i is preirrigation (antecedent) soil wetness.

depth is small and the underlying soil is relatively dry the suction gradients, augmenting the gravitational gradient, are likely to be strong and hence induce a rapid rate of redistribution. On the other hand, when the initial wetting depth is considerable and the underlying soil itself is wet the suction gradients are small and redistribution occurs primarily under the influence of gravity, as in the case of internal drainage described above.

At first the decrease of soil wetness in the initially wetted zone can be expected to occur more rapidly during the redistribution of moisture in profiles which had been subjected to shallow wetting than in the internal drainage of profiles which had been wetted deeply. Sooner or later, however, the redistribution process "spends itself out," so to speak, and the flux slows down for two reasons: (1) the suction gradients between the wet and dry zones decrease as the former loses, and the latter gains, moisture; (2) as the initially wetted zone quickly desorbs, its hydraulic conductivity decreases correspondingly. With both gradient and conductivity decreasing simultaneously, the flux falls rapidly. The rate of advance of the wetting front decreases accordingly, and this front, which was relatively sharp during infiltration, gradually flattens out and dissipates during redistribution. This is illustrated in Fig. 3.4.

The figure shows that the upper, initially wetted zone drains monotonically, though at a decreasing rate. On the other hand, the sublayer at first wets up but eventually begins also to drain. The time dependence of soil wetness in the upper zone is illustrated in Fig. 3.5 for a sandy soil, in

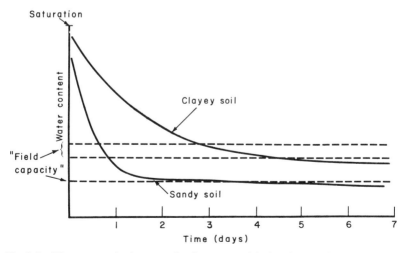

Fig. 3.5. The monotonic decrease of soil wetness with time in the initially wetted zone during redistribution.

which the unsaturated conductivity falls off rapidly with increasing suction, and for a clayey soil, in which the decrease of conductivity is more gradual and hence redistribution tends to persist longer.

D. Hysteretic Phenomena in Redistribution

Unlike the internal drainage process, in which an initially wet soil desorbs monotonically, the redistribution process necessarily involves hysteresis, as the upper part of the profile desorbs (drains) while the lower part absorbs water. Moreover, the zone just below the infiltration-wetted zone first absorbs water and then begins to desorb as redistribution progresses deeper into the profile. This sorption–desorption pattern is portrayed schematically in Fig. 3.6. We recall from Chapter 7 of Hillel (1980) that the relation between wetness and suction is not unique but depends on the history of wetting and drying that take place at each point in the soil. This relationship, when plotted, exhibits two limiting curves which apply when wetting or drying starts from the extreme conditions of dryness or saturation, respectively. Between the wetting and drying branches there lies an infinite number of possible "scanning curves" which describe wetting or drying between different intermediate values of wetness. Such scanning curves tend to be "flatter" than the primary desorption curve; hence it requires a greater increase in suction to effect a unit decrease in wetness during desorption after partial wetting than in desorption after complete wetting.

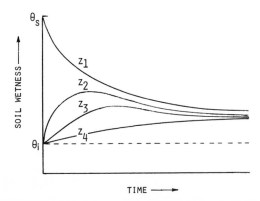

Fig. 3.6. Pattern of change of soil wetness at different depths. Z_1 represents an infiltration-wetted zone, draining monotonically. Z_2–Z_4 represent progressively greater depths below the infiltration wetting front.

It is useful to ponder, if only hypothetically, the expectable pattern of redistribution in an initially dry horizontal soil column of finite length wetted from one end. If infiltration is terminated when the wetting front has moved only part way into the column, redistribution will take place in the horizontal column under the influence of matric suction gradients only. After a time (a *long* time, to be sure, as the process becomes progressively slower owing to decrease of gradient and conductivity) we might expect the column to approach a state of static equilibrium, in which the matric suction becomes equal throughout. At this time, if we measure the distribution of moisture along the column, we are likely to find it distinctly nonuniform. Why? Have we failed to wait long enough for complete attainment of static equilibrium? Not necessarily. Even at equilibrium, with the same matric potential, the soil having undergone desorption will tend to be wetter than the soil having undergone sorption. Clearly, therefore, the effect of the hysteresis phenomenon is to retard redistribution and to retain more water than would otherwise remain in the initially wetted zone.

Horizontal redistribution is of lesser interest to us, quite obviously, than vertical redistribution. If hysteresis were nonexistent, we would expect a draining vertical column under the influence of the gravity gradient to tend toward an equilibrium such that soil wetness would increase with depth. With hysteresis, however, more of the moisture remains in the infiltration-wetted part of the profile, where it might be available for subsequent plant use, rather than flow downward beyond the reach of plant roots. Evidence for this was provided by Rubin (1967), as shown in Fig. 3.7. An intriguing hypothesis was advanced in this regard by Youngs (1958a,b, 1960b). After infiltration is stopped, further downward movement of water out of the infiltration-saturated zone requires intrusion of air, either at the soil surface or at the wetting front. On the basis of hysteresis, Youngs reasoned that for air to enter the surface the length of the wetted zone must be greater than the difference between the water-entry and air-entry values of the soil. If the depth (thickness) of the wetted zone is sufficiently small, as Youngs put it, the water ought to remain stationary near the soil surface until extracted by evapotranspiration.

The steep reduction in hydraulic conductivity in the draining part of the profile, as well as the overall decrease in matric suction gradients, along with hysteresis, all act in concert to increase the efficiency of soil moisture storage. If we did not know of these synergistic effects, we might never understand how the soil, being a bottomless barrel, nevertheless manages to retain enough water in its upper part long enough to support the growth of vegetation even in arid regions, where water supply is meager and infrequent. The

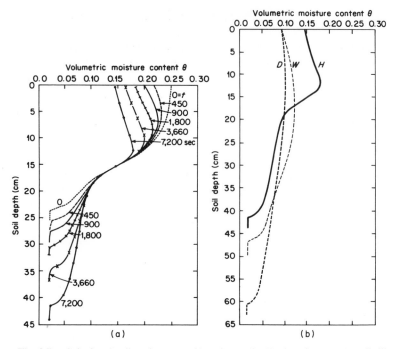

Fig. 3.7. Calculated soil moisture profiles after redistribution, in a sandy soil, illustrating the effect of hysteresis on retention of moisture in the upper layers. Rehovot sand: $n = 0.387$, $\theta_{fc} = 0.066$. H denotes hysteretic redistribution, W the nonhysteretic, wetting soil characteristic curve, and D the nonhysteretic, drying soil characteristic curve. In (b), $t = 2$ hours. (After Rubin, 1967.)

first soil physicist to have gained an insight into the principles governing this phenomenon must have felt that he had discovered a great secret of nature. Indeed he had.

The hysteresis phenomenon complicates the redistribution process and makes it difficult to describe and analyze mathematically. To do this, it is necessary to know the wetting history and the wetness value when desorption commences at each depth within the profile. Furthermore, it is necessary to know the slopes of all possible scanning curves (these slopes, in general, differ considerably from those originating at the primary soil-moisture characteristic curves). Hence, it appears at present that the best way to handle hysteresis is by numerical analysis (e.g., Rubin, 1967), while analytical or quasianalytical attempts to treat the redistribution process have so far tended to avoid hysteresis or to assume rather gross simplifications (e.g., Youngs, 1958a,b; Gardner *et al.*, 1970a,b).

E. Analysis of Redistribution Processes

Once again we return to the general equation for flow in a vertical profile, Eq. (3.12):

$$\frac{\partial \theta}{\partial t} = -\frac{\partial}{\partial z}\left(K \frac{\partial \psi}{\partial z} + K\right)$$

In the case of redistribution involving hysteresis, this equation may be written (Miller and Klute, 1967)

$$\left(\frac{d\theta}{d\psi}\right)_h \frac{\partial \psi}{\partial t} = -\frac{\partial}{\partial z}\left[K_h(\psi) \frac{\partial \psi}{\partial z}\right] - \frac{\partial K_h(\psi)}{\partial z} \tag{3.14}$$

In the above, θ is volumetric wetness, t time, K conductivity, z depth, ψ suction head, and the subscript h indicates a hysteretic function. After the cessation of infiltration and in the absence of evaporation flux through the soil surface is zero, and hence the hydraulic gradient at the surface must also be zero. Conservation of matter requires that

$$\int_{z=0}^{z=\infty} \theta \, dz = \text{const} \tag{3.15}$$

for all time, provided no sinks are present (e.g., no extraction of water by plant roots).

When redistribution begins, the upper portion of the profile, which was wetted to near saturation during the preceding infiltration process, begins to desorb monotonically. Below a certain depth, however, the soil first wets during redistribution, then begins to drain, and the value of wetness at which this turnabout takes place decreases with depth. Each point in the soil thus follows a different scanning curve, and the conductivity and water-capacity functions vary with position. An approximate analysis of these phenomena was given by Youngs (1958a, b; 1960a,b). He showed that in coarse materials with uniform pore sizes and hence with "abrupt" $\theta(\psi)$ curves redistribution can be slower for shallow than for deep wettings.

The flow equation describing redistribution was analyzed numerically by Staple (1969), Remson *et al.* (1965, 1967), and Rubin (1967). Rubin's analysis was based on the assumptions that, despite hysteresis, there exists at any time and depth in the soil a unique, single-valued relation between wetness and suction, that a soil element that has begun to desorb will not wet up again, and that hysteresis in the relation of conductivity to wetness (but not to suction) can be neglected.

To solve Eq. (3.12), one of the two dependent variables (namely, ψ) was eliminated:

$$\frac{\partial \theta}{\partial t} = -\frac{\partial}{\partial z}\left(\frac{K}{c}\frac{\partial \theta}{\partial z} + K\right) \qquad (3.16)$$

The solution was then sought, subject to the following initial and boundary conditions:

$$
\begin{array}{llll}
t = 0, & z > 0, & \theta = \theta(z) & \\
t > 0, & z = 0, & q = \dfrac{K}{c}\dfrac{\partial \theta}{\partial z} + K = 0 & (3.17) \\
t > 0, & z = \infty, & \theta = \theta_i &
\end{array}
$$

where q is the flux, $\theta(z)$ is a function of soil depth describing the postinfiltration (preredistribution) moisture profile, θ_i is the soil's preinfiltration wetness, c is the differential water capacity $d\theta/d\psi$, and the other variables are as defined above. The ratio K/c is the diffusivity D.

Equation (3.16) was approximated by an implicit difference equation. In order to take hysteresis into account, empirical equations were used to describe the dependence of ψ on θ in primary wetting in primary drying and in intermediate (transitional) scanning from wetting to drying. Another empirical equation was used to describe the relation of K to θ. Empirical equations cannot be expected to pertain to any but the particular soil considered. The results of the analysis, however, are thought to be generally valid in principle, provided the basic assumptions are met in the real situation, at least approximately.

Rubin's findings (Fig. 3.7) indicate that the hysteretic moisture profile is not bounded by the two possible nonhysteretic profiles (one assuming the desorbing branch of the soil-moisture characteristic, the other assuming the sorbing branch). The hysteretic redistribution was shown to be clearly slower than the nonhysteretic one. These results demonstrate the importance of hysteresis in redistribution processes, particularly in coarse-textured soils in which the hysteresis phenomenon is often more pronounced.

A different approach to the analysis of redistribution was taken by Gardner *et al.* (1970a,b). Equation (3.16) can be solved analytically by separation of variables (Gardner, 1962b) in the special case where the empirical relation $D = C\theta^n$ applies (with C and n constants). In this procedure, it is assumed that the solution is of the form $\theta = T(t)Z(z)$, where T is a function of t alone and Z is a function of z alone. It is further assumed that $K = B\theta^m$ (with B and m constants). Their analysis suggests that both where matric suction forces predominate and where, alternatively, the gravity force predominates the time dependence of soil wetness (and of total water content in any depth zone) obeys a relation of the type

$$W, \bar{\theta}, \text{ or } \int \theta \, dz \approx a(b + t)^{-c} \qquad (3.18)$$

where W is the total water content and $\bar{\theta}$ is the mean wetness of the initially wetted zone at time t during redistribution. This theory accords with experimentally measured patterns of redistribution in a sandy loam soil irrigated with different quantities of water (Fig. 3.8 and 3.9). The constants in Eq. (3.18) were shown to be related to the soil's hydraulic conductivity and diffusivity. Of these, the value of b could be neglected after a day or so, so that the simplified equation could be cast into logarithmic form:

$$\log W = \log a - c \log t \qquad (3.19)$$

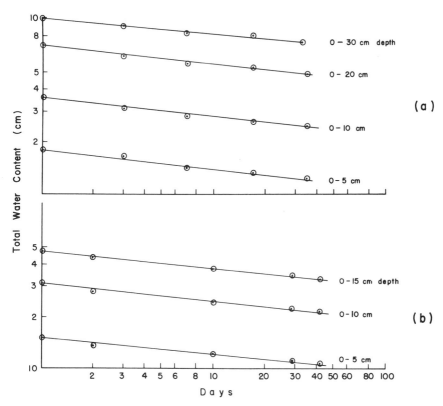

Fig. 3.8. Total amount of water retained in various depth layers within the initially wetted zone of a fine sandy loam during redistribution following irrigations of (a) 10 and (b) 5 cm of water. (After Gardner *et al.*, 1970a, b.)

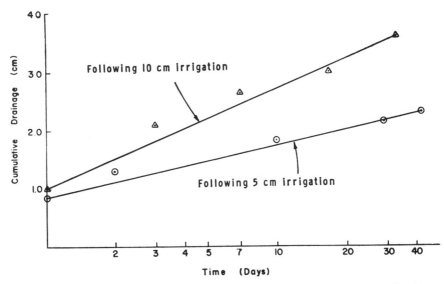

Fig. 3.9. Cumulative downward flow through the initial (end of infiltration) wetting front during redistribution following irrigations of 5 and 10 cm of water. (After Gardner *et al.*, 1970a, b.)

The values of a and c can be obtained from a logarithmic graph of W versus t if the data do indeed indicate a straight line, as shown in Fig. 3.9.

Equation (3.18) can also be differentiated with respect to time to give the rate of decrease of water content in the initially wetted zone $(-dW/dt)$, equal to the flux through the initial (end of infiltration) wetting front:

$$-dW/dt = ac/(b + t)^{c+1} \tag{3.20}$$

This relationship is shown in Figure 3.8.

An alternative approximate analysis which provides insight into the redistribution process and permits useful predictions was carried out by Peck (1970). He introduced two assumptions:

(1)   The envelope of all instantaneous moisture profiles can be represented by a rectangular hyperbola for soil wetness values lower than the maximum wetness attained during infiltration.

(2)   Soil wetness at the interface between the desorbing and the sorbing parts of the profile is approximately a linear function of the desorbing zone above this interface.

With these assumptions, a first-order ordinary differential equation was derived which can be integrated numerically to relate the mean wetness of the desorbing zone to the redistribution time.

Peck defined the mean wetness in the desorbing zone $0 \leq z < z_*$ by

$$\bar{\theta} = \frac{1}{z_*} \int_0^{z_*} \theta \, dz \tag{3.21}$$

where z_* is the depth of the transition plane at any particular time. He showed that with increasing infiltration quantity I both the depth of the transition plane and its rate of downward movement are increased. However, the flux of water through the transition plane is reduced. Neglecting the influence of gravity, the time needed for mean soil wetness to decrease to any specified value was shown to increase with I^2. Analysis of the possible effect of antecedent (preinfiltration) soil wetness indicated that a higher antecedent wetness leads immediately to greater penetration of the infiltrating water, but since the overall gradient is reduced, subsequent movement through the transition plane occurs less rapidly. Finally, the greater the antecedent soil wetness, the greater the wetness of the infiltration-wetted part of the profile at all times during redistribution. However, the *storage increment* (wetness after a specified period of redistribution minus the antecedent wetness) of the same zone is reduced at higher antecedent moisture values.

The foregoing discussion is based on several assumptions which may not accord with reality in any particular case. For instance, the pattern of redistribution in layered profiles, and consequently soil moisture storage, are likely to be very different from those in uniform profiles. No less important are the possible differences between redistribution in the absence of evaporation and the same process when it occurs simultaneously with soil moisture evaporation and with plant activity (i.e., moisture extraction by roots). These processes and conditions can now be analyzed by means of computer-based dynamic simulation (Hillel, 1977), and it appears that the dynamics of soil moisture storage are indeed strongly dependent on the textural and structural composition of the soil profile (Hillel and Talpaz, 1977) and on evaporation processes (Gardner et al., 1970).

F. "Field Capacity"

Early observations that the rate of flow and water-content changes decrease in time (Alway and McDole, 1917; Richards and Moore, 1952; Veihmeyer and Hendrickson, 1931) have been construed to indicate that the flow rate generally becomes negligible within a few days, or even that the flow ceases entirely. The presumed water content at which internal drainage allegedly ceases, termed the *field capacity*, had for a long time been accepted almost universally as an actual physical property, characteristic of and constant for each soil.

Though the field capacity concept originally derived from rather crude measurements of water content in the field (where sampling and measurement errors necessarily limited the accuracy and validity of the results), some workers have sought to explain it somehow in terms of a static equilibrium value or a discontinuity in the capillary water. It was commonly assumed that the application of a certain quantity of water to the soil will fill the deficit to field capacity to a certain and definite depth, beyond which this quantity of water will not penetrate. It became an almost universal practice to calculate the amount of irrigation to be applied at any particular time on the basis of the "deficit to the field capacity" of the soil depth zone to be wetted.

In recent years, with the development of theory and more precise experimental techniques in the study of unsaturated flow processes, the field capacity concept, as originally defined,[2] has been recognized as arbitrary and not an *intrinsic* physical property independent of the way it is measured. When and how can one determine that redistribution has virtually ceased, or that its rate has become negligible or practically zero? Obviously, the criteria for such a determination are subjective, depending as often as not upon the frequency and accuracy with which the soil water content is measured. The common working definition of field capacity (namely, the wetness of the initially wetted zone, say, two days after infiltration) takes no account of such factors as the antecedent (preinfiltration) wetness of the soil, the depth of wetting, the possible presence and depth of a groundwater table, etc.

The redistribution process is in fact continuous and exhibits no abrupt "breaks" or static levels. Although its rate decreases constantly, in the absence of a water table the process continues and equilibrium is approached, if at all, only after very long periods.

The soils for which the field capacity concept is most tenable are the coarse-textured ones, in which internal drainage is initially most rapid but soon slows down owing to the relatively steep decrease of hydraulic conductivity with increasing matric suction. In medium- or fine-textured soils, however, redistribution can persist at an appreciable rate for many days. As

[2] According to Veihmeyer and Hendrickson (1949), the field capacity is "the amount of water held in soil after excess water has drained away and the rate of downward movement has materially decreased, which usually takes place within 2–3 days after a rain or irrigation in pervious soils of uniform structure and texture." By "amount of water" they obviously meant amount per unit volume, or mass, of soil. Apart from that, however, this definition raises more qeustions than it answers. What is "excess water?" How can we be sure just when it has "drained away" or just when downward movement has "materially decreased?" Can we depend universally on the statement that this "usually takes place within 2–3 days after a rain or irrigation," particularly if the amount of water supplied and antecedent soil conditions are left unspecified? And what are "pervious soils?" Finally, we must ask, what about all the world's soils which are not "of uniform structure and texture?"

an example, we can cite the case of a loessial silt loam in the Negev region of Israel, in which the changes in water content shown in Table 3.1 were observed in the 60–90 cm depth zone following a wetting to a depth exceeding 150 cm (with evaporation effectively prevented by means of a paper mulch). It is seen that the water loss continued incessantly for over five months. The rate of decrease of water content accorded with the function given by Richards *et al.* (1956):

$$-dW/dt = at^{-b} \tag{3.22}$$

where W is the water content, t the time, and a and b constants related to the boundary conditions and conductance properties of the soil. (The exponential constant b, which is related to soil diffusivity, is obviously most important, and the greater its value, the steeper the decrease in water content.)

An agriculturist engaged in irrigated farming and accustomed to frequent irrigations is interested mainly in the short-run storage "capacity" of his soil. For him, the field capacity of the loessial soil cited can be taken at about 18%. By way of contrast, an agriculturist engaged in dryland farming is sometimes interested in storing soil water from one season, or even from one year, to the next. For the dry-land farmer, therefore, the field capacity of the same loessial soil cannot be taken at 18% (since the soil does not retain this content beyond a few days) but at 14%, or even less.

A commonly, if tacitly, held assumption is that the redistribution process occurs by itself, independently of other processes of water extraction from the soil. This assumption is seldom, if ever, realistic. When several processes of soil-water extraction (e.g., internal drainage, evaporation, and uptake by plants) occur simultaneously, the reduction in soil water content is obviously more rapid and less apt to cease at any point such as "field capacity."

Various laboratory methods have been proposed for the estimation of field capacity. These include equilibration of presaturated samples with a centrifugal force 1000 times the gravity force (the so-called moisture equiv-

Table 3.1

	Moisture (percent by mass)
At the end of infiltration	29
After 1 day	20.2
2 days	18.7
7 days	17.5
30 days	15.9
60 days	14.7
156 days	13.6

alent) or with a matric suction value of $\frac{1}{10}$ or $\frac{1}{3}$ bar. Although the results of such tests may correlate with measurements of soil moisture storage in the field in certain circumstances, it is a fundamental mistake to expect criteria of this sort to apply universally, since they are solely static in nature while the process they purport to represent is highly dynamic. For example, two texturally and structurally identical profiles will drain quite differently if one of them is uniform throughout all depths whereas the other is underlain by a clay pan. The former profile will tend to drain readily while the latter may remain nearly saturated for long periods.

These and other shortcomings of the field capacity concept were recognized some twenty years ago by L. A. Richards, who proclaimed in a keynote address to the International Congress of Soil Science that "the field capacity concept may have done more harm than good" (Richards, 1960). Alas, the concept has not faded away. Notwithstanding the fundamental impossibility of fixing unequivocally any unique point in time at which internal drainage or redistribution ceases, there remains a universal need for a simple criterion to characterize the ability of soils to retain, or store, moisture (i.e., the upper limit of soil water content which can be depended on, more or less, in the field). However indefinite the field capacity concept, the fact is that the rapid slowing of internal drainage or redistribution is an exceedingly important property of soils, responsible for retaining (albeit temporarily) water available for plant use during periods of no rain or irrigation.

Granted that the field capacity concept is subjective, it is nevertheless considered by many to be necessary.[3] If so, can it be improved, in principle and in practice? In the first place, since no laboratory system yet devised is capable of duplicating soil-water dynamics in the field, it should be obvious that, to be realistic, field capacity must be measured directly in the field. Too many practitioners still ignore this simple truism, preferring instead to assume that, say, "moisture retention at $\frac{1}{3}$ bar *is* field capacity" when in fact such a value can, at best, be *correlated* with it (a correlation which must be verified in each case and never taken for granted a priori).

Second, the field determination itself must be made reproducible by standardizing a consistent procedure. Such vague specifications as "wet the soil to the depth of interest," and "allow the soil to drain for approximately 2 days" (Peters, 1965) are not good enough. Wetting depth is extremely important, and the preferred depth is the maximal one—considerably beyond the "depth of interest" (presumably, the rooting zone of crops, which itself is too variable to specify). The internal drainage of a profile initally wetted to its *entire* depth is a much more reproducible and reliable criterion

---

[3] A wag once remarked that if field capacity did not exist, someone (perhaps even a soil physicist, though one can never be certain) would have had to invent it.

than redistribution in a profile wetted to some unspecified partial depth without regard to antecedent conditions.

Third, the measurement of soil-moisture content and depth distribution should be made repeatedly rather than only once at an arbitrary time such as 2 days. Periodically repeated measurements, preferably by a nondestructive method such as neutron gauging, will provide information on the dynamic pattern of internal drainage and allow evaluation of whether any single value of soil moisture at any specifiable characteristic time can be designated as the field capacity. To make this judgment objectively, one must decide at the outset what drainage rate one is willing to consider negligible.[4] For different soils, the characteristic time may vary from a few hours to a few weeks, depending on soil hydraulic properties and on the flux criterion used.

A still better approach is to characterize, for each field to be considered, the internal drainage process (starting from well-defined initial and boundary conditions) as a complete function of time. One way to do this would be to specify the half life of stored moisture, as explained in Section B of this chapter. Another way is to fit the measured function to an empirical equation, such as that proposed by Richards et al. (1956) and by Ogata and Richards (1957), as given above in Eq. (13.22):

$$q = -dW/dt = at^{-b}$$

It would then be up to the user to calculate what value of soil moisture gives the best estimate of soil moisture storage to suit his purposes. Assuming, for the sake of argument, that the data obtained in the field accords with the above equation, then, for whatever limiting value of flux q_n, one wishes to regard as negligible, one gets

$$t_n = (a/q_n)^{1/b} \tag{3.23}$$

where t_n is the time period required from the cessation of infiltration for the drainage rate to fall to a negligible value. The constants a and b are to be calculated in each case from the complete internal drainage curve as measured in the field.

Ultimately, of course, soil moisture storage could be monitored continuously by solution of the flow equation, taking into account all processes of water entry and extraction from the soil volume of interest. It is already possible to do this with a properly programmed computer, given the necessary parameters and the current data on rainfall, evapotranspiration, etc. It will be some time, however, before computers become accessible and

---

[4] A likely criterion might be $\frac{1}{10}$ or less of the daily potential evapotranspiration (PET). Thus, if PET is 5 mm/day, one might designate a drainage flux of 0.5 mm/day through the bottom of the root zone as small enough to be disregarded.

economical enough for all those who wish to estimate soil moisture storage. Until then, we shall probably have to live with the field capacity concept and strive to make it more meaningful.

G. Summary of Factors Affecting Field Capacity

The field capacity, as commonly measured, may vary between about 4% (by mass) in sands, to about 45% in heavy clay soils, and up to 100% or even more in certain organic soils. Among the factors affecting redistribution and the apparent field capacity are the following:

(1) Soil texture. Clayey soils retain more water, and retain it longer, than sandy ones. Hence, the finer the texture the higher the apparent field capcity, the slower its attainment, and the less distinct and stable its value. Soil structure may also affect water retention.

(2) Type of clay present. The higher the content of montmorillonite is, the greater is the content of water adsorbed and retained at any time.

(3) Organic matter content. Soil organic matter can help retain more water, though the amount of organic matter normally present in mineral soils is too low to retain any significant amounts of water. The effect of organic matter on soil structure can, however, be significant.

(4) Depth of wetting and antecedent moisture. In general (but not always) the wetter the profile is at the outset, and the greater the depth of wetting during infiltration, the slower the rate of redistribution and the greater the apparent field capacity.

(5) The presence of impeding layers in the profile, such as layers of clay, sand, or gravel, can inhibit redistribution and increase the observable field capacity. The rate of outflow from any given layer in the soil thus depends not only on the texture or hydraulic characteristics of that layer, but also on the composition and structure of the entire profile, since the presence at any depth of an impeding layer can retard the movement of water out of the layers above it.

(6) Evapotranspiration. The rate and pattern of the upward extraction of water from the soil can affect the gradients and flow directions in the profile and thus modify the redistribution or internal drainage process.

Sample Problems

1. Assume gravity drainage out the bottom of a uniformly wetted soil profile, 100 cm deep, with an initial wetness of 45%. Further assume the hydraulic conductivity K to be an exponential function of the soil's volume wetness θ, namely, $K = ae^{b\theta}$, where $a = 1.5 \times 10^{-6}$ cm/day, and $b = 35$. Estimate the remaining soil wetness when the drainage flux decreases to $\frac{1}{2}$, then to $\frac{1}{10}$, then to $\frac{1}{100}$ of the evapotranspiration rate (ET), which is 0.5 cm/day.

With gravity alone (no suction gradients), vertical drainage occurs at a rate q equal to the hydraulic conductivity, which is, in turn, a function of the soil's wetness at the depth considered (e.g., at the bottom of the root zone):

$$q = K(\theta) = ae^{b\theta} = 1.5 \times 10^{-6}e^{35\theta}$$

We can now calculate the flux at different assigned wetness values to obtain a plot of log q versus θ, as follows:

When $\theta = 0.45$, $q = 1.5 \times 10^{-6}e^{35 \times 0.45} = 10.38$ cm/day;
when $\theta = 0.40$, $q = 1.5 \times 10^{-6}e^{35 \times 0.40} = 1.80$ cm/day;
when $\theta = 0.35$, $q = 1.5 \times 10^{-6}e^{35 \times 0.35} = 3.15 \times 10^{-1}$ cm/day;
when $\theta = 0.30$, $q = 1.5 \times 10^{-6}e^{35 \times 0.30} = 5.4 \times 10^{-2}$ cm/day;
when $\theta = 0.25$, $q = 1.5 \times 10^{-6}e^{35 \times 0.25} = 9.45 \times 10^{-3}$ cm/day;
when $\theta = 0.20$, $q = 1.5 \times 10^{-6}e^{35 \times 0.20} = 1.65 \times 10^{-3}$ cm/day.

From the q–θ plot we can determine the θ values at which $q = 2.5 \times 10^{-1}$ cm/day ($\frac{1}{2}$ET), 5×10^{-2} cm/day ($\frac{1}{10}$ET), and 5×10^{-3} cm/day. The θ values turn out to be approximately 43, 30, and 23.2%, respectively. Which of the q values is to be considered negligible is a matter of individual decision. Alternatively, we can calculate the θ remaining at any flux value, as follows: $q = ae^{b\theta}$. Hence ln $q = $ ln $a + b\theta$. Therefore $\theta = (\ln q - \ln a)/b$. Substituting, for example, $q = 5 \times 10^{-2}$ cm/day ($\frac{1}{10}$ET), we get

$$\theta = [-3 - (-13.4)]/35 = 10.4/35 = 0.297$$

which is close to the 30% estimated above. The actual plot of q versus θ is again left as an instructive exercise to the enterprising student.

2. A uniform soil profile, 150 cm deep, was wetted to saturation ($\theta = 48\%$) and then gravity drained while evaporation was prevented. The "drainable" fraction was 70% of the water initially present, while the "residual" 30% of the water is assumed to remain even at very long times. The profile water content was monitored continuously and the volume of drainable water was found to decrease to half its initial value within 20 hours ($\frac{5}{6}$ of a day). Estimate the profile's water content after 1, 2, 3, 4, 5, 6, and 7 days of internal drainage, using the exponential decay equation.

The exponential decay process is described by Eq. (3.6):

$$-dW_{\rm d}/dt = \lambda W_{\rm d}$$

from which we can derive [see Eqs. (3.10) and (3.11)]:

$$W_d = W_{di}e^{-\lambda t} \quad \text{and} \quad t_{1/2} = 0.693/\lambda$$

In this equation W_{di} is the initial amount of drainable water present in the profile, W_d is the content of drainable water content at any time t during the internal drainage process, $t_{1/2}$ is the half life of drainable soil moisture (i.e., the time at which W_d diminishes to half of W_{di}), and λ is the exponential "decay" coefficient.

From the last equation and the data given, we obtain

$$\lambda = 0.693/t_{1/2} = 0.693/(5/6) = 0.8316$$

Using this value, we can estimate total profile water content at any time, according to

$$W = W_r + W_{di}/e^{0.8316t}$$

where $W_{di} = 0.7 \times 0.48 \times 150 \text{ cm} = 50.4 \text{ cm}$, W_r (the residual content) $= 0.3 \times 0.48 \times 150 \text{ cm} = 21.6 \text{ cm}$, and W is the total profile water content (initially 72.0 cm).

> After 1 day, $W = 21.6 + 50.4/e^{0.8316 \times 1} = 43.54 \text{ cm}$.
> After 2 days, $W = 21.6 + 50.4/e^{0.8316 \times 2} = 31.15 \text{ cm}$.
> After 3 days, $W = 21.6 + 50.4/e^{0.8316 \times 3} = 25.76 \text{ cm}$.
> After 4 days, $W = 21.6 + 50.4/e^{0.8316 \times 4} = 23.41 \text{ cm}$.
> After 5 days, $W = 21.6 + 50.4/e^{0.8316 \times 5} = 22.39 \text{ cm}$.
> After 6 days, $W = 21.6 + 50.4/e^{0.8316 \times 6} = 21.94 \text{ cm}$.
> After 7 days, $W = 21.6 + 50.4/e^{0.8316 \times 7} = 21.75 \text{ cm}$.

Note that the daily loss of water from the profile diminished rapidly as follows:

Days:	1	2	3	4	5	6	7
Drainage:	28.46	12.39	5.39	2.35	1.02	0.45	0.19 cm

3. Use the equation of Richards *et al.* (1956) to estimate the water content of a 1 m deep profile at the end of 1, 2, 3, 4, 4, 5, 6, 7 days of redistribution if the initial water content is 50 cm, factor $a = 4$, and exponent $b = 0.8$. Estimate the time necessary for the downward flux at a depth of 1 m to diminish to 1 cm/day.

Recalling Eq. (3.22), with W and t as in the preceding problem,

$$dW/dt = -at^{-b}$$

from which we obtain

$$dW = -at^{-b} \, dt$$

Integrating,

$$W = -[a/(-b + 1)]t^{-b+1} + c$$

where c, the constant of integration, equals the profile water content at $t = 0$. We can now calculate the water content at various times, as follows:

At zero time, $W = 50 - [4/(-0.8 + 1)] \times 0^{0.2} = 50.0$ cm.
After 1 day, $W = 50 - (4/0.2) \times 1^{0.2} = 30.0$ cm.
After 2 days, $W = 50 - (4/0.2) \times 2^{0.2} = 27.0$ cm.
After 3 days, $W = 50 - (4/0.2) \times 3^{0.2} = 25.1$ cm.
After 4 days, $W = 50 - (4/0.2) \times 4^{0.2} = 23.6$ cm.
After 5 days, $W = 50 - (4/0.2) \times 5^{0.2} = 22.4$ cm.
After 6 days, $W = 50 - (4/0.2) \times 6^{0.2} = 21.4$ cm.
After 7 days, $W = 50 - (4/0.2) \times 7^{0.2} = 20.5$ cm.

To estimate the time t_n needed for the flux to diminish to any specified value q_n, we use Eq. (3.23):

$$t_n = (a/q_n)^{1/b}$$

Accordingly,

$$t_n = (4/1)^{1/0.8} = 5.66 \text{ days}$$

4 *Groundwater Drainage*

A. Introduction: Some Basic Concepts of Groundwater Hydrology

It is an interesting fact that some 75% of the earth's fresh water is locked away in polar ice caps and glaciers, while less than 2% is in *surface waters*, such as lakes and streams, and a relatively minute amount is contained in the generally unsaturated topsoil. That leaves nearly 22% of our planet's fresh water, which permeates and saturates porous rocks and subsoils, generally at some depth below the ground surface. It is this amount which we call *groundwater.*

In many regions, groundwater constitutes an important source of fresh water for domestic, agricultural, or industrial use. Since an adequate water supply is a prerequisite for the development of settlements and industries, including the agricultural industry, and since injudicious exploitation of groundwater can deplete the *groundwater reservoir* and diminish its quality, it is important to acquire and disseminate knowledge pertaining to the behavior of groundwater and to methods of managing it. Extraction of groundwater in excess of the *annual recharge* (including the natural percolation of precipitation water, seepage from reservoirs and streams, and artificial injection through wells and by surface ponding) will cause depletion, whereas an excess of recharge over extraction will cause a buildup of groundwater.

We now wish to define a number of terms used in the science of groundwater hydrology. An *aquifer* (literally, a water carrier) is a porous geological formation which accumulates and transmits water in sufficient quantities to serve as a source of supply for human use (Todd, 1967). Aquifers vary greatly in their characteristics. Some consist of unconsolidated coarse sediments such as sand and gravel, and some are pervious bedrock such as sandstone, conglomerate, limestone, or even fissured volcanic formations (e.g., basalt deposits). A formation which, in contrast with an aquifer, neither contains nor transmits significant amounts of water is called an *aquifuge*. An example is a formation of tight, impermeable rock such as granite. Between the two extremes, aquifer and aquifuge, hydrologists recognize an intermediate type of formation which contains water but, owing to low permeability, cannot transmit enough to serve as a source of supply (e.g., clay layers and shales). Such a formation is called an *aquiclude*.

A distinction can be made between two major types of aquifers: confined and unconfined. A *confined aquifer* (also called *artesian*) occurs when a permeable stratum is overlain by an impermeable stratum (i.e., when an aquifer is "confined" by an aquiclude). A confined aquifer is generally not recharged through the aquiclude, but from areas (which may be some distance away) where the permeable formation outcrops, as shown in Fig. 4.1. Often, the water contained in a confined aquifer is under pressure, which may be sufficient to bring water up to ground surface whenever a well is drilled through the aquiclude into the aquifer. The equilibrium height to which water rises in an unpumped well, called the *piezometric*[1] *level*, is a measure of the hydrostatic pressure of water in the confined aquifer. Layers which are sufficiently permeable to transmit water vertically to or from the confined aquifer but not permeable enough to laterally transport water like an aquifer are called *aquitards* (Bouwer, 1978). An aquifer bound by one or two aquitards is called a *leaky* or *semiconfined* aquifer.

An *unconfined aquifer* (also called a *phreatic*[2] or *water-table aquifer*) is one in which the water level is free to fluctuate up and down, to rise and fall periodically, and thus to seek an equilibrium level. Phreatic aquifers may reach levels close to the ground surface, and at times and places even rise above it. Typically, however, the soil remains nearly always unsaturated to some depth (called the *unsaturated zone*, or the *zone* of *aeration*), below which a saturated condition prevails in what is known as the *zone of saturation*. If we dig a hole through the unsaturated zone into the saturated zone, water will tend to seep into the hole, and at equilibrium this free water will come to rest at an elevation called the *water table*,[3] defined as the surface at which the pressure of groundwater equals atmospheric pressure. Taking

[1] From the Greek *piezein*, to press or squeeze. Hence a piezometer is a device to measure pressure.

[2] From the Greek *phrear*, a well.

[3] The term "water table" is something of a misnomer, since the surface it describes can be very irregular.

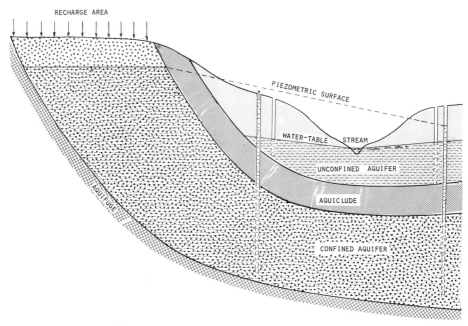

Fig. 4.1. A confined (artesian) and an unconfined (phreatic) aquifer. The former rests on an aquifuge, and the latter is perched on an aquiclude. Note that the deep wells are artesian.

atmospheric pressure as our reference, we can think of the water table as a surface of *zero hydrostatic pressure.*

In this chapter, we shall be dealing mainly with shallow unconfined groundwater, which occasionally encroaches upon the root zone and must then be drained if optimal growing conditions are to be maintained. For more fundamental expositions of groundwater theory, the reader is referred to publications by Harr (1962), Chow (1964), Todd (1967), Remson *et al.* (1971), Bouwer (1978), and Freeze and Cherry (1979). Problems of groundwater pollution were described by Fried (1976).

B. Flow of Confined Groundwater

We now wish to describe a number of simple cases illustrating flow phenomena in confined aquifers. The simplest example is that of flow through a horizontal layer of uniform thickness and properties sandwiched between impervious layers above and below it. As illustrated in Fig. 4.2, flow takes place from the channel on the left to the channel on the right. If the water level in each of the two channels is maintained at a constant level, the flow

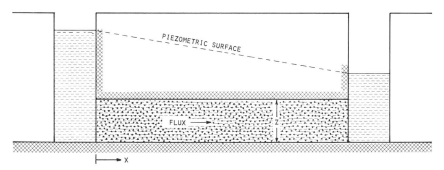

Fig. 4.2. Flow through a horizontal confined aquifer of uniform thickness and hydraulic properties.

will take place at a steady state, i.e., a state in which the flux and gradient are time invariant.

The process can be formulated as follows. Begin with the Darcy equation (with q being flux, K conductivity, H hydraulic head, and x horizontal distance).

$$q = -K \, dH/dx \qquad (4.1)$$

Now separate differential variables:

$$dH = -(q/K) \, dx$$

Next, integrate

$$\int_{H_1}^{H_2} dH = -\frac{q}{K} \int_0^x dx \qquad (4.2)$$

to obtain

$$H_2 - H_1 = -(q/K)x \qquad \text{or} \qquad H_2 = H_1 - (q/K)x \qquad (4.3)$$

which indicates that, as long as the flux and conductivity are constant, the hydraulic head decreases linearly with distance along the direction of flow.

To calculate the *discharge* Q, that is to say, the total volume of flow per unit time per unit thickness (perpendicular to the plane of the drawing), we multiply the flux by the height of the aquifer Z:

$$Q = qZ = -KZ \, dH/dx = -\tau \, dH/dx \qquad (4.4)$$

Here $\tau = KZ$ is a composite parameter called the *transmissivity*, and it characterizes the amount of water obtainable from an aquifer under a unit hydraulic gradient. In evaluating the potential usefulness of an aquifer as a source of water, the transmissivity rather than the conductivity is the parameter of interest, since a less permeable aquifer can deliver more water than a highly permeable one if the dimensions of the former are sufficiently larger than those of the latter.

Now let us consider the slightly more complex case of steady radial flow toward a single well. Suppose the initial piezometric head at the well was H_i and this head was uniform for an infinite distance in the region surrounding the well, as shown in Fig. 4.3. As we begin to pump water out of the well, we notice that the piezometric head drops. A hydraulic head gradient is thereby created, and water flows radially inward toward the well from the surroundings, creating a *cone of depression* in the piezometric head. If we continue pumping water at a steady rate, and if the head remains constant at some distance from the well, the cone of depression will stabilize at the exact configuration designed to supply the well with the amount of water withdrawn per unit time. Henceforth the depression of piezometric head (from the original level H_i) at each point, called the *drawdown*, will be a function of radial distance and of discharge, but not of time.

To formulate this process in mathematical terms, we cast Darcy's law into the radial form (with r, the radius, as the space coordinate):

$$q = -K\,dH/dr \tag{4.5}$$

To obtain the discharge through any concentric cylindrical surface at any radial distance r from the well, we multiply the flux q by the area through which flow takes place, namely $2\pi r Z$ (where $2\pi r$ is the circumference of

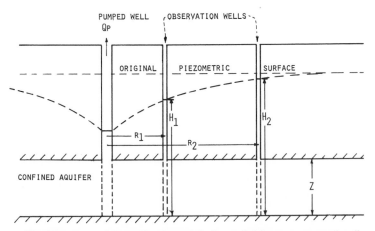

Fig. 4.3. Cone of depression formed during radial flow to a pumped well.

the circle and Z, once again, is the height, or thickness, of the aquifer):

$$Q = -2\pi rZK \, dH/dr \tag{4.6}$$

At steady state, the discharge through all concentric cylindrical surfaces must be equal to the discharge from the well, i.e., to the pumping rate $-Q_p$ (volume withdrawn per unit time), so that in the above $Q = -Q_p$. We can now separate the differential variables:

$$dH = \frac{Q_p}{2\pi ZK} \frac{dr}{r}$$

and integrate between any arbitrary radial distances r_1 and r_2:

$$\int_{H_1}^{H_2} dH = \frac{Q_p}{2\pi ZK} \int_{r_1}^{r_2} \frac{dr}{r} \tag{4.7}$$

to obtain

$$H_2 - H_1 = (Q_p/2\pi ZK)\ln(r_2/r_1) \tag{4.8}$$

If we wish to measure the aquifer transmissivity $\tau = ZK$, we can drill two observation wells at radial distances r_1 and r_2 from the pumping well, begin pumping at a steady rate, and wait until the piezometric levels in the two observation wells stabilize. Then, knowing the values of r_1, r_2, H_1, H_2, and Q_p, we can calculate the value of τ from the well-known *Thiem equation*:

$$\tau = ZK = [Q_p/2\pi(H_2 - H_1)]\ln(r_2/r_1) \tag{4.9}$$

An alternative derivation is to begin with the cylindrical form of the Laplace equation:

$$\frac{1}{r}\frac{\partial}{\partial r}\left(r\frac{\partial H}{\partial r}\right) + \frac{1}{r^2}\frac{\partial^2 H}{\partial \alpha^2} + \frac{\partial^2 H}{\partial z^2} = 0 \tag{4.10}$$

The last two terms of this equation are zero if the aquifer is uniform throughout its height and the flow is cylindrically symmetrical (i.e., the flux varies neither with elevation nor with angle α). Therefore,

$$\frac{1}{r}\frac{\partial}{\partial r}\left(r\frac{\partial H}{\partial r}\right) = 0 \quad \text{or} \quad \frac{1}{r}\frac{\partial}{\partial r}(r\,q_r) = 0 \tag{4.11}$$

where q_r is the flux in the radial direction. Hence

$$\frac{\partial}{\partial r}(r\,q_r) = 0 \tag{4.12}$$

Integrating, we obtain

$$r\, q_r + C = 0 \tag{4.13}$$

or

$$r\, q_r - r_0 q_0 = 0 \tag{4.14}$$

The constant of integration C has been evaluated by setting $r = r_0$ (the radius of the perforated casing of the well), for which $C = -r_0 q_0$. The discharge from the well Q_p (i.e., the volume extracted per unit time) can be equated to the product of the flux at the well casing q_0 and the area of that casing $(2\pi r_0 Z)$:

$$Q_p = -2\pi r_0 Z q_0 \tag{4.15}$$

in which the negative sign is introduced to account for the inward direction of flow, which is negative with respect to r (since the latter increases outward). Thus

$$q_0 = -Q_p/2\pi r_0 Z \tag{4.16}$$

Substituting this equation into (4.14) then gives

$$r q_r + Q_p/2\pi Z = 0 \tag{4.17}$$

or

$$Q_p = -2\pi Z r q_r \tag{4.18}$$

Recalling the radial form of Darcy's equation,

$$q_r = -K\, dH/dr \tag{4.19}$$

and combining this with the preceding equation, we get

$$Q_p = 2\pi Z r K\, dH/dr \tag{4.20}$$

This can be integrated as follows:

$$\int_{H_0}^{H} dH = \frac{Q_p}{2\pi ZK} \int_{r_0}^{r} \frac{dr}{r} \tag{4.21}$$

where H_0 is the hydraulic head at r_0 (the surface of the wall cavity), and H is the hydraulic head at any radial distance r. The solution is

$$H - H_0 = (Q_p/2\pi ZK) \ln(r/r_0) \tag{4.22}$$

which is analogous to the Thiem equation and indicates that the hydraulic head varies logarithmically with radial distance from the well throughout the cone of depression.

Equation (4.9) presupposes steady-state conditions. But what if such conditions take too long to be established, and we lose patience while waiting? Is there some way to obtain the necessary solution even during the transient-state, falling-head phase of the process? To do this, we must first consider the change in aquifer storage resulting from a change in piezometric head. It is easy to perceive that in an unconfined aquifer, a fall in piezometric head means a fall in the water table and consequent desaturation of a corresponding depth increment in the soil, so that the effective height (or thickness) of the aquifer is visibly reduced. But a confined aquifer remains saturated even while the piezometric head declines, so it is difficult to see where the water comes from. However, a decline of piezometric head naturally means a decrease of hydrostatic pressure, which causes decompression (expansion) of water. The volume of water remaining in the aquifer can thus remain constant although some water is withdrawn. But strictly speaking, the volume does not remain constant, since the overburden (i.e., the weight of the strata overlying the confined aquifer), formerly supported in part by the hydrostatic pressure in the aquifer, now causes an incremental compression of the aquifer as the hydrostatic pressure is partially relieved. Consequently, because of these two effects acting in concert, a decline in piezometric head even in a confined aquifer releases water, and the volume released per unit area per unit decline in piezometric level is generally called *storativity*.

Now we can return to our problem of formulating the transient-state phase of the process under consideration. In the past [see Chapter 9 of Hillel (1980)] we have used the continuity equation embodying the conservation of mass principle (inflow rate minus outflow rate must equal rate of change of storage) as a basis for handling transient-state water flow. With storativity $\mathscr{S}$ taken into account, the time rate of change of storage per unit width of aquifer is $\mathscr{S} \, \partial H/\partial t$, and the rate of change of discharge with distance is $Z \, \partial q/\partial x$. (Note that for the sake of simplicity we have returned to the x–y rectangular system depicted in Fig. 4.2). The continuity equation therefore becomes

$$\mathscr{S} \frac{\partial H}{\partial t} = -Z \frac{\partial q}{\partial x} \tag{4.23}$$

Introducing the Darcy equation ($q = -K \, dH/dx$), gives

$$\mathscr{S} \frac{\partial H}{\partial t} = Z \frac{\partial}{\partial x} \left(K \frac{\partial H}{\partial x} \right)$$

which, on the assumption that the saturated conductivity remains constant, becomes

$$\mathscr{S}\frac{\partial H}{\partial t} = ZK\frac{\partial^2 H}{\partial x^2} \tag{4.24}$$

or

$$\frac{\partial^2 H}{\partial x^2} = \frac{\mathscr{S}}{\tau}\frac{\partial H}{\partial t} \tag{4.25}$$

The following transient-state solution for radial geometry is known as the *Theis equation*:

$$d(r, t) = H_i - H(r, t) = (Q_p/4\pi\tau)W(v) \tag{4.26}$$

wherein d is the drawdown at any radial distance r from the pumping well at any time t, H_i is the initial piezometric head, $H(r, t)$ is the piezometric head at any distance and time, Q_p as before is the pumping discharge, τ is aquifer transmissivity, v is a composite variable equal to $r^2\mathscr{S}/4t\tau$, and $W(v)$ is known as the well function.

C. Flow of Unconfined Groundwater

While water in unsaturated soil is strongly affected by suction gradients and its movement is subject to very considerable temporal variations in conductivity resulting from changes in soil wetness, groundwater is always under positive hydrostatic pressure and hence saturates the soil. Thus, no suction gradients and no variations in wetness or conductivity normally occur below the water table and the hydraulic conductivity is maximal and fairly constant in time, though it may vary in space and direction. (We are disregarding here possible effects due to overburden pressures and swelling phenomena.)

Despite the differences between the saturated and unsaturated zones, the two are not independent realms but parts of a continuous flow system. Groundwater is sustained (recharged) by percolation through the unsaturated zone, and the position of its surface, the water table, is determined by the relative rates of recharge versus outflow. Reciprocally, the position of the water table affects the moisture profile and flow conditions above it. One problem encountered in attempting to distinguish between the unsaturated and saturated zones is that the boundary between them may not be exactly at the water table but at some elevation above it, corresponding to the upper extent of the *capillary fringe* (at which the suction is equal to the air-entry value for the soil). Frequently, this boundary is diffuse and scarcely definable, particularly when affected by hysteresis.

The various circumstances, or boundary conditions, under which the movement of unconfined groundwater can occur were elucidated by Kirkham (1957), van Schilfgarrde (1957, 1974), and Childs (1969).

Water may seep into or out of the saturated zone through the soil surface or by lateral flow through the sides of a channel or a porous drainage tube. The groundwater table is hardly ever entirely level and may exhibit steep gradients in the drawdown regions near drainage channels, tubes, or wells. Where the land surface elevation varies, as well as where the amount of infiltration water supply varies areally, the water table can change in depth and may in places and at times even intersect the soil surface and emerge as free (ponding) water above it.

If the depth of the water table remains constant, the indication is that the rate of inflow to the groundwater and the rate of outflow are equal. In other words, where there is downward seepage out of the unsaturated zone, this must be offset by downward or horizontal outflow of the groundwater if the water table is to remain stationary. On the other hand, a rise or fall of the water table indicates a net recharge or discharge of groundwater, respectively. Such vertical displacements of the water table can occur periodically, as under a seasonally fluctuating regimen of rainfall or irrigation. The rise and fall of the water table can also be affected by barometric pressure changes, though generally to a minor degree.

Groundwater flow can be geometrically complex where the profile is heterogeneous or anisotropic (Maasland, 1957) or where sources and sinks of water are distributed unevenly. If the profile above the water table consists of a sequence of layers such that a highly conductive one overlies one of low conductivity, then it is possible for the flow rate into the top layer to exceed the transmission rate through the lower layer. In such circumstances, the accumulation of water over the interlayer boundary can result, temporarily at least, in the development of a "perched" (or secondary) water table with positive hydrostatic pressures. If infiltration persists at a relatively high rate, the two bodies of water will eventually merge. If infil-

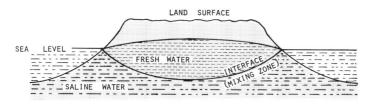

Fig. 4.4. Idealized representation of a Ghyben–Herzberg lens of fresh water floating on saline water under an island or narrow peninsula.

tration ceases, the perched water table will eventually tend to disappear by downward seepage into the primary water table or by evapotranspiration.

In some cases, groundwaters of different density (caused by salinity or temperature differences) may come into contact. One such case, of common occurrence along sea coasts, is that of a body of fresh water (called a *Ghyben–Herzberg lens*; see Fig. 4.4) floating over a body of saline water. At the contact plane, or interface, of the two bodies of water, miscible displacement phenomena can be observed (Bear, 1972).

D. Analysis of Falling Water Table

Truly steady-state flow processes are rare in unconfined aquifers (as elsewhere). More typical are transient-state processes which involve a change in water-table height. Transient flow under such conditions has been described in terms of the *specific-yield* concept, generally defined as the volume of water extracted from the groundwater per unit area when the water table is lowered a unit distance (or as the ratio of drainage flux to rate of fall of the water table). The assumption that there exists anything like a fixed value of *drainable porosity* and that this fraction of the soil volume drains instantaneously as the water table descends is a gross approximation. In actual fact, the volume of water drained increases gradually with the increasing suction which accompanies the progressive descent of the water table.

According to Luthin (1966), the drainable porosity f_d is not a constant but a function of the capillary pressure h (i.e., the negative pressure head, also called the suction head) and can be written as $f_d(h)$. As the water table drops from h_1 to h_2, the volume of water V_w drained out of a unit column will be

$$V_w = \int_{h_2}^{h_1} f_d(h)\, dh \tag{4.27}$$

The function $f_d(h)$ is related to the soil moisture characteristic and is generally a complex function. However, it is sometimes possible to write an approximate expression for the relation of drainable porosity to capillary pressure and still end up with a reasonable prediction of the amount of water that drains out of a profile of soil in which the water table is dropping. The simplest equation to use is that of a straight line $f_d = ah$, where a is the slope of the line. The quantity of water drained is now given by

$$V_w = \int_{h_2}^{h_1} ah\, dh = \frac{a}{2}(h_1^2 - h_2^2) \tag{4.28}$$

Problems involving falling water tables have been handled by representing the transient process as a succession of steady states (Childs, 1947; Kirkham and Gaskell, 1951; Collis-George and Youngs, 1958; Isherwood, 1959; Bouwer and van Schilfgaarde, 1963). Childs and Poulovassilis (1962) showed that the soil-moisture profile above a descending water table depends upon the time rate of that descent.

The time-dependent vertical drainage of a soil column following a drop of the water table has been studied by Day and Luthin (1956), Youngs (1960), Gardner (1962a), Jensen and Hanks (1967), and Jackson and Whisler (1970).

Gardner assumed a constant mean diffusivity and a linear relation between hydraulic head and soil wetness. His equation is

$$\frac{V_w}{V_{w\infty}} = 1 - \frac{8}{\pi^2} \exp\left(-\frac{\bar{D}\pi^2 t}{4L^2}\right) \tag{4.29}$$

where V_w is the volume of water per unit area removed during time t, $V_{w\infty}$ is the total drainage after infinite time, $\bar{D}$ is the weighted mean diffusivity of the soil, and L is the column length (height).

Young's derivation was based on the capillary-tube model and on the assumption of a downward moving *drainage front*, i.e., the assumption that a distinct demarcation of constant suction exists between the still-saturated and the already drained zones. This implies that an equal quantity of water drains for an equal downward advance of the drainage front. The above assumption is analogous to the Green and Ampt assumption pertaining to infiltration (namely, that there exists a distinct wetting front which remains at constant suction as it advances down the profile). The approximate equation obtained is

$$V_w/V_{w\infty} = 1 - \exp(-q_0 t/V_{w\infty}) \tag{4.30}$$

where q_0 is the initial flux of drainage at the base of the column (the water table) and the other variables are as above.

Youngs' approach was modified by Jackson and Whisler (1970), who assumed that hydraulic conductivity is linearly related to average soil wetness during drainage and that the latter is related to the average length of soil column yet to be drained. They also assumed a constant drainable porosity. They claimed that their theory accords with experimental data over a much longer time than previously published theories.

In general, the applicability of such approximate theories depends on the validity of the assumptions made in their derivation. Some approximate solutions which provide reasonably accurate predictions of drainage outflow do not give an accurate description of the changing moisture and suction profiles above the water table (Fig. 4.5). More exact solutions of

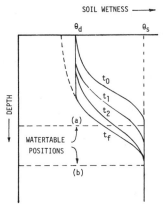

Fig. 4.5. Idealized succession of soil moisture profiles following a rapid drop of the water table from position (a) to (b) (schematic). t_0 and t_f represent the equilibrium moisture profiles at time zero and at final completion of vertical drainage, respectively. t_1 and t_2 represent intermediate times during vertical drainage. θ_s indicates saturation and θ_d the presumed wetness after drainage; i.e., $\theta_s - \theta_d =$ "drainable" porosity. Note that in reality θ_d is generally not a constant, but itself depends on proximity to the water table (as shown by the dashed line extension of the t_f profile), on soil profile characteristics, and on the fact that true equilibrium is practically never attained. Moreover, the soil moisture profile above a fluctuating water table also depends on hysteresis and, if the water table is close to the soil surface, on plant activity as well.

nonsteady drainage problems can be obtained by numerical techniques (e.g., Jensen and Hanks, 1967), but such procedures require the use of a computer for each case considered.

Most of the equations available for the estimation of drainage from soils with falling water tables disregard the possible effect of evapotranspiration. If evapotranspiration occurs while the soil is draining, the amount of drainage will obviously be lessened, and equations such as those cited will give overestimations.

E. Review of Equations Pertaining to Flow of Unconfined Groundwater

As stated in Chapter 8 of Hillel (1980), Darcy's law alone is sufficient to describe only steady flow processes. In general, however, Darcy's law must be combined with the mass-conservation law to obtain the general flow equation for homogeneous isotropic media (Hubbert, 1940; de Wiest, 1969).

$$
\frac{\partial \theta}{\partial t} = -\left(\frac{\partial q_x}{\partial x} + \frac{\partial q_y}{\partial y} + \frac{\partial q_z}{\partial z} \right)
$$

$$
\frac{\partial \theta}{\partial t} = K\left(\frac{\partial^2 H}{\partial x^2} + \frac{\partial^2 H}{\partial y^2} + \frac{\partial^2 H}{\partial z^2} \right)
$$

$$(4.31)$$

where q_x, q_y, and q_z are the fluxes in the x, y, and z directions, respectively, θ is wetness, t time, K conductivity, and H hydraulic head. In a saturated, stable medium, there is no change of wetness (water content) with time, and we obtain

$$K_s\left(\frac{\partial^2 H}{\partial x^2} + \frac{\partial^2 H}{\partial y^2} + \frac{\partial^2 H}{\partial z^2}\right) = 0 \tag{4.32}$$

where K_s is the saturated conductivity. Since K_s is not zero, it follows that

$$\frac{\partial^2 H}{\partial x^2} + \frac{\partial^2 H}{\partial y^2} + \frac{\partial^2 H}{\partial z^2} = 0 \tag{4.33}$$

This equation is known as the *Laplace equation*. The expression $\partial^2/\partial x^2 + \partial^2/\partial y^2 + \partial^2/\partial z^2$, or in vector notation ∇^2, is known as the Laplacian operator. Accordingly, we can write Laplace's equation $\nabla^2 H = 0$.

If, instead of using Cartesian coordinates (x, y, z), we cast Eq. (4.33) into cylindrical coordinates (r, α, z), we obtain

$$\frac{1}{r}\frac{\partial}{\partial r}\left(r\frac{\partial H}{\partial r}\right) + \frac{1}{r^2}\frac{\partial^2 H}{\partial \alpha^2} + \frac{\partial^2 H}{\partial z^2} = 0 \tag{4.34}$$

Laplace's equation also applies to systems other than fluid flow in porous media, namely to the flow of heat in solids and of electricity in electrical conductors. Solutions for boundary values appropriate to the latter systems, some of which are also applicable to soil-water flow, are given by Smythe (1950) and by Carslaw and Jaeger (1959).

The direct analytical solution of Laplace's equation for conditions pertinent to groundwater flow is not generally possible. Therefore, it is often necessary to resort to approximate or indirect methods of analysis. Where flow is restricted to two dimensions, the equation becomes

$$\frac{\partial^2 H}{\partial x^2} + \frac{\partial^2 H}{\partial y^2} = 0 \tag{4.35}$$

which is more readily soluble (Childs, 1969; Domenico, 1972).

A well-known numerical technique is the *method of relaxation* (Southwell, 1946), in which the potentials and flow rates throughout the system are calculated successively and reiteratively in reference to known boundary values. Newer techniques are those of *successive overrelaxation* (Smith, 1965), and *alternating-direction implicit methods* (Peaceman and Rachford, 1955; Varga, 1962).

Approximate solutions for unconfined groundwater flow problems were reviewed by van Schilfgaarde (1957), who stressed that the simplifications possible in a less than rigorous approach can be of considerable value in practice, but that they require a constant awareness of the limitations which are inherent in the use of such simplifications. His discussion encompasses both steady-state processes (in which the potentials in the system do not change with time) and nonsteady, or transient-state, processes (during which the potentials and fluxes vary).

In the solution of problems relating to unconfined groundwater flow toward a shallow sink (a drainage tube or ditch), it is often convenient to employ the *Dupuit–Forchheimer assumptions* (Forchheimer, 1930) that, in a system of gravity flow toward a shallow sink, all the flow is horizontal and that the velocity at each point is proportional to the slope of the water table but independent of depth. Though these assumptions are obviously not correct in the strict sense and can in some cases lead to anomalous results (Muskat, 1946), they often provide feasible solutions in a form simpler than obtainable by rigorous analysis. They are most suitable where the flow region is of large horizontal extent relative to its depth. Kirkham (1967) showed that the Dupuit–Forchheimer assumptions give exact results if they are applied to a soil having infinite conductivity in the vertical direction. For real soils the results are only approximate.

The equation for the free-water surface can be written (van Schilfgaarde, 1957)

$$\frac{\partial^2 h}{\partial x^2} + \frac{\partial^2 h}{\partial y^2} = 0 \qquad (4.36)$$

where h is the height of the saturated soil above an impervious stratum and x and y are the horizontal coordinates. This equation permits determination of the shape of the free-water surface and the velocity at any point for a shallow gravity-flow system in steady state.

Computer-based models and numerical techniques for solving groundwater flow problems were described by Remson *et al.* (1971), Fleming (1972), and Domenico (1972).

F. Flow Nets, Models, and Analogs

One interesting (though often laborious) method for describing groundwater flow is the use of a *flow net* (Cedergren, 1967) which is a graphical representation of the distribution of hydraulic potentials, gradients, flow directions, and fluxes within the soil. The flow net includes lines, or curves, of two types: *equipotentials*, which are curves joining points of equal hydraulic head, and *streamlines*, which are perpendicular to the equipotentials and represent the direction of flow from higher to lower potential. The space between each pair of adjacent equipotentials represents a constant potential drop, while the space between each pair of adjacent streamlines represents a constant discharge. When the complete flow net is known for

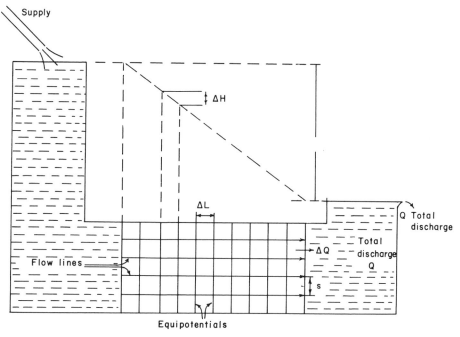

Fig. 4.6. Flow net in a constant-head permeameter.

any system, and the hydraulic conductivity is also known, it is possible to calculate the exact fluxes and their directions in different parts of the soil.

The simplest example of a flow net is the one describing a *constant-head permeameter*, as shown in Fig. 4.6, in which water flows through a soil model of unit thickness (in the plane perpendicular to the page) of height S and of length L. The same model also represents the horizontal flow in a confined aquifer of uniform thickness, depicted in Fig. 4.2. In this hypothetical case, all of the flow lines are straight and parallel. The number of flow lines drawn in the net is arbitrary, but their spacing is such that the discharge rates between each adjacent pair are equal.

In the simple case shown, there is actually no need to draw a flow net, since the geometry is one dimensional and one can use Darcy's law simply and directly. However, in most cases in the field the lines are neither straight nor parallel. In any case, however, flow nets are so drawn that the streamlines intersect the equipotentials perpendicularly,[4] and the discharge rate

---

[4] Strictly speaking, this principle holds only in an isotropic soil. A nonisotropic soil (one with directionally variable conductivity, such as might be caused by a particular pattern of fissures or thin layers) may exhibit streamlines which are not orthogonal to the equipotential surfaces. In such cases, the flow direction is not necessarily that of the steepest potential gradient.

Q in the section bounded by each pair of streamlines is equal to the total discharge divided by the number of such sections in the flow net. Similarly, the head drop in the sections between pairs of equipotentials ΔH is also constant.

In the field, flow conditions can be determined by the measurement of hydraulic head at different points in the soil. This is generally done by means of piezometers, which are borings made to below the water table, to allow repeated measurements of the hydraulic head at various sites (Fig. 4.7). The water levels in piezometers do not indicate the water-table position if there is a vertical component to the flow below the water table, as shown in Fig. 4.8. In this case, the water level in a piezometer depends on the piezometer's depth of penetration. The position of the water table can be estimated from the water level in the shallowest piezometer. The water level difference between any pair of adjacent piezometers inserted to different depths will be proportional directly to the downward (or upward) flux

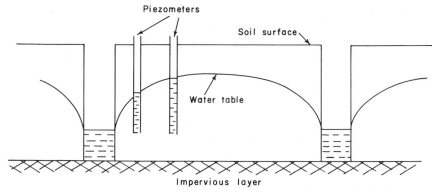

Fig. 4.7.   Piezometers in a system of groundwater drainage by means of open ditches.

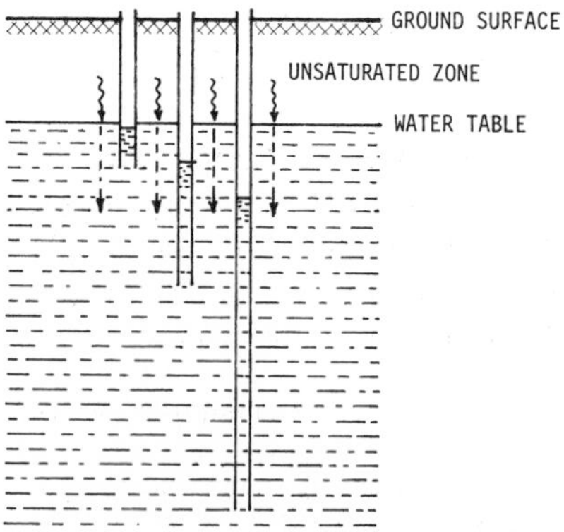

Fig. 4.8.   Battery of piezometers in an unconfined aquifer with downward flow.

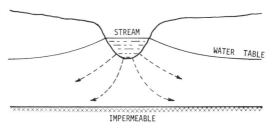

Fig. 4.9. Seepage from a stream into a shallow unconfined aquifer overlying an impermeable layer.

and inversely to the hydraulic conductivity of the intervening soil layer. An example of downward flow below a water table is shown in Fig. 4.9.

To obtain the flow net in the field, the *equipotentials* must first be mapped out by making numerous measurements of the spatial variation of pressure head under the water table. This can be done by means of a grid of regularly spaced batteries of piezometers inserted to different depths and designed to measure the pressure head profiles over the area of interest. A grid of regularly spaced points is then drawn, representing the cross section of the flow domain. The gravity head (relative to a convenient datum) is added to the pressure head to give the total hydraulic head for each point in the grid. Finally, equipotential lines (or curves), analogous to contour lines on a topographic map, are drawn by interpolation between the hydraulic head values of adjacent points in the grid.

After equipotentials of equal interval spacing (say, increments of 10% of the total hydraulic potential range) have been drawn, *streamlines* are sketched perpendicularly to the equipotentials. To achieve a true flow net, the streamlines too must be drawn so that the region bound between each pair of adjacent streamlines, called a *streamtube*, carry an equal discharge (say, 10% of the total flow). The flow Q_i in each streamtube can be calculated by applying Darcy's equation to a segment of the tube between two adjacent equipotentials, yielding

$$Q_i = \tau W \, \Delta H / L \qquad (4.37)$$

where W and L are the width and length of the segment, respectively, ΔH is the total-head difference or equipotential drop of the segment, and τ is the aquifer transmissivity (Bouwer, 1978). The process of constructing a flow net is thus seen to be a rather tedious task, but often worthwhile. An example of how a flow net can provide us with insight into the nature of a complex flow regime in the field is given in Fig. 4.10. A two-dimensional profile is shown of groundwater flow toward a central stream. The curved lines without arrows are equipotentials (connecting points of equal

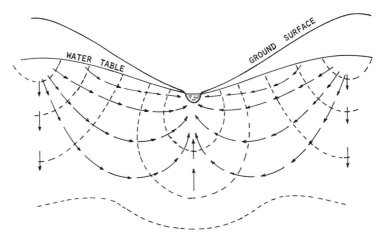

Fig. 4.10. Two-dimensional groundwater flow toward a river showing equipotentials and streamlines.

hydraulic potential). The arrowed lines perpendicular to the equipotentials indicate "downhill'' directions of flow from higher to lower potential and thus represent streamlines. Note that no two equipotentials or two stream-lines, intersect, but streamlines can diverge or converge in different regions of the overall flow system. Note that the water table is an isobar and, unless perfectly horizontal, *not* an equipotential. Significantly, the flow toward the stream is not horizontal but convergent, with a strong upward component for the inflow through the stream bottom. The commonly held notion that groundwater flow is limited to the region just under the water table, or that it is parallel to the water table, while the greater body of groundwater beneath lies dormant and stagnant is seen to be in gross error. This fact has important consequences in connection with the disposal of liquid wastes from so-called septic tank leaching fields or "*sanitary*" *landfills*. The lengths of streamlines in a flow net indicates the path lengths for "average" parcels of water flowing in different regions of the system and are therefore indicative of the residence times for water (and for solutes carried by the water) in the soil. In the example given (Fig. 4.10) the migration time of a water-borne contaminant to the stream is seen to depend greatly on the location of the point of entry into the aquifer.

Hypothetical flow nets for different systems are given in Figs. 4.11 and 4.12. Where the streamlines converge, as they do close to the drains, the flux is greatest.

In a layered soil, the streamlines are often refracted in passage from one layer to the next. This has the effect of changing the flow direction in all

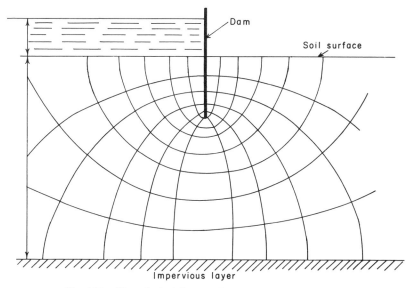

Fig. 4.11. Hypothetical flow net for seepage out of reservoir.

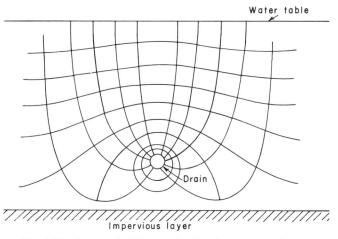

Fig. 4.12. Hypothetical flow net for flow into a drain tube.

cases except where this direction is perpendicular to the interlayer boundary (or where the two layers happen to have equal conductivities). The following relationship holds (Casagrande, 1937):

$$K_1 \cot \alpha_1 = K_2 \cot \alpha_2 \qquad (4.38)$$

where K_1 and K_2 are the conductivities of the two adjacent layers, and

α_1 and α_2 are the incident and emergence angles, respectively, of the streamlines as they pass from layer 1 to layer 2.

Hydraulic models are often used to help visualize groundwater flow systems. Such models generally consist of containers of various dimensions filled with uniform or layered soil in which the boundary and flow conditions of a given flow problem are simulated. The side wall of such a container is generally made of transparent glass or plastic, allowing visual observation of flow. Dyes are sometimes injected at various points to help in tracing the streamlines. One hydraulic model that has often been used in groundwater hydrology is the *Hele–Shaw model*, in which a fluid or fluids (generally oils of higher viscosity than water) are caused to flow between parallel glass plates. The effective hydraulic conductivity at various sections of the system can be varied by changing the separation width between the plates.

Electrical analogs are also used widely in simulating groundwater flow. Their applicability is based on the formal similarity between Ohm's and Darcy's laws.[5] Numerous types of electrical analogs have been proposed to aid in the solution of drainage and soil-water flow problems, from simple conducting sheets to complex arrays of variable resistors and capacitors (e.g., Hanks and Bowers, 1960; Bouwer, 1962, 1964; Luthin, 1974).

G. Groundwater Drainage

The term "drainage" can be used in a general sense to denote outflow of water from soil. More specifically, it can serve to describe the *artificial* removal of excess water, or the set of management practices designed to prevent the occurrence of excess water. The removal of free water tending to accumulate over the soil surface by appropriately shaping the land is termed *surface drainage* and is outside the scope of our present discussion. Finally, *groundwater drainage* refers to the outflow or artificial removal of excess water from within the soil, generally by lowering the water table or by preventing its rise.

Soil saturation *per se* is not necessarily harmful to plants. The roots of very many plants can, in fact, thrive in water, provided it is free of toxic

---

[5] Ohm's law states the linear relation of electric current to the electric potential gradient. It is usually written $I = E/R$ wherein I is current in amperes (a flow rate of one coulomb per second), E is the potential drop (volts) and R is resistance (ohms). Ohm's law can also be written in terms of conductivity K_c. Since the total resistance of a conductor varies directly as the length L, and inversely as the area A and the conductivity K_c, we can write $I = AK_cE/L$ or $i = K_cE/L$ where i is the flow rate per unit area (the flux) and E/L is the potential gradient. This form is analogous to the way Darcy's law is usually written, namely, $q = K \Delta H/L$.

substances and contains sufficient oxygen to allow normal respiration. As is well known, plant roots must respire constantly, since most terrestrial plants are unable to transfer the required oxygen from their canopies to their roots. The problem is that water in a saturated soil seldom can provide sufficient oxygen for root respiration. Excess water in the soil tends to block soil pores and thus retard aeration and in effect strangulate the roots. In *water-logged* soils, gas exchange with the atmosphere is restricted to the surface zone of the soil, while within the profile proper, oxygen may be almost totally absent and carbon dioxide may accumulate. Under anaerobic conditions, various substances are reduced from their normally oxidized states. Toxic concentrations of ferrous, sulfide, and manganous ions can develop. These, in combination with products of the anaerobic decomposition of organic matter (e.g., methane) can greatly inhibit plant growth. At the same time, nitrification is prevented, and various plant and root diseases (especially fungal) are more prevalent.

The occurrence of a high-water-table condition may not always be clearly evident at the very surface, which may be deceptively dry even while the soil is completely water-logged just below the surface zone. Where the effective rooting depth is thus restricted, plants may suffer not only from lack of oxygen in the soil, but also from lack of nutrients. If the water table drops periodically, plants growing in water-logged soils may even, para- doxically, suffer from occasional lack of water, especially when the trans- pirational demand is very high. High moisture conditions at or near the soil surface cause the soil to be susceptible to compaction by animal and ma- chinery traffic. Necessary operations (e.g., tillage, planting, spraying, and harvesting) are thwarted by poor trafficability (i.e., the ability of the ground to support vehicular traffic). Tractors are bogged down and cultivation tools are clogged by the soft, sticky, wet soil. Furthermore, the surface zone of a wet soil does not warm up readily at springtime, owing to greater thermal inertia and downward conduction and to loss of latent heat by the higher evaporation rate. Consequently germination and early seedling growth are retarded.

Plant sensitivity to restricted drainage is itself affected by temperature. A rise in temperature is associated with a decrease in the solubility of oxy- gen in water and with an increase in the respiration rate of both plant roots and soil microorganisms. The damage caused by excessive soil moisture is therefore likely to be greater in a warm climate than in a cold one. Moreover, in a warm climate, the evaporation rate and, hence, the hazard of salinity are likely to be greater than in a cool climate. The process of evaporation inevitably results in the deposition of salts at or near the soil surface, and these salts can be removed and prevented from accumulating only if the water table remains deep enough to permit leaching without subsequent resalinization through capillary rise of the groundwater. Irrigated lands, even in arid regions, frequently require drainage. In fact, irrigation without drainage can be disastrous. Once-thriving civilizations based on irrigated agriculture in river valleys (as in Mesopotamia, for instance) have been destroyed through the insidious, and for a time invisible, process of salt accumulation caused by poor drainage.[6]

In many regions, such as coastal plains and river valleys, large tracts of land that are potentially highly productive lie waste, or are of restricted use, because of excessive moisture. Wherever topographic conditions, soil imperviousness, or the presence of shallow groundwater prevent the profile from draining itself adequately, the soil may become an unsuitable medium for plant growth unless drained artificially. Artificial drainage of such lands can result in the reclamation of millions of acres for the production of food and fiber for the world's growing population. In large areas, likewise, drainage is required for the long-term maintenance of soil productivity. Irrigated agriculture cannot long be sustained in many arid regions unless drainage is provided for salinity control as well as for effective soil aeration (Richards, 1954; Fireman, 1957).

On the other hand, the presence of a shallow water table in the soil profile (provided it is not too shallow) can in certain circumstances be beneficial. Where precipitation or irrigation water is scarce, the availability of groundwater within reach of the roots can supplement the water requirements of crops. However, to obtain any lasting benefit from the presence of a water table in the soil, its level and fluctuation must be controlled.

Numerous investigations of groundwater flow and drainage have resulted in a very extensive body of literature on this subject. Reference should be made particularly to the books by Luthin (1957) and by van Schilfgaarde (1974).

H. Factors Influencing Drainage

The artificial drainage of groundwater is generally carried out by means of *drains*, which may be ditches, pipes, or "mole channels," into which groundwater flows as a result of the hydraulic gradients existing in the soil. The drains themselves are made to flow, by gravity or by pumping, to the *drainage outlet*, which may be a stream, a lake, an evaporation pond, or the sea. In some places, drainage water may be recycled, or reused, for agricultural, industrial, or residential purposes. Because drainage water may contain potentially harmful concentrations of salts, fertilizer nutrients, and

---

[6] This process is not simply a thing of the past, as some irrigation farmers in the United States southwest have lately, and belatedly, discovered.

pesticide residues, it is not enough to provide means to "get rid" of it; nowadays one must be concerned with the quality of the water to be disposed of and with the long-term consequences of its disposal.

The flow rate from soil to drains depends upon the following factors (among others):

(1)   Hydraulic conductivity of the soil, which may differ from one layer or place to another if the soil is heterogeneous and may also vary directionally if the soil is anisotropic. Some soils can be drained easily; others are extremely difficult to drain. Generally, coarse-textured soils drain more readily than fine-textured ones, though of course permeability does not depend on texture alone. The presence of profile layers of low conductivity can greatly retard drainage by affecting the flow pattern in the soil.

(2)   Configuration of the water table and hydraulic pressure of the groundwater. The water table is seldom horizontal or of constant depth. Furthermore, in some cases, the groundwater may be confined and exhibit artesian pressure.

(3)   Depths of the channel or tube drains, relative to the groundwater table and to the soil surface, as well as the slopes of these drains and their outlet elevation.

(4)   Horizontal spacing between drains. According to Kirkham (1949), the discharge per drain becomes constant when the drains are spaced more than about 7 m apart. The total drainage discharged from a field then becomes proportional to the number (i.e., density) of drains installed.

(5)   Character of the drains, whether open ditches or tubes. Open ditches allow a greater seepage surface and are easier to monitor than underground tubes, which are also more expensive to install. However, open ditches break the continuity of the field and are difficult to traverse. They constrain field operations, might scour and collapse, and often allow the proliferation of weeds and pests. Tubes seem therefore to be the preferred mode of drainage.

(6)   Inlet openings in the drain tubes. If segmented tubes are used, small gaps are left between the sections of the tubes to allow inflow from the soil. If continuous tubes are used, they are perforated.

(7)   Envelope materials. Drainage tubes are commonly embedded in gravel to increase the seepage surface and therefore the discharge and to prevent scouring or collapse of the soil at the inlets and possible clogging of the drains by penetration of soil material.

(8)   Diameters of the drains. These must be sufficient to convey the necessary drainage discharge. Installation of wide enough tubes at the outset is no permanent guarantee: drainage tubes tend to clog within a few years, owing to penetration and deposition of sediments and to precipitation of salts (e.g., gypsum) and oxides (e.g., reduced iron and manganese

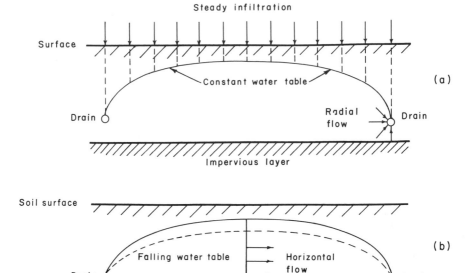

Fig. 4.13. Groundwater drainage (a) under steady flow conditions (infiltration rate equals drainage rate and the water table remains at constant depth), and (b) under unsteady flow resulting in a falling water table.

carried in solution may oxidize and precipitate in the drains). Hence the tubes must be flushed out periodically.

(9) The rate at which water is added to the groundwater (Freeze, 1969) by the excess of infiltration over evapotranspiration or by lateral flow from an external source of water (i.e., outside the field being drained). A steady flow condition, in which the rate of inflow by infiltration is equal to the rate of outflow by drainage, is depicted in Fig. 4.13, in comparison with an unsteady flow condition resulting in a falling water table.

It is a truism which nevertheless bears repeating that water will not spontaneously flow out of the soil into a large cavity or drain unless the pressure of soil water is greater than atmospheric. Drains must be located below the water table to draw water, and the water table cannot be lowered below the drains. Hence the depth and spacing of drains is of crucial importance. Insufficient depth of placement will prevent a set of drains from lowering the water table to the extent necessary. Too great a depth might, on the other hand, deprive the plants of a possibly important source of water during drought periods or whenever evapotranspiration exceeds precipitation.

I. Drainage Design Equations

Various equations, empirically or theoretically based, have been proposed for the purpose of determining the desirable depths and spacings of drain pipes or ditches in different soil and groundwater conditions. Since field conditions are often complex and highly variable, these equations are generally based upon assumptions which idealize and simplify the flow system. The available equations are therefore approximations which should not be applied blindly. Rather, the assumptions must be examined in the light of all information obtainable concerning the circumstances at hand.

One of the most widely applied equations is that of Hooghoudt (1937), designed to predict the height of the water table which will prevail under a given rainfall or irrigation regime when the conductivity of the soil and the depth and horizontal spacing of the drains are known. This equation, like others of its type, oversimplifies the real field situation, as it disregards additional factors which may have a bearing upon groundwater movement, such as the variable rate of evapotranspiration, soil layering, etc. It is based on the following tacit assumptions:

(1) the soil is homogeneous and of constant hydraulic conductivity;
(2) the drains are parallel and equally spaced;
(3) the hydraulic gradient at each point beneath the water table is equal to the slope of the water table above that point (this gradient is generally directed toward the nearest drain);
(4) Darcy's law applies;
(5) an impervious layer exists at a finite depth below the drain;
(6) the supply of water from above, due to rain or irrigation (presumably, after taking evapotranspiration into account) is at a constant flux q.

The shape of the water table between parallel drains is generally described as elliptical.

To derive the Hooghoudt equation, let us examine flow in a profile section of a field having a width of one unit and bounded at its sides by two adjacent drains (tubes or ditches) a distance S apart (Fig. 4.14). Assuming symmetry, we can draw a vertical midplane between the drains which will divide flow toward one drain from flow to the other. Now let us consider flow toward one of the drains through any arbitrary vertical plane located a distance x from that drain. The quantity of water passing through this plane per unit time must be equal to the percolation flux q multiplied by

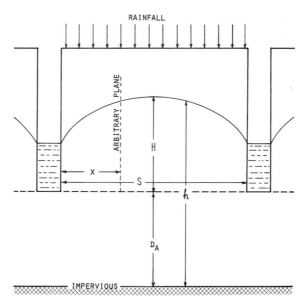

Fig. 4.14. Model used in derivation of Hooghoudt's equation.

the width from the arbitrary plane to the midplane between the drains. This width is $\frac{1}{2}S - x$. Accordingly, the horizontal flow per unit time through the arbitrary plane is

$$Q = -q(\tfrac{1}{2}S - x) \tag{4.39}$$

At the same time, Q can be obtained from Darcy's law. If we assume the effective gradient to be equal to the slope of the water table (dh/dx) at the arbitrary vertical plane, we get

$$Q = -Kh\,dh/dx \tag{4.40}$$

where K is the hydraulic conductivity and h is the height of the water table above an impervious layer which is assumed to form the "floor" of the flow system. Now we can equate the two equations:

$$q(\tfrac{1}{2}S - x) = Kh\,dh/dx \tag{4.41}$$

or

$$\tfrac{1}{2}qS\,dx - qx\,dx = Kh\,dh$$

which can be integrated to yield

$$qSx - qx^2 = Kh^2 \tag{4.42}$$

Assuming that at $x = 0$ (i.e., at the drain) $h = d_a$ (the height of the drain above the impervious floor), while at $x = \frac{1}{2}S$ (the midplane) $h = H + d_a$ (where H is the maximal height of the water table above the drains), we obtain Hooghoudt's equation for the elliptical shape of the water table between drains:

$$S^2 = (4KH/q)(2d_a + H) \tag{4.43}$$

This equation has been used widely for determining the desirable spacing and depth of drains needed to maintain the water table below a certain level. That level, as well as the average infiltration flux and hydraulic conductivity, must be known a priori. A depth must also be known, or assumed, for the impervious layer.

In the event that H is negligible compared to d_a, we can write

$$Q = -Kd_a \, dH/dx \tag{4.44}$$

Equating this with (4.39), we obtain

$$q(\tfrac{1}{2}S - x) = Kd_a \, dH/dx, \qquad \tfrac{1}{2}qS \, dx - qx \, dx = Kd_a \, dH$$

Integration yields

$$qSx - qx^2 = 2KHd_a + \text{const} \tag{4.45}$$

Since at $x = 0$, $H = 0$, the constant of integration is also equal to 0. Therefore

$$H = qx(S - x)/2Kd_a \tag{4.46}$$

At the midpoint between drains, where $x = \frac{1}{2}S$, we get for the maximum height $H_{\max}$ of the water-table mound

$$H_{\max} = qS^2/8Kd_a \tag{4.47}$$

We thus see that the height of rise of the water table between drains is related directly to the recharging flux q and to the square of the distance S between drains and inversely related to the soil's hydraulic conductivity K.

An equation to describe the time rate of water-table drop at the midpoint between drains following an abrupt drop of the water table at the drains, known as the *Glover equation*, is as follows:

$$S^2 = (\pi^2 K \bar{h}_i t/f_d) \ln(4H_i/\pi H) \tag{4.48}$$

wherein $\bar{h}_i$ is the average initial depth of the water-bearing stratum, f_d the assumed drainable porosity, H_i the initial height of the midpoint water table above the drains, and H the height at any time t.

Among the most serious weaknesses of the approach described are the assumptions of an impervious layer at some definable shallow depth and

Table 4.1
PREVALENT DEPTHS AND SPACINGS OF DRAINAGE TUBES
IN VARIOUS SOIL TYPES

Soil type	Hydraulic conductivity (cm/day)	Spacing of drains (m)	Depth of drains (m)
Clay	0.15	10–20	1–1.5
Clay loam	0.15–0.5	15–25	1–1.5
Loam	0.5–2.0	20–35	1–1.5
Fine, sandy loam	2.0–6.5	30–40	1–1.5
Sandy loam	6.5–12.5	30–70	1–2
Peat	12.5–25	30–100	1–2

the disregard of that portion of the total flow which occurs above the water table (Donnan, 1947; Bouwer, 1959). Corrections to account for that flow were described by van Schilfgaarde (1974).

Other equations, derived by alternative and in some cases more rigorous procedures, have been offered by, among others, Kirkham (1958), Ernst (1962), and the U.S. Bureau of Reclamation (Luthin, 1966). The ranges of depth and spacing generally used for the placement of drains in field practice are shown in Table 4.1.

It is of interest to note that in Holland, the country with the most experience in drainage, a common criterion for drainage is to provide for the removal of about 7 mm/day and to prevent a water-table rise above 50 cm from the soil surface. In more arid regions, because of the greater evaporation rate and groundwater salinity, the water table must generally be kept very much deeper. In the Imperial Valley of California, for instance, the drain depth ranges from about 150 to 300 cm and the water-table depth midway between drains is about 120 cm. For medium- and fine-textured soils the depth should be greater still where the salinity risk is high. Since there is a practical limit to the depth of drain placement, it is the density of drain spacing which must be increased under such circumstances. By setting adjacent lines closer together, we can ensure that their drawdown curves will intersect at a lower midpoint level.

Sample Problems

1. Two confined aquifers, geometrically similar to the one depicted in Fig. 4.2, were compared, as shown in the accompanying table.

		Aquifer A	Aquifer B
(a)	Horizontal distance between channels	200 m	90 m
(b)	Hydraulic head difference between channels	10 m	15 m
(c)	Discharge per unit length of channel	0.02 m³/m hr	0.08 m³/m hr
(d)	Vertical thickness of conducting layer	8 m	6 m

Calculate the transmissivity and hydraulic conductivity values.
The system is described by a form of Eq. (4.4):

$$Q = KZ \, \Delta H/\Delta x = \tau \, \Delta H/\Delta x$$

where Q is discharge, K hydraulic conductivity, Z vertical thickness of the aquifer, H hydraulic head difference, Δx distance ($\Delta H/\Delta x$ being the hydraulic gradient), and τ transmissivity ($= KZ$).

To calculate transmissivity, we set

$$\tau = Q/(\Delta H/\Delta x)$$

and to calculate conductivity

$$K = Q/Z(\Delta H/\Delta x) = \tau/Z$$

For Aquifer A,

$$\tau = (0.02 \text{ m}^3/\text{m hr})/(10 \text{ m}/200 \text{ m}) = 0.4 \text{ m}^2/\text{hr}$$

$$K = (0.4 \text{ m}^2 \text{ hr})/8 \text{ m} = 0.05 \text{ m/hr} = 1.39 \times 10^{-3} \text{ cm/sec}$$

For Aquifer B,

$$\tau = (0.08 \text{ m}^3/\text{m hr})/(15 \text{ m}/90 \text{ m}) = 0.48 \text{ m}^2/\text{hr}$$

$$K = (0.48 \text{ m}^2/\text{hr})/6 \text{ m} = 0.08 \text{ m/hr} = 2.22 \times 10^{-3} \text{ cm/sec}$$

2. Consider a confined aquifer of the type depicted in Fig. 4.3, with a single pumping well and two piezometers located at 50 m intervals along a radial line from the well. Water is pumped out continuously at the rate of 100 m³/hr. After the "cone of depression" (drawdown) stabilizes and steady-state conditions are established, the hydraulic head difference between the two piezometers is found to be 2 m. Calculate the aquifer's transmissivity. If its thickness is 60 m, calculate the aquifer's hydraulic conductivity.

This steady-state radial-flow case is described by Eq. (4.9), known as the Thiem equation:

$$\tau = ZK = [Q_p/2\pi(H_2 - H_1)] \ln(r_2/r_1)$$

where τ is transmissivity, Z aquifer thickness, K hydraulic conductivity,

Q_p pump discharge, and H_1 and H_2 are hydraulic head values at the piezometers located at radial distances r_1 and r_2, respectively, from the well.

Substituting the appropriate values, we can write

$$\tau = 60 \text{ m} \times K = \frac{100 \text{ m}^3/\text{hr}}{6.28 \times 2 \text{ m}} \ln \frac{100 \text{ m}}{50 \text{ m}} = 5.52 \text{ m}^2/\text{hr}$$

$$K = \frac{5.52 \text{ m}^2/\text{hr}}{60 \text{ m}} = 2.55 \times 10^{-3} \text{ cm/sec}$$

3. An initially saturated vertical soil column is drained by dropping the water table abruptly to a level 1 m below its original height, as depicted in Figure 4.15. Plot the fractional amount of drainable water removed as a function of time if the weighted mean diffusivity of the draining soil is $10^{-2} \text{ cm}^2/\text{sec}$. Use Eq. (4.29).

According to Gardner (1962a), the volume of water V_w drained per unit area, as a fraction of the total drainable water $V_{w\infty}$ (i.e., the cumulative volume per unit area which drains after infinite time), is a function of time t, weighted mean diffusivity $\bar{D}$ and column length L, according to

$$V_w/V_{w\infty} = 1 - (8/\pi^2) \exp(-\bar{D}\pi^2 t/4L^2)$$

We can now substitute the appropriate values of $\bar{D}$ and L and assign successive values of t to obtain actual solutions, as follows:

At $t = 1 \text{ hr} = 3600 \text{ sec}$,

$$V_w/V_{w\infty} = 1 - 8/(9.87 \times e^\varepsilon)$$

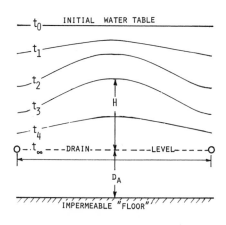

Fig. 4.15. Idealized succession of water table positions during drainage of an initially saturated profile.

where $\varepsilon = 10^{-2} \times 9.87 \times 3600/(4 \times 100^2) = 8.88 \times 10^{-3}$. Hence

$$V_w/V_{w\infty} = 0.197 = 19.7\%$$

At $t = 4$ hr $= 14400$ sec,

$$\varepsilon = 10^{-2} \times 9.87 \times 14400/(4 \times 100^2) = 3.55 \times 10^{-2}$$

$$V_w/V_{w\infty} = 1 - 8/(9.87 \times e^{3.55 > 10^{-2}}) = 0.218 = 21.8\%$$

At $t = 1$ day $= 86400$ sec,

$$\varepsilon = 10^{-2} \times 9.87 \times 86400/(4 \times 100^2) = 0.213$$

$$V_w/V_{w\infty} = 1 - 8/(9.87 \times e^{0.213}) = 0.345 = 34.5\%$$

At $t = 2$ days $= 172800$ sec,

$$\varepsilon = 10^{-2} \times 9.87 \times 172800/(4 \times 100^2) = 0.426$$

$$V_w/V_{w\infty} = 1 - 8/(9.87 \times e^{0.426}) = 0.471 = 47.1\%$$

At $t = 4$ days $= 345600$ sec,

$$\varepsilon = 0.852 \qquad V_w/V_{w\infty} = 1 - 8/(9.87 \times e^{0.852}) = 0.654 = 65.4\%$$

At $t = 8$ days $= 691200$ sec,

$$\varepsilon = 1.704 \qquad V_w/V_{w\infty} = 1 - 8/(9.87 \times e^{1.704}) = 0.853 = 85.3\%$$

Note: The actual plotting of these data is left as an exercise for students, who may also wish to compare Gardner's theory with one or more of the other theories extant. Note that the equation used is not suitable for very small values of t, for which the fraction of water drained is overestimated (since $V_w/V_{w\infty}$ does not tend to zero as t approaches zero).

4. Use the Hooghoudt equation to compare the necessary drain spacings S for two soils with hydraulic conductivity K values of 10^{-4} and 10^{-5} cm/sec. Assume the allowable maximum water-table mound H_{max} to be 1 m above the drains, which are 2 m (d_a) above an impervious stratum. Total rainfall is 1200 mm and total evapotranspiration 1000 mm during a 6 month growing season.

Recall Eq. (4.47):

$$H_{max} = qS^2/8Kd_a$$

which we can solve for S:

$$S = (8Kd_a H_{max}/q)^{1/2}$$

where q is the average percolation flux. Substituting the given values, we have

for Soil A,

$$S = \left[\frac{8 \times 10^{-4} \text{ cm/sec} \times 200 \text{ cm} \times 100 \text{ cm}}{(120 \text{ cm} - 100 \text{ cm})/15{,}760{,}000 \text{ sec}} \right]^{1/2}$$

in which the number of seconds by which we divide the denominator represents a 6 month period. Hence,

$$S = 3550.8 \text{ cm} = 35.5 \text{ m}$$

for Soil B,

$$S = \left[\frac{8 \times 10^{-5} \text{ cm/sec} \times 200 \text{ cm} \times 100 \text{ cm}}{(120 - 100) \text{ cm}/15{,}760{,}000 \text{ sec}} \right]^{1/2} = 1122.6 \text{ cm} = 11.23 \text{ m}$$

Vapor of vapors, saith Koheleth;
vapor of vapors, all is vapor.
Ecclesiastes 1:2

5   *Evaporation from Bare-Surface Soils*

A. Introduction

Evaporation in the field can take place from plant canopies, from the soil surface, or, more rarely, from a free-water surface. Evaporation from plants, called *transpiration*, is the principal mechanism of soil-water transfer to the atmosphere when the soil surface is covered with vegetation. Soil-water uptake and transpiration by plants is, however, the subject of our next chapter. When the surface is at least partly bare, evaporation can take place from the soil as well as from plants. Since it is generally difficult to separate these two interdependent processes, they are commonly lumped together and treated as if they were a single process, called *evapotranspiration*.

In the absence of vegetation, and when the soil surface is subject to radiation and wind effects, evaporation occurs directly and entirely from the soil. This process is the subject of our present chapter. It is a process which, if uncontrolled, can involve very considerable losses of water in both irrigated and unirrigated agriculture. Under annual field crops, the soil surface may remain largely bare throughout the periods of tilllage, planting, germination, and early seedling growth, periods in which evaporation can deplete the moisture of the surface soil and thus hamper the growth of young plants during their most vulnerable stage. Rapid drying of a seedbed can thwart germination and thus doom an entire crop from the start. The problem can also be acute in young orchards, where the soil surface is often kept bare continuously for several years, and in dryland farming in arid zones, where the land is regularly fallowed for several months to collect and conserve rainwater from one season to the next.

109

Evaporation of soil water involves not only loss of water but also the danger of soil salinization. This danger is felt most in regions where irrigation water is scarce and possibly brackish and where annual rainfall is low, as well as in regions with a high groundwater table.

B. Physical Conditions

Three conditions are necessary if the evaporation process from a given body is to persist. First, there must be a continual supply of heat to meet the latent heat requirement (which is about 590 cal/gm of water evaporated at 15°C). This heat can come from the body itself, thus causing it to cool, or as is more commonly the case, it can come from the outside in the form of radiated or advected energy. Second, the vapor pressure in the atmosphere over the evaporating body must remain lower than the vapor pressure at the surface of that body[1] (i.e., there must be a vapor-pressure gradient between the body and the atmosphere), and the vapor must be transported away, by diffusion or convection, or both. These two conditions—namely, supply of energy and removal of vapor—are generally external to the evaporating body and are influenced by meteorological factors such as air temperature, humidity, wind velocity, and radiation, which together determine the *atmospheric evaporativity* (the maximal flux at which the atmosphere can vaporize water from a free-water surface).[2]

The third condition is that there be a continual supply of water from or through the interior of the body to the site of evaporation. This condition depends upon the content and potential of water in the body as well as upon its conductive properties, which together determine the maximal rate at which the body can transmit water to the evaporation site. Accordingly, the actual evaporation rate is determined either by external evaporativity or by the soil's own ability to deliver water, whichever is the lesser (and hence the limiting factor).

If the top layer of soil is initially quite wet, as it typically is at the end of an infiltration episode, the process of evaporation into the atmosphere will

---

[1] Even in a humid region, the atmosphere on a clear day is likely to be at a relative humidity equivalent to a negative water potential of many score or even hundreds of bars.

[2] Atmospheric evaporativity, also called "the evaporative demand of the atmosphere," is not entirely independent of the properties of the evaporating surface. For instance, the net supply of energy for evaporation is affected by the reflectivity, emissivity, and thermal conductivity of the surface soil. Hence, the evaporative demand acting on a soil will not be exactly equal to evaporation from a free-water surface. The latter itself depends on the size and depth of the water body considered.

generally reduce soil wetness and thus increase matric suction at the surface. This, in turn, will generally cause soil water to be drawn upward from the layers below, provided they are sufficiently moist.

Among the various sets of conditions under which evaporation may occur are the following:

(1)   A shallow groundwater table may be present at a constant or variable depth—or it may be absent (or too deep to affect evaporation). Where a groundwater table occurs close to the surface, continual flow may take place from the saturated zone beneath through the unsaturated soil to the surface. If this flow is more or less steady, continued evaporation can occur without materially changing the soil-moisture content (though cumulative salinization may take place at the surface). In the absence of shallow groundwater, on the other hand, the loss of water at the surface and the resulting upward flow of water in the profile will necessarily be a transient-state process causing the soil to dry.

(2)   The soil profile may be uniform (homogeneous and isotropic). Alternatively, soil properties may change gradually in various directions, or the profile may consist of distinct layers differing in texture or structure.

(3)   The profile may be shallow (of *finite depth*), resting on bedrock or some other impervious floor, or it may be deep (*semi-infinite*). In intermediate cases, the soil profile may be effectively semi-infinite for a time, then become finite as the downward-propagating effect of the evaporation process reaches the bottom boundary.

(4)   The flow pattern may be one dimensional (vertical), or it may be two or three dimensional, as in the presence of vertical cracks which form secondary evaporation planes inside the profile.

(5)   Conditions may be nearly isothermal or strongly nonisothermal. In the latter case, temperature gradients and conduction of heat and vapor through the system may interact with liquid water flow.

(6)   External environmental conditions may remain constant or fluctuate. Such fluctuation, furthermore, can be predictably periodic and regular (e.g., diurnal or seasonal) or it can be highly irregular (e.g., spells of cool or warm weather).

(7)   Soil moisture flow may be governed by evaporation alone or by both evaporation (at the top of the profile) and internal drainage, or redistribution, down below.

(8)   The soil may be stable or unstable. For instance, the surface zone may become denser under traffic or under raindrop impact and subsequent shrinkage. Additionally, the soil surface may become encrusted or infused with salt, which then precipitates as the soil solution evaporates. This is quite

apart from the fact that, as a soil dries, its thermal properties inevitably change, including its thermal conductivity and reflectivity.

(9)   The surface may or may not be covered by a layer of mulch (e.g., plant residues) differing from the soil in hydraulic, thermal, and diffusive properties.

(10)   Finally, the evaporation process may be continuous over a prolonged period of time or it may be interrupted by regularly recurrent or sporadic episodes of rewetting (e.g., intermittent rainfall or scheduled irrigation).

To be studied systematically, each of the listed circumstances, as well as others not listed but perhaps equally relevant, must be formulated in terms of a specific set of initial and boundary conditions. A proper formulation of an evaporation process should account for spatial and temporal variability, as well as for interactions with the above-ground and below-ground environment. We shall now proceed to describe a few of the circumstances under which evaporation of soil moisture may occur. We begin, however, with a description of capillary rise, which is often a precursor and contributor to the process of evaporation in the presence of a high water table.

C. Capillary Rise from a Water Table

The rise of water in the soil from a free-water surface (i.e., a water table) has been termed *capillary rise*. This term derives from the *capillary model*, which regards the soil as analogous to a bundle of capillary tubes, predominantly wide in the case of a sandy soil and narrow in the case of a clay soil. Accordingly, the equation relating the equilibrium height of capillary rise h_c to the radii of the pores is

$$h_c = (2\gamma \cos \alpha)/r\rho_w g \qquad (5.1)$$

where γ is the surface tension, r the capillary radius, ρ_w the water density, g the gravitational acceleration, and α the wetting angle, normally (though not always justifiably) taken as zero. This equation predicts that water will rise higher, albeit less rapidly, in a clay than in a sand. However, soil pores are not capillary tubes of uniform or constant radii, and hence the height of capillary rise will differ in different pores. Above the water table, matric suction will generally increase with height. Consequently, the number of water-filled pores, and hence wetness, will decrease in each soil as a function of height. The rate of capillary rise, i.e., the flux, generally decreases with time as the soil is wetted to greater height and as equilibrium is approached.

The wetting of an initially uniformly dry soil by upward capillary flow,

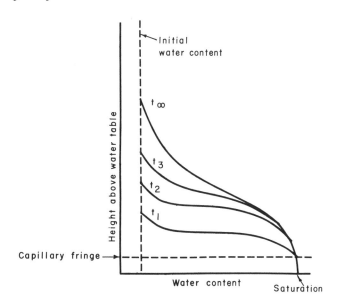

Fig. 5.1. The upward infiltration of water from a water table into a dry soil: water-content distribution curves (moisture profiles) for various times ($t_1 < t_2 < t_3 < t_\infty$). t_∞ is the profile after infinitely long time (equilibrium). Note that the equilibrium curve is in effect the wetting branch of the *soil moisture characteristic*. Note also that what is at first a sharp wetting front, representing the limit of the advancing water, gradually becomes diffuse and ends up as a smooth curve which is characteristic of the particular soil's pore-size distribution.

illustrated in Figure 5.1, is a rare occurrence in the field. In its initial stages, this process is similar to infiltration, except that it takes place in the opposite direction. At later stages of the process, the flux does not tend to a constant value, as in downward infiltration, but to zero. The reason is that the direction of the gravitational gradient is opposite to the direction of the matric suction gradient, and when the latter (which is large at first but decreases with time) approaches the magnitude of the former, the overall hydraulic gradient approaches zero.

Such an ideal state of static equilibrium between the gravitational head and the suction head is the exception rather than the rule under field conditions. In general, the condition of soil water is not static but dynamic—that is to say, constantly in a state of flux rather than at rest. Where a water table is present, soil water generally does not attain equilibrium even in the absence of vegetation, since the soil surface is subject to solar radiation and the evaporative demand of the ambient atmosphere. However, if soil and external conditions are constant, that is, if the soil is of stable structure, the water table is stationary, and atmospheric evaporativity also remains constant (at least approximately)—then, in time, a steady-state flow situation can develop from water table to atmosphere via the soil. To be sure, we must hasten to qualify this statement by noting that in the field the flow regime will at best be a *quasi*-steady-state, since diurnal fluctuations and other perturbations will prevent attainment of truly stable flow conditions. Nevertheless, the representation of this process as a steady-state flow is a useful approximation from the analytical point of view.

D. Steady Evaporation in the Presence of a Water Table

The steady-state upward flow of water from a water table through the soil profile to an evaporation zone at the soil surface was first studied by Moore (1939). Theoretical solutions of the flow equation for this process were given by several workers, including Philip (1957d), Gardner (1958), Anat *et al.* (1965), and Ripple *et al.* (1972).

The equation describing steady upward flow is

$$q = K(\psi)(d\psi/dz - 1) \tag{5.2}$$

or

$$q = D(\theta)\, d\theta/dz - K(\psi) \tag{5.3}$$

where q is flux (equal to the evaporation rate under steady-state conditions), ψ suction head, K hydraulic conductivity, D hydraulic diffusivity, θ volumetric wetness, and z height above the water table. The equation shows that flow stops ($q = 0$) when $d\psi/dz = 1$. Another form of Eq. (5.2) is

$$q/K(\psi) + 1 = d\psi/dz \tag{5.4}$$

Integration should give the relation between depth and suction or wetness:

$$z = \int \frac{d\psi}{1 + q/K(\psi)} = \int \frac{K(\psi)}{K(\psi) + q}\, d\psi \tag{5.5}$$

or

$$z = \int \frac{D(\theta)}{K(\theta) + q}\, d\theta \tag{5.6}$$

In order to perform the integration in Eq. (5.5), we must know the functional relation between K and ψ, i.e., $K(\psi)$. Similarly, the functions $D(\theta)$ and $K(\theta)$ must be known if Eq. (18.6) is to be integrated. An empirical equation for $K(\psi)$, given by Gardner (1958), is

$$K(\psi) = a(\psi^n + b)^{-1} \tag{5.7}$$

where the parameters a,b, and n are constants which must be determined for each soil. Accordingly, Eq. (5.2) becomes

$$e = q = \frac{a}{\psi^n + b}\left(\frac{d\psi}{dz} - 1\right) \tag{5.8}$$

where e is the evaporation rate.

With Eq. (5.7), Eq. (5.5) can be used to obtain suction distributions with height for different fluxes, as well as fluxes for different surface-suction values. The theoretical solution is shown graphically in Fig. 5.2 for a fine sandy loam soil with an n value of 3.

The curves show that the steady rate of capillary rise and evaporation depend on the depth of the water table and on the suction at the soil surface. This suction is dictated largely by the external conditions, since the greater the atmospheric evaporativity, the greater the suction at the soil surface upon which the atmosphere is acting. However, increasing the suction at the soil surface, even to the extent of making it infinite, can increase the flux through the soil only up to an asymptotic maximal rate which depends on

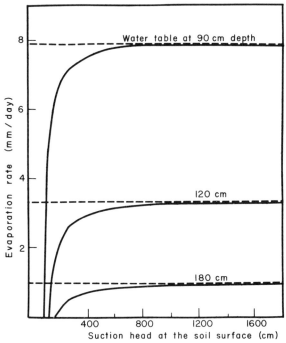

Fig. 5.2. Steady rate of upward flow and evaporation from a water table as function of the suction prevailing at the soil surface. The soil is a fine sandy loam, with $n = 3$. (After Gardner, 1958.)

the depth of the water table. Even the driest and most evaporative atmosphere cannot steadily extract water from the surface any faster than the soil profile can transmit from the water table to that surface. The fact that the soil profile can limit the rate of evaporation is a remarkable and useful feature of the unsaturated flow system. The maximal transmitting ability of the profile depends on the hydraulic conductivity of the soil in relation to the suction.

Disregarding the constant b of Eq. (5.7), Gardner (1958) obtained the function

$$q_{max} = Aa/d^n \tag{5.9}$$

where d is the depth of the water table below the soil surface, a and n are constants from Eq. (5.7), A is a constant which depends on n, and q_{max} is the limiting (maximal) rate at which the soil can transmit water from the water table to the evaporation zone at the surface.

We can now see how the actual steady evaporation rate is determined either by the external evaporativity or by the water-transmitting properties of the soil, depending on which of the two is lower, and therefore limiting. Where the water table is near the surface, the suction at the soil surface is low and the evaporation rate is determined by external conditions. However, as the water table becomes deeper and the suction at the soil surface increases, the evaporation rate approaches a limiting value regardless of how high external evaporativity may be.

Equation (5.9) suggests that the maximal evaporation rate decreases with water-table depth more steeply in coarse-textured soils (in which n is greater)[3] than in fine-textured soils. Nevertheless, a sandy loam soil can still evaporate water at an appreciable rate (as shown in Fig. 5.2) even when the water table is as deep as 180 cm. Figure 5.3 illustrates the effect of texture on the limiting evaporation rate.

The subsequent contributions of a number of workers (Visser, 1959; Wind, 1959; Talsma, 1963) have generally accorded with the above theory.[4] Anat *et al.* (1965) developed a modified set of equations employing dimensionless variables. For basic explanations of the principles and criteria involved in the dimensionless approach, see Corey (1977) and Chapter 12 in this book (by E. E. Miller). Their theory also leads to a maximal evaporation rate e_{max} varying inversely with water-table depth d to the power of n:

$$e_{max} = [1 + 1.886/(n^2 + 1)] d^{-n} \tag{5.10}$$

[3] That is, because in sandy soils the conductivity generally falls off more steeply with increasing suction than in a clayey soil.

[4] Hadas and Hillel (1968), however, found that experimental soil columns deviated from predicted behavior, apparently owing to spontaneous changes of soil properties, particularly at the surface.

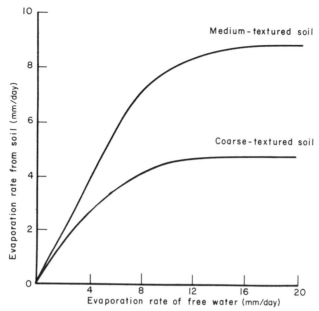

Fig. 5.3. Theoretical relation between the rate of evaporation from coarse- and medium-textured soils (water table depth, 60 cm) and the rate of evaporation from a free-water surface. (After Gardner, 1958.)

A theoretical analysis of steady evaporation from a two-layered soil profile was carried out by Willis (1960), with the following assumptions: (a) the steady flow through the layered profile is governed only by the transmission properties of the profile (external evaporativity taken to be infinite); (b) matric suction is continuous at and through the interlayer boundary, though wetness and conductivity may be discontinuous (i.e., change abruptly); (c) the same empirical $K(\psi)$ function given by Eq. (5.7) holds for both layers, but the values of parameters $a, b,$ and n are different; and (d) each soil layer is internally homogenous. With these assumptions, Eq. (18.2) leads to

$$\int_0^L dz + \int_L^{d+L} dz = \int_{\psi 0}^{\psi L} \frac{d\psi}{1 + e/K_1(\psi)} + \int_{\psi L}^{\psi L + d} \frac{d\psi}{1 + e/K_2(\psi)} \quad (5.11)$$

where L and d are the thicknesses of the bottom and top layers, respectively. The integral in this equation relates water-table depth $L + d$ to the suction at the soil surface for any given evaporation rate. By assuming that the suction at the soil surface is infinite, one can calculate the limiting (maximal) evaporation rate for any given water-table depth and profile layering sequence. Willis developed a graphical method for obtaining the necessary solution.

All of the above treatments apply to cases in which soil properties are the sole factor determining the evaporation rate. A more realistic approach should include cases in which meteorological conditions can also play a role. A more flexible treatment of steady-state evaporation from multilayer profiles might also be based on numerical, rather than analytical or graphical, methods of solution. Such an approach was indeed developed by Ripple *et al.* (1972). Their procedure makes it possible to estimate the steady-state evaporation from bare soils (including layered ones) with a high water table. The field data required include soil-moisture characteristic curves, water-

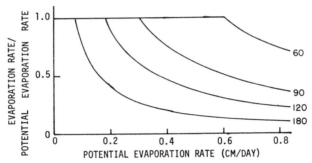

Fig. 5.4. Dependence of relative evaporation rates e/e_{pot} upon the potential evaporation rate (evaporativity), for a clay soil. Numbers labeling the curves indicate the depth to water table (cm). (After Ripple *et al.*, 1972.)

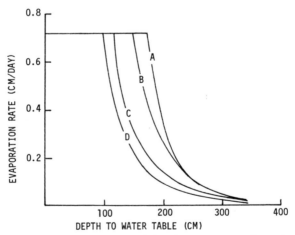

Fig. 5.5. Influence of layering on the relation between evaporation rate and depth to water table. Limiting curves of soil-water evaporation are shown for the homogeneous case (A), a two-layer soil with the upper layer thickness of either 3 (B) or 10 cm (C), and a three-layer soil with the thickness of the intermediate and uppermost layers equal to 10 cm each (D). (After Ripple *et al.*, 1972.)

table depth, and standard-elevation records of air temperature, air humidity, and wind velocity. The theory takes into account both the relevant atmospheric factors and the soil's capability to transmit water in liquid and vapor forms. The possible effects of thermal transfer (except in the vapor phase) and of salt accumulation are still neglected, however. The results of their work are illustrated in Fig. 5.4 and 5.5.

E. Hazard of Salinization Due to High Water Table

The rise of water from a shallow water table can in some cases serve the useful purpose of supplying water to the root zone of crops. On the other hand, this process also entails the hazard of salinization, especially where the groundwaters are brackish and potential evaporativity is high. In fine-textured soils, the danger of salinization can be appreciable even where the water table is several meters deep. The tendency for water to be drawn from the water table toward the soil surface will persist as long as the suction head prevailing at the surface is greater than the depth of the water table. The gradual and irreversible salinization of the soil may have been the process responsible for the destruction of once-thriving agricultural civilizations based on the irrigation of river valleys with high-water-table conditions. Excessive irrigation tends to raise the water table and thus aggravate the salinization problem.

Lowering the water table by drainage can decisively reduce the rate of capillary rise and evaporation. Drainage is a costly operation, however, and it is therefore necessary, ahead of time, to determine the optimal depth to which the water table should be lowered. Among the important considerations in this regard is the necessity to limit the rate of capillary rise to the surface (Gardner, 1958). In the soil described by Fig. 5.2, for example, the maximal rate of profile transmission to the surface is 8 mm/day when the water table is at a depth of 90 cm. Since potential evaporativity is seldom greater than this, it follows that a water-table rise above the depth of 90 cm would not be likely to increase the evaporation rate. On the other hand, a lowering of the water table to a depth of 180 cm can decrease the evaporation rate to 1 mm/day.[5] An additional lowering of the water table to a depth of 360 cm will reduce the maximal evaporation rate to 0.12 mm/day, while any further lowering of the water table can cause only a negligible reduction of evaporation and might in any case be prohibitively expensive to carry out.

Uniform profiles occur only rarely. As mentioned in our preceding section,

[5] This is a consequence of the equation $q_{max} = Aad^{-n}$. When $n = 3$, the doubling of d (from 90 to 180 cm) will reduce q_{max} by 2^3, i.e., by a factor of 8.

the soil often consists of several more or less distinct layers. Layered conditions may in fact exist even when the soil is texturally uniform, owing to structural differentiation. Holmes *et al.* (1960) pointed out the possible effects of surface layers with different tilths upon evaporation in the presence of a water table. Talsma (1963) found that tillage reduced the evaporation rate by half and decreased suction below the tilled zone. He attributed these effects to the formation of layered conditions and to the resulting shift in the actual evaporation zone. Gardner (1958) and Gardner and Fireman (1958) showed that a dry layer at the surface reduces the steady evaporation rate in hyperbolic relation to its thickness. Water movement through such a layer occurs mainly by vapor diffusion.

The problem of salt movement and accumulation in soils is treated by Hillel (1980, Chapter 9). A specific elucidation of soil salinity as related to high groundwater conditions and to drainage can be found in the recent book edited by van Schilfgaarde (1974).

F. Evaporation in the Absence of a Water Table (Drying)

Steady evaporation from soils is not a widespread occurrence, since, even where high-water-table conditions exist, water-table depths (as well as external conditions) seldom remain constant for very long. More commonly, soil-moisture evaporation occurs under unsteady conditions and results in a net loss of water from the soil, i.e., it results in drying. The process of drying involves considerable losses of water, especially in arid regions, where these losses can amount to 50% or more of total precipitation (Hide, 1954, 1958). In this section, we shall consider the drying of initially wetted soil profiles which do not contain a water table anywhere near enough to the soil surface to have any bearing on the evaporation process.

We begin once again by assuming that external conditions, and hence atmospheric evaporativity, are constant. Under such conditions (which, incidentally, happen to be the easiest to set up in laboratory experiments), the soil drying process has been observed to occur in three recognizable stages (Fisher, 1923; Pearce *et al.*, 1949):

(1) An initial *constant-rate stage*, which occurs early in the process, while the soil is wet and conductive enough to supply water to the site of evaporation at a rate commensurate with the evaporative demand. During this stage, the evaporation rate is limited by, and hence also controlled by, external meteorological conditions (i.e., radiation, wind, air humidity, etc.) rather than by the properties of the soil profile. As such, this stage, being *weather controlled*, is analogous to the *flux-controlled stage* of infiltration (in

contrast with the *profile-controlled stage*; see Sections B and J, on rain infiltration, in Chapter 2). The evaporation rate during this stage might also be influenced by soil surface conditions, including surface reflectivity and the possible presence of a mulch, insofar as these can modify the effect of the meterorological factors acting on the soil. In a dry climate, this stage of evaporation is generally brief and may last only a few hours to a few days.

(2) An intermediate *falling-rate stage*, during which the evaporation rate falls progressively below the potential rate (the evaporativity). At this stage, the evaporation rate is limited or dictated by the rate at which the gradually drying soil profile can deliver moisture toward the evaporation zone. Hence it can also be called the *soil profile-controlled stage*. This stage may persist for a much longer period than the first stage.

(3) A residual *slow-rate stage*, which is established eventually and which may persist at a nearly steady rate for many days, weeks, or even months. This stage apparently comes about after the surface-zone has become so desiccated that further liquid-water conduction through it effectively ceases. Water transmission through the desiccated layer thereafter occurs primarily by the slow process of vapor diffusion, and it is affected by the vapor diffusivity of the dried surface zone and by the adsorptive forces acting over molecular distances at the particle surfaces. This stage is often called the *vapor diffusion stage* and can be important where the surface layer is such that it becomes quickly desiccated (e.g., a loose assemblage of clods).

Whereas the transition from the first to the second stage is generally a sharp one, the second stage generally blends into the third stage so gradually that the last two cannot be separated so easily. A qualitative explanation for the occurrence of these stages follows.

During the initial stage, the soil surface gradually dries out and soil moisture is drawn upward in response to steepening evaporation-induced gradients. The rate of evaporation can remain nearly constant as long as the moisture gradients toward the surface compensate for the decreasing hydraulic conductivity (resulting from the decrease in water content). In terms of Darcy's law, $q = K(d\psi/dz)$, we can restate the above principle by noting that the flux q remains constant because the gradient $d\psi/dz$ increases sufficiently to offset the decrease of K. Sooner or later, however, the soil surface approaches equilibrium with the overlying atmosphere (i.e., becomes approximately air dry). From this moment on, the moisture gradients toward the surface cannot increase any more, and in fact, must tend to decrease as the soil in depth loses more and more moisture. Since, as the evaporation process continues, both the gradients and the conductivities at each depth near the surface are decreasing at the same time, it follows that the flux toward the surface and the evaporation rate inevitably decrease as well. As

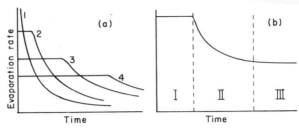

Fig. 5.6. (a) Relation of evaporation rate (flux) to time under different evaporativities (curves 1–4 are in order of decreasing initial evaporation rate). (b) Relation of relative evaporation rate (actual rate as a fraction of the potential rate) to time, indicating the three stages of the drying process.

shown in Fig. 5.6, the end of the first, i.e., the beginning of the second, stage of drying can occur rather abruptly. The pattern of evaporation under different evaporative conditions is also shown in terms of cumulative evaporation in Fig. 5.7.

The tendency of the moisture gradients toward the soil surface to become steeper during the first stage of the process, as the surface becomes progressively drier, and their tendency to become less and less steep during the second stage, after the surface has dried to its final "air-dry" value, is depicted in Fig. 5.8. Continuation of the evaporation process for a prolonged period is sometimes accompanied by the downward movement into the profile of a "drying front" and the development of a distinct desiccated zone, through which water can move from the still-moist underlying layers only by vapor diffusion.

The length of time the initial stage of drying lasts depends upon the intensity of the meteorological factors that determine atmospheric evaporativity, as well as upon the conductive properties of the soil itself. Under similar external conditions, the first stage of drying will be sustained longer in a clayey than in a sandy soil, since clayey soils retain higher wetness and conductivity values as suction develops in the upper zone of the profile.

When external evaporativity is low, the initial, constant-rate stage of drying can persist longer. This fact has led to the hypothesis that an initially

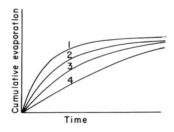

Fig. 5.7. Relation of cumulative evaporation to time (curves 1–4 are in order of decreasing initial evaporation rate).

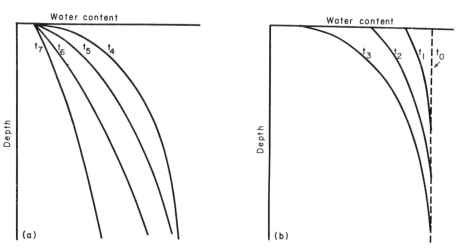

Fig. 5.8. The changing moisture profile in a drying soil. (a) The second stage of drying, in which the moisture gradients decrease as the deeper layers lose moisture by continued upward movement. (b) The first stage, during which the gradients toward the surface become steeper.

high evaporation rate may in the long run reduce cumulative moisture loss to the atmosphere. This hypothesis was raised in a number of Russian papers and cited by Lemon (1956). Gardner and Hillel (1962), on the other hand, concluded that the higher initial drying rate will in fact result in a higher cumulative loss at any time. The total water loss resulting from any finite initial evaporation rate will gradually approach (but never surpass) the total loss in the extreme case where the initial evaporation rate is infinitely high. It is at present debatable whether this pattern holds, strictly speaking, for all profile and environmental conditions or whether the presence of nonuniformities (profile layers, mulches, hysteresis effects, etc.) may modify it.

G. Analysis of the First and Second Stages of Drying

Since the gravitational effect is in general relatively negligible for evaporation,[6] it is possible to base an analysis on the hydraulic diffusivity and water-content gradient relationships. Both stages of drying depend upon the diffusivity: the first stage for its duration, the second stage for its rate.

[6] As we have already pointed out, the equivalent suction exercised by the atmosphere can amount to many (often hundreds of) bars. This suction, distributed over a few centimeters of soil depth, generally constitutes a force many times greater than gravity. For instance, a suction difference of 1 bar over 1 cm of soil is a head gradient of 1000, three orders of magnitude greater than the gravitational head gradient.

A mathematical study of the constant-rate stage of drying of both finite-length and semi-infinite soil columns was carried out by Covey (1963), who neglected gravity and used an exponential dependence of diffusivity upon wetness. For homogeneous soil columns, initially uniformly wet, Covey devised a criterion for determining when the column behaves as though it were infinitely long and when finiteness of length becomes important. When the drying rate is slow, a condition is soon reached in which all depths lose water at about the same rate per unit volume, and the profile dries nearly uniformly with depth. If e is the rate of evaporation (volume of water lost per unit area per unit time) from a soil initially uniformly wet to a depth L, then after a short time, which depends on the diffusivity, the rate of water loss per unit volume will be e/L, and we can then write for the flow equation (Gardner and Hillel, 1962)

$$-\frac{e}{L} = \frac{\partial \theta}{\partial t} = \frac{\partial}{\partial z}\left(D \frac{\partial \theta}{\partial z}\right) \qquad (5.12)$$

where z is height above the bottom of the profile, D diffusivity, θ volume wetness, and t time. The first integral of this equation can be written directly:

$$-ez/L = D \, d\theta/dz + const \qquad (5.13)$$

The constant of integration is zero if there is no flow through the bottom boundary of the profile ($z = 0$), since at that location $d\theta/dz$ is zero. To perform the second integration, D must be a known function of θ. Using an exponential function ($D = D_f \exp aC$, where a is a constant, and C is the soil wetness minus the final air-dry wetness at which $D = D_f$) Gardner and Hillel evaluated the time and remaining water content at the end of the first stage of drying, when the actual evaporation rate ceases being equal to the potential rate. This occurs when C falls to zero at $z = L$. The total water content W of the profile is then approximated by

$$W = (L/a) \ln(1 + eaL/2D_f) \qquad (5.14)$$

At the end of the first stage of drying, the soil surface is more or less at its constant, "final" moisture and the profile in depth has a moisture distribution which depends upon the preceding evaporation rate. In early studies of the second (falling-rate) stage of drying, the initial evaporation rate was assumed to be infinitely high. Subject to an infinite evaporativity, the soil surface is brought instantly to its final state of dryness, so that the first stage of drying ends essentially as soon as it begins. Under such conditions, the profile moisture distribution is obviously not the same as it would be at the onset of the second stage after a prolonged period of constant evaporation at a moderate rate, but the effect of this initial distribution tends to become small after a time, beyond which the evaporation rate depends only on the total remaining water content of the soil. Hence an analysis of the second stage of drying based on the assumption of an initially infinite evaporation rate can be a valid approximation in many cases (Gardner and Hillel, 1962).

For semi-infinite soil columns subjected to infinite evaporativity at the surface, a solution of the flow equation, neglecting gravity, indicates that the cumulative evaporation E is related linearly to the square root of time according to the equation (Gardner, 1959)

$$E = 2(\theta_i - \theta_f)\sqrt{\bar{D}t/\pi} \qquad (5.15)$$

The evaporative flux e, being the time derivative of E, is thus inversely proportional to the square root of time:

$$e = dE/dt = (\theta_i - \theta_f)\sqrt{\bar{D}/\pi t} \qquad (5.16)$$

In these equations, θ_i is the initial profile wetness, θ_f is the final (surface) wetness, and $\bar{D}$ is the weighted mean diffusivity which can be used to characterize the drying process.[7]

Using the same exponential diffusivity function as mentioned above (namely, $D = D_0 \exp[\beta(\theta - \theta_f)/(\theta_i - \theta_f)]$), Gardner obtained an approximate solution for the flow equation by means of an iterative procedure (based on Crank, 1956, p. 152). He also reported experimental results indicating that cumulative evaporation from semi-infinite columns is indeed linear with the square root of time, as predicted by Eq. (5.15). Accordingly, the time required for soil moisture at a specified depth to fall to a given level of dryness is proportional to that depth squared.

Gardner (1959), as well as Klute *et al.* (1965), studied falling-rate evaporation from finite-length columns. In this case, the flux, though initially proportional to $t^{-1/2}$, decreases more steeply with time as the wetness at the

---

[7] Equations (5.14) and (5.15) are similar to those given for horizontal infiltration. The major difference is, however, that $\bar{D}$ must be weighted differently for drying than for wetting. According to Crank (1956), the weighted mean diffusivity for desorption processes is

$$\bar{D} = \frac{1.85}{(\theta_i - \theta_0)^{1.85}} \int_{\theta_0}^{\theta_i} D(\theta)(\theta_i - \theta)^{0.85} \, d\theta$$

This weighting function yields lower values of $\bar{D}$ for corresponding θ values than the approximate weighting function for sorption processes. The need to weight the diffusivity differently can be ascribed to the fact that in infiltration the maximal flux occurs at the wet end of the column, where diffusivity is highest, whereas in drying the greatest flux is through the dry end, where diffusivity is lowest. This fact makes sorption processes inherently faster than desorption processes and contributes to soil-moisture conservation.

bottom of the column is reduced. The evaporation rate thereafter becomes roughly proportional to the product of the average diffusivity and the water content. A field study of evaporation from a sandy soil was reported by Black et al. (1969), who also found that the evaporation rate declined as the square root of time.

Gardner and Hillel (1962) used the following equation to predict the rate of evaporation from finite-length columns during the decreasing-rate stage of drying:

$$e = -dW/dt = D(\bar{\theta})W\pi^2/4L^2 \tag{5.17}$$

where $\bar{\theta}$ is the average volumetric wetness obtained by dividing the total water content of the soil W by the depth of wetting L and $D(\theta)$ is the known diffusivity function. The cumulative evaporation can be obtained by integrating (5.17) with respect to time.

Rose (1966), using the sorptivity concept of Philip (1957d), described evaporation in terms of the following equation:

$$E = st^{1/2} + bt \tag{5.18}$$

where s, the sorptivity (or desorptivity, since we are dealing with a drying process) is positive and b is negative. This approach also applies only to the falling-rate stage of drying.

The numerical procedure of Hanks and Bowers (1962), first developed for infiltration, has been applied to drying by Hanks and Gardner (1965), who studied the effects of different $D-\theta$ relations as well as of layering. The use of computer-based simulation techniques in the analysis of drying processes under various initial and boundary contitions was demonstrated by Hillel (1977).

H. The "Drying-Front" Phenomenon

An interesting phenomenon which deserves attention is the possible development of what might be called a "drying-front," which moves downward into the soil as the drying process progresses. Such a front may be discerned from the inflection of the soil moisture profile. This curve (Fig. 5.9) is usually convex along its entire length at first, but may eventually become concave in its upper part. It appears that this inflection may relate to the diffusivity versus wetness function of the surface-zone soil.

The hydraulic diffusivity for liquid water is known to decrease exponentially with a decrease in the soil's volumetric wetness. However, though the liquid diffusivity decreases, the vapor diffusivity increases as the soil dries, so that the overall diffusivity often indicates a minimum at some low value of

soil wetness

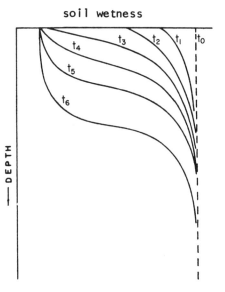

Fig. 5.9. The development of a drying front and its movement into the soil during the course of soil moisture evaporation (hypothetical). Note that t_i represents the initially uniform soil moisture profile at progressively later stages of the drying process.

wetness, beyond which further drying results in an increase, rather than continued decrease, of diffusivity. This phenomenon was pointed out by Philip (1955a,b; 1974; see Fig. 5.10) and studied by Jackson (1964a–c). Its possible effect upon the drying process, and the shape of the soil moisture profile in the surface zone, was postulated by Hillel (1968) and analyzed by van Keulen and Hillel (1974). A description of their work follows.

To consider only the profile-controlled, falling-rate stage of drying, as explained in the preceding section, we can again assume that the soil surface

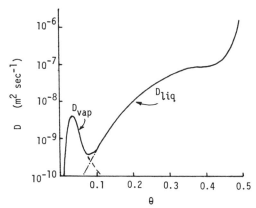

Fig. 5.10. Relation between moisture diffusivity D and moisture content θ for Yolo light clay (18). For $\theta < 0.06$, D includes dominant contribution in vapor phase. (After Philip, 1974.)

becomes instantly air-dry at the start of the process. The one-dimensional, diffusion-type flow equation is once again

$$\frac{\partial \theta}{\partial t} = \frac{\partial}{\partial z}\left[D(\theta)\frac{\partial \theta}{\partial z}\right] \tag{5.19}$$

in which $D(\theta)$ is the overall diffusivity for both liquid and vapor flow. For a uniform soil, and disregarding thermal and gravity effects, it is possible to respresent the drying process by the following conditions:

$$\begin{aligned} \theta &= \theta_i, & z \geq 0, & \quad t = 0 \\ \theta &= \theta_f, & z = 0, & \quad t > 0 \end{aligned} \tag{5.20}$$

For these conditions, the moisture flow equation can be simplified by means of the *Boltzmann transformation* (see Hillel, 1980, Chapter 9). We recall that this consists of substituting a composite variable of space and time, which transforms the partial differential equation into an ordinary one. Introducing the composite variable $B = zt^{-1/2}$, we obtain instead of Eq. (5.19)

$$\frac{d\theta}{dB} = -\frac{2}{B}\frac{d}{dB}\left[D(\theta)\frac{d\theta}{dB}\right] \tag{5.21}$$

which is an ordinary differential equation with B as the only independent variable. This equation can be rewritten in the form

$$\frac{d^2\theta}{dB^2} = -\frac{1}{D}\left(\frac{B}{2}\frac{d\theta}{dB} + \frac{dD}{dB}\frac{d\theta}{dB}\right) \tag{5.22}$$

with the following transformed conditions:

$$\begin{aligned} \theta &= \theta_f, & B = 0 \\ \theta &= \theta_i, & B = \infty \end{aligned} \tag{5.23}$$

Using the procedure described by de Wit and van Keulen (1972), Eq. (5.22), subject to (5.23), was solved numerically for various $D(\theta)$ relations by van Keulen and Hillel (1974), who showed that a "hooked" diffusivity function does indeed result in an inflected soil-moisture profile and in the development of what might be called a drying front.

I. Diurnal Fluctuations of Surface-Zone Moisture and Hysteresis Effects

The foregoing treatments of the evaporation process were based on the precept that the soil surface is subjected to a constant, meterorologically induced, *evaporativity*, which determines the *potential evaporation rate*. In nature, however, evaporativity is obviously not constant but intermittent, as it fluctuates diurnally and varies from day to day, so it may become difficult or even impossible to discern or distinguish between the stages defined in our preceding sections. The resulting course of evaporation may not be described accurately by a simplistic theory based on the assumption of constant evaporativity.

Recently, detailed experimental observations by Jackson (1973) and Jackson *et al.* (1973) showed that the surface-zone soil moisture content fluctuates in a manner corresponding to the diurnal fluctuation of evaporativity; that is to say, the soil surface dries during daytime and tends to rewet during nighttime, apparently by sorption from the moister layers beneath. The phenomenon is illustrated in Fig. 5.11. A similar pattern was found throughout a layer of soil several centimeters thick. The amplitude of the diurnal fluctuation decreased with depth and time, and the daily maxima and minima exhibited an increasing phase lag at greater depths. Because of the high evaporativity occurring in midday, the soil surface desiccated much sooner than it would if the evaporativity had remained constant at its average value. However, Jackson and his co-workers were unable on the basis of their evidence to resolve conclusively the question as to whether the early formation of a dry surface layer might serve to reduce cumulative evaporation below that which would occur had evaporativity remained steady. They noted, in any case, that the concept of three stages of drying appears to have little meaning under natural field conditions.

More recently, Hillel (1975, 1976a) developed a dynamic simulation model capable of monitoring the evaporation process continuously through repeated cycles of increasing and decreasing evaporativity. The model was used in an attempt to clarify the extent to which the diurnal pattern of evaporativity may influence the overall quantity of evaporation and the resulting soil moisture distribution in space and time. Of particular interest is the fluctuating wetness of the near-surface zone, where germination takes place. Computations carried out for a 10 day period accord with experimental findings that the diurnal cycle of evaporativity causes nighttime resorption of moisture and hence a somewhat higher average wetness in the soil surface zone.

The alternating desorption and resorption of moisture by the soil surface zone inevitably involves *soil moisture hysteresis*. Definable as the dependence of the equilibrium state of soil moisture (namely, the relation between wetness and suction) upon the direction of the antecedent process (whether sorption or desorption), soil moisture hysteresis was first studied by Haines (1930), and later by Miller and Miller (1956), Youngs (1960b), Poulovassilis (1962), and many others. More recently, several investigators (e.g., Rubin,

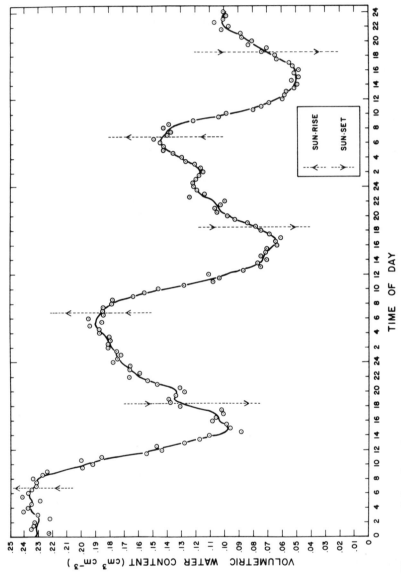

Fig. 5.11. Time course of changing volumetric wetness at the surface (0–0.5 cm) of a loam soil during three days of drying, 5–7 days after irrigation (March 1971). (After Jackson, 1973.)

1966; Bresler *et al.*, 1969; Vachaud and Thony, 1971) have studied the effect of hysteresis on soil-water dynamics, particularly during the postinfiltration redistribution phase, in which hysteresis appears to inhibit the downward drainage of water from the infiltration-wetted top layer.

A priori reasoning leads us to suppose that the hysteresis effect might tend to retard evaporation in a regime of fluctuating evaporativity. After its strong daytime desiccation, the surface zone of the soil draws moisture from below during the night, so that the top layer is in a process of sorption while the underlying donor layer is in a process of desorption. In principle, the hysteresis phenomenon causes a sorbing zone of soil to approach potential equilibrium with a desorbing zone of the same soil while the former is at a lower moisture content, and hence at a lower value of hydraulic conductivity. It would seem to follow that the hysteresis effect can contribute to the self-arresting tendency of the evaporation process by causing it to fall below the potential rate earlier than it would if hysteresis were nonexistent.

A test of this hypothesis was carried out by Hillel (1975), who modified the the earlier simulation model to account for hysteresis. He reported a system-atic reduction of cumulative evaporation with increasing magnitude of the hysteresis range (as indicated by the displacement between the primary desorption and sorption curves). At its greatest, the reduction of evaporation due to hysteresis, as calculated, amounted to about 33% of the nonhysteretic evaporation.

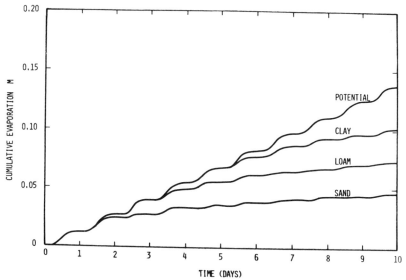

Fig. 5.12. Cumulative evaporation during simultaneous drainage and evaporation from initially saturated uniform profiles of sand, silt, and clay. (After Hillel and van Bavel, 1976.)

A similar study of evaporation from soils of various textures (Hillel and van Bavel, 1976) showed that differences in soil hydraulic properties can strongly influence cumulative evaporation, with coarse-textured soils (sand) evaporating the least and fine-textured soils (clay) evaporating the most, under both steady and cyclic evaporativity regimes (Fig. 5.12).

J. Nonisothermal Evaporation

The discussion so far has dealt with isothermal conditions only. In recent years, it has become increasingly evident that in many cases the effect of temperature gradients and heat flow should not be neglected. The heat required to vaporize water must be transported to the evaporation site, which requires a nonzero temperature gradient. Nevertheless, it has generally been assumed (e.g., Miller and Klute, 1967) that the isothermal flow equation $(\partial\theta/\partial t = \partial(D\,\partial\theta/\partial z)/\partial z)$ predicts many of the essential features of the constant-rate and falling-rate stages of evaporation. According to Philip (1957d), the isothermal model can be expected to represent evaporation processes so long as the surface soil is not extremely dry. Recent work (Fritton *et al.*, 1970) has indicated that the isothermal flow equation can describe cumulative evaporation quantities for both wind and radiation treatments, but that this equation cannot describe the soil water distribution pattern adequately in the case of nonisothermal flow. They concluded that where temperature gradients are important (e.g., near the soil surface) simultaneous heat and mass transfer analysis gives better fit with experimental data than the isothermal diffusion equation.[8]

Vapor movement occurs primarily in the surface zone of the soil. Where conditions are nearly isothermal, the square root of time dependence of

---

[8] Simultaneous equations for analyzing nonisothermal transfer of vapor and liquid water under combined temperature and moisture gradients were offered by Philip and de Vries (1957):

$$\frac{q_w}{\rho_w} = -D_T \frac{dT}{dz} - D_\theta \frac{d\theta}{dz} - K, \qquad q_h = -\lambda \frac{dT}{dz} - \rho_w L D_{\theta\,vap} \frac{dz}{d\theta}$$

where q_w is water flux density, ρ_w is density of water, D_T is thermal water diffusivity, T is temperature, z is depth, D_θ is diffusivity of soil moisture, θ is volumetric wetness, K is hydraulic conductivity, q_h is soil heat flux density, λ is thermal conductivity, which includes the contribution of vapor movement, L is latent heat of evaporation, and $D_{\theta\,vap}$ is vapor diffusivity (where $D_\theta = D_{\theta\,liq} + D_{\theta\,vap}$ and $D_{\theta\,liq}$ is liquid water diffusivity). Philip (1957d) eliminated the dT/dz term from the two simultaneous equations and obtained

$$Z = -\int_{\theta(0)}^{\theta(z)} \{(\lambda D_\theta + \rho_w L D_T D_{\theta vap})/[\lambda(K + E) - D_T q_h]\}\, d\theta$$

where E is evaporation rate (q_w/ρ_w in the first of the two equations above).

evaporation will still hold even if much of the water movement is in the vapor phase, since isothermal vapor movement can be described by a nonlinear diffusion equation of the same form as the diffusivity equation for liquid movement (Jackson, 1964a–c).

A study of the interaction of heat and moisture flow may involve the application of the thermodynamics of irreversible processes (Wiegand and Taylor, 1961). Fundamental studies in this field have been carried out by Taylor and Cary (1960) and by Cary (1963, 1964). Rose (1968) attempted to establish the magnitude of vapor versus liquid water movement during evaporation under nonisothermal conditions. The effect of warming the soil is to lower the suction and to raise the vapor pressure of soil water. Hence the effect of a thermal gradient is to induce flow and distillation from warmer to cooler regions. When the soil surface is warmed by radiation, this effect would tend to counter the tendency to upward flow of water in response to evaporation-induced moisture gradients. For this reason the evaporation rate might be lower when the surface is dried by radiation than when it is dried by wind of comparable evaporation potential (Hanks *et al.*, 1967).

In isothermal models of evaporation, the potential evaporation rate (evaporativity), whether constant or time variable, is generally introduced as an external forcing function. No account is taken of the surface radiation balance or of the soil's energy balance, including such processes as absorption, emission, and conduction of heat by the soil and exchange of heat between soil and atmosphere, nor of thermally induced transport of vapor in the soil's surface zone. Recently, van Bavel and Hillel (1975, 1976) published a comprehensive simulation model of nonisothermal evaporation based on the use of actual weather data (radiation, air temperature and humidity, and wind speed) and soil physical properties as inputs. The model permits the *calculation* of evaporation, both potential and actual, as a resultant of the interaction between weather and soil factors, rather than its *imposition* as a forcing function. In addition to the storage and disposition of water in the soil profile, this model can predict the transformation and partitioning of radiant energy into sensible and latent heat and the resulting pattern of soil temperature.

The reported results obtained with this model suggest that the representation of potential evaporation as a purely external variable is a gross approximation which can lead to appreciable error in the estimation of evaporation even during the early stage of the process. The soil does not simply act in passive obedience to an external dictate and in fact is an active participant in determining its own energy balance and hence its potential and actual daily evaporation rates. Almost from the start, the emittence and long-wave radiance decline with time as the soil loses water by drainage and surface drying. Concurrently, the albedo rises and less shortwave energy is retained. Further, as the soil drains and dries, its thermal conductivity and heat capacity decrease, affecting the soil heat flux. Surface temperature and air stability (affected by the bouyancy of the lower air as it is heated by the ground) might also vary. Because of all these factors, the concept of potential evaporation, even from so "simple" a surface as a bare soil, escapes precise definition. Useful as the idea may be, its use as a quantitative predictor of daily evaporation from a wet soil may lead to errors of 10% or more. The reported study indicates that when the relative humidity at the soil surface drops below 100%, as generally happens within a few days, the diurnal amplitude of surface temperature becomes very much greater, suggesting a practical criterion for defining the transition from the first stage of evaporation to the second stage. The study shows that the daily evaporation during the second stage is indeed related inversely to the square root of time, as predicted by the simple isothermal model elucidated in Section G of this chapter.

K. Effect of Albedo Changes on Nonisothermal Evaporation

Recent work by Jackson *et al.* (1974) has drawn attention to the changes of albedo which occur naturally during evaporation. Since albedo is one soil surface property which can readily be modified artificially, it is of interest to establish how and to what extent its modification might influence evaporation. Preliminary experimental work along these lines has been reported by Stanhill (1965) and by Hillel (1968). In principle, an increase in albedo can reduce the energy load on the surface and hence the temperature of the evaporation zone. On the other hand, a decrease of albedo, while causing a greater warming of the soil surface, might also help to drive water vapor down into the profile and to arrest the evaporation process earlier by the more rapid desiccation of the surface. To answer the question, a comprehensive model, capable of handling soil water and energy dynamics simultaneously, is needed.

Using a modified version of the van Bavel and Hillel (1976) model described in our preceding section, Hillel (1977) conducted a simulation study in which three treatments affecting soil surface albedo were compared:

(1)   leaving the soil surface in its "natural" state, with the albedo varying from a minimum of 0.1 to a maximum of 0.35 as the volumetric wetness of the surface soil decreased from 35% to 10% or less,

(2)   darkening the soil surface, as if by the application of charcoal powder, to produce a low albedo value of 0.05,

(3)   whitening the soil surface, as if by the application of chalk powder, to effect a high albedo of 0.50.

The soil used in the simulation was assigned properties similar to those of a fine sandy loam, and the climatic inputs were arbitrarily chosen to represent a typical late spring or early summer period in a semiarid region.

The results of this study showed that a low albedo condition can cause about 30% more evaporation in comparison with a high albedo condition during the first three days. The difference between the two albedo conditions later appeared to decrease, so that after about ten days the cumulative evaporation value of the low-albedo condition is only about 8% greater than that of the high-albedo condition. It thus appears that increasing albedo can result in reduction of evaporation only in the short run. It is noteworthy that the evaporation curve for the variable albedo condition (designed to simulate the pattern for a natural soil) at first resembled that of the low albedo condition, but as the soil surface dried and albedo increased, the pattern of evaporation tended to approach that of the high albedo condition.

Considering these results, we can conjecture that the principal potential benefit to be derived from modification of surface albedo is likely to be in affecting seedbed conditions during a germination period lasting several days. Where the weather is warm enough so that low temperatures are not likely to inhibit germination but soil moisture may be limiting, whitening the surface may conserve enough water in the seedbed to enhance germination appreciably. On the other hand, under cool weather conditions, where not moisture but low temperature may be the limiting factor, darkening the soil surface may be beneficial in hastening germination despite the increase of evaporation.

L. Evaporation from Irregular Surfaces and Shrinkage Cracks

Thus far we have discussed the drying of soils as if it were entirely a one-dimensional process, i.e., vertically downward from a smooth horizontal surface plane. In many cases, the surface is far from smooth and horizontal: the land as a whole may slope in various directions and degrees of inclination, and the surface may exhibit the sort of geometrically complex microrelief (ridges and furrows, large and small clods) that would cause the drying process to be uneven and three dimensional, at least in its early stages. Differences in temperature and drying pattern between soil surfaces of differing roughness characteristics, e.g., different directions and elevations of ridges, have long been recognized in principle and even utilized in practice in attempts to enhance germination and crop-stand establishment in unfavorable climates (Adams, 1970; Hillel and Rawitz, 1972). As examples we might mention the practice of planting on ridge tops or on sun-facing ridge slopes in cool and wet climates and the opposite practice of planting in furrows or depressions in warm and dry climates. However, a fundamental definition of the role of surface configuration in evaporation and drying is difficult to achieve, as it involves not only three dimensionality in the soil realm but also complex effects in the above-ground realm involving radiation, air motion, and vapor transfer processes.

A particular example of multidimensional evaporation is that of soils which tend to form vertical fissures, commonly called *cracks*, as they dry. Especially prone to cracking are the soils known as *grumusols*, or *vertisols*, which have a high content of montmorillonitic clay and occur typically in an environment of alternating wet and dry seasons. They appear in every continent and are estimated to cover about 40 million hectares in all. These soils are characterized by an extreme tendency to swell upon wetting and to shrink upon drying. Unconfined clods of a typical grumusol have been reported to decrease as much as 40% in specific volume in drying from a suction of $\frac{1}{3}$ to 15 bar (Johnston and Hill, 1945). The cracks which form in a grumusol can be as wide as 10 cm and as deep as 1 meter and form a characteristic polygonal pattern. Such cracks can tear and desiccate plant roots and cause the soil to dry to an extreme degree down to considerable depths. The wide cracks permit the entry of turbulent air, which sweeps out water vapor from the deeper reaches of the soil profile (Adams and Hanks, 1964). The exposed vertical sides of the cracks become secondary evaporation planes, which increase the evaporating surface threefold or fourfold over the evaporating surface of a noncracking soil (Adams and Hanks, 1964). Lateral movement of water toward crack walls can take place from a distance exceeding 10 cm and cracks can increase evaporation by 12–30% (Selim and Kirkham, 1970; Ritchie and Adams, 1974).

The process of crack formation is self-generating in the sense that once a crack begins preferential drying and shrinkage inside it can cause the crack to extend into the soil. However, in the field the cracking pattern appears to remain consistent through repeated cycles of wetting and drying. A thorough wetting causes the soil blocks between the cracks to expand and close the cracks; gradual drying opens them up again. Surface desiccation often results in loose aggregates falling into the cracks, so that a typical grumusol in effect churns or mixes itself as it alternately wets and dries. The large cracks, incidentally, have a very considerable effect on the infiltrability of what might otherwise be very impermeable soils. This effect is most notable during the initial stage of infiltration, before the cracks are closed by soil swelling.

The formation of cracks does not increase the evaporation rate during the

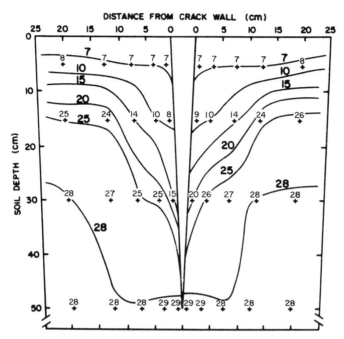

Fig. 5.13. Soil moisture distribution in the vicinity of a shrinkage crack in a vertisol at the end of a summer growing season. The numbers refer to mass wetness values. (After Ritchie and Adams, 1974.)

early stage of the drying process, when that rate is dictated mainly by external weather factors. However, it can extend the duration of that stage and greatly increase the rate at which evaporation continues during the profile-controlled second stage, and perhaps the third stage as well. After the soil surface becomes air dry, evaporation from shrinkage cracks becomes the dominant component of the drying process, as deeper zones in the profile become the main source of water for evaporation.

A two-dimensional representation of soil moisture distribution pattern as it develops in the vicinity of a shrinkage crack is shown in Fig. 5.13. It is at present unclear whether timely cultivation to fill in the cracks can in fact reduce evaporation to any significant degree.

M. Reduction of Evaporation from Bare Soils

In principle, the evaporation flux from the soil surface can be modified in three basic ways: by controlling energy supply to the site of evaporation (e.g., modifying the albedo through color or structure of the soil surface, shading the surface) by reducing the potential gradient, or the force driving water upward through the profile (e.g., lowering the water table, if present, or warming the surface so as to set up a downward-acting thermal gradient), or by decreasing the conductivity or diffusivity of the profile, particularly of the surface zone (e.g., tillage and mulching practices).

The actual choice of means for reduction of evaporation depends on the stage of the process one wishes to regulate: whether it be the first stage, in which the effect of meteorological conditions on the soil surface dominates the process, or the second stage, in which the rate of water supply to the surface, determined by the transmitting properties of the profile, becomes the rate-limiting factor. Methods designed to affect the first stage do not necessarily serve during the second stage, and vice versa.

Covering or *mulching* the surface with vapor barriers or with reflective materials can reduce the intensity with which external factors, such as radiation and wind, act upon the surface (Hanks *et al.*, 1961; Bond and Willis, 1969). Thus, such surface treatments can retard evaporation during the initial stage of drying. A similar effect can result from application of materials which lower the vapor pressure of water (Law, 1964). Retardation of evaporation during the first stage can provide the plants with a greater opportunity to utilize the moisture of the uppermost soil layers, an effect which can be vital during the germination and establishment phases of plant growth. The retardation of initial evaporation can also enhance the process of internal drainage, and thus allow more water to migrate downward into the deeper parts of the profile, where it is conserved longer and is less likely to be lost by evaporation (Hillel, 1968). We shall discuss the practice of mulching in greater detail later in this section.

An opposite approach to evaporation control during the first stage is to induce a temporarily higher evaporation rate so as to rapidly desiccate the surface, thus hastening the end of the first stage and using the hysteresis effect to help arrest or retard subsequent outflow. Heating the surface by darkening its color or by means of flaming, microwave, or plasma-jetting devices has been suggested but not proven in practice. Shallow cultivation designed to pulverize the soil at the surface often has the immediate effect of causing the loosened layer to dry faster and more completely but may over a period of time help conserve the moisture of the soil underneath. The problem here is how to obtain a favorable balance between the short-run loss and the long-run gain.

During the second stage of drying, the effect of surface treatments is likely to be only slight and reduction of the evaporation rate and of eventual water loss will depend on decreasing the diffusivity or conductivity of the soil profile in depth. Deep tillage, for instance, by possibly increasing the range of variation of diffusivity with changing water content, may reduce the rate at which the soil can transmit water toward the surface during the second stage of the drying process. However, the evaporation rate is usually much lower during the second stage than during the first, and it is questionable whether it might be worthwhile to invest in the control of evaporation in this stage.

An irrigation regime having an excessively high irrigation frequency can cause the soil surface to remain wet and the first stage of evaporation to persist most of the time, resulting in a maximum rate of water loss. Water loss by evaporation from a single deep irrigation is generally smaller than from several shallow ones with the same total amount of water. However, water losses due to percolation are likely to be greater from deep irrigations than from shallow ones. New water application methods such as *drip* (or *trickle*) *irrigation*, which concentrate the water in a small fraction of the area while maintaining the greater part of the soil surface in a dry state, are likely to reduce the direct evaporation of soil moisture very significantly.

We shall now proceed to discuss two of the major practices most commonly applied in the effort to reduce evaporation losses: tillage and mulching.

1. TILLAGE AND EVAPORATION

Tillage for seedbed preparation, weed control, or other purposes is perhaps the most ubiquitous of soil management practices. Soil manipulation by tillage implements generally results in opening and loosening of the tilled layer. As mentioned, this often enhances drying of the tilled layer, but might also reduce water movement from the layers below. Tillage might thus produce either a larger or smaller total loss than would have occurred from the undisturbed soil, the net effect depending on duration of the process, as well as on depth, degree, and frequency of tillage. The long-term effect also depends on subsequent rainfall and its influence in reconsolidating the tilled layer. In many cases, failure to conserve water by shallow cultivation is due to the simple fact that most soils cannot be cultivated when wet. Once the soil surface becomes dry enough to permit cultivation, the first stage of evaporation is generally over, so that surface modification can no longer help (since it is the profile in depth which now begins to control the process).

Marshall and Gurr (1966) reported that the zone of evaporation from a moist soil is a layer about 1 cm thick at the top of the moist zone. This zone may thicken somewhat as the drying effect progresses more deeply into the soil, though H. R. Gardner and Hanks (1966) found that it remains about 1 cm thick even as its location descends into the soil. Work of W. R. Gardner (1959) and of Hanks and Gardner (1965) suggests that the total water loss during the first stage of drying may be reduced if the water diffusivity in the wet range is reduced. Only if this is achieved to an appreciable depth does it affect the second stage of drying as well. Accordingly, a shallow aggregated layer or other mulch at the soil surface can be expected to affect the evaporation process primarily during its initial stage. An important qualification is that the profile and surface conditions desired must be achieved prior to water application, since after wetting no mechanical treatments can be performed in practice until the surface dries, by which time much of the water loss has already taken place.

Due to indiscriminate application of experimental findings by F. H. King in Wisconsin during the latter part of the nineteenth century there arose the misconception that soil water could always be conserved by cultivation after every rain. It was assumed that working the soil surface layer into a finely pulverized condition could help to break the "continuity" of capillary "tubes" through which water was presumed to flow from lower layers to the soil surface. As a consequence, expensive operations were adopted which in many cases not only failed to conserve water, but also damaged soil structure instead of improving it. The finely pulverized soil tended to slake and crust after each cultivation because of raindrop impact. In recent years, the trend seems to have been reversed, and the search is on for appropriate methods of soil management based on *minimum* tillage. This trend has been reinforced by the rising cost of energy. Problems remain, however, with our collective tendency to over-generalize specific experiences obtained in given sets of circumstances and extend our conclusions to situations in which they do not really apply.

There are two situations in which cultivation can be beneficial in the context of water conservation. The first is in controlling the growth of weeds, which represent an uneconomical use of soil moisture. In the last decade, increasing use of chemical herbicides, which can be applied by spraying from the air, seemed to obviate the need for cultivating wet soils for weed control. However, recent recognition of the danger of environmental pollution may curtail the widespread application of some of these chemicals.

The second situation requiring tillage is the case of soils which, because of appreciable swelling and shrinking, as a consequence of wetting and drying, tend to crack markedly during the dry season. These cracks become secondary evaporation surfaces penetrating inside the soil, causing it to desiccate deeply and to an extreme degree. Timely cultivations may prevent the development of such cracks or at least help obliterate cracks after they have begun to form. The quality of tillage, i.e., the degree of pulverization and clod size distribution and porosity of the tilled layer are also important. Coarse tillage leaves large cavities among the clods into which air currents can penetrate and increase vapor outflow (Hillel, 1968). Tillage that is too fine may pulverize the soil unduly and result in a compact surface that would defeat the aim of reducing diffusivity. Each set of soil and meteorological conditions may require its own optimum depth, timing, and quality of tillage to effect maximal reduction of evaporation losses.

In principle, the best condition from the standpoint of both evaporation and infiltration control is to have a coarse layer with large pores at the surface, overlying a finer textured or structured profile. The coarse surface layer would conduct the water readily at saturation during infiltration but would restrict liquid flow as suction develops during evaporation and as the soil with the large pores desaturates and its hydraulic conductivity decreases drastically. Since vapor diffusion in dry soils is generally much slower than liquid flow in moist soils (typically by several orders of magnitude) rapid drying of the surface few millimeters may reduce subsequent evaporation.

2. MULCHING

A *mulch* is a cover of the soil surface, and *mulching* is the artificial application of a mulch, practiced to obtain beneficial changes in the soil environment. Mulches can consist of many different types of materials, ranging from natural plant residues to gravel to various industrial products such as paper or plastic sheets. A soil cover may change the soil's radiation balance, water balance, temperature regime, and nutrient status, among other factors. The possible effects of a mulch are in fact so complex that they cannot in general be predicted universally. However, various mulches have commonly been observed to reduce evaporation and to alter the temperature regime of the uppermost soil layers (van Bavel and Hillel, 1975).

The effectiveness of vegetative mulches may be limited unless they are sufficiently thick, because their high porosity permits rapid diffusion and air currents (Hanks and Woodruff, 1958). The initial evaporation rate under a mulch is usually reduced (Greb, 1966), so water is saved if rains are frequent. But for extended rainless periods a mulch may keep the soil surface moister and thus produce no net saving of water. Since mulches of crop residues tend to be more reflective than most soils, they may result in cooler surface temperatures. Consequently, early plant growth under mulches is often retarded.

Gravel mulching is an age-old method and can be very effective in water conservation (both in enhancing infiltration and in suppressing evaporation) even in layers as thin as 5–10 mm (Faribourn and Cluff, 1974). Such mulches can also help reduce erosion by water and wind. If light colored, they cool, and if dark (e.g., volcanic ash) they warm the soil. The main problem here is that gravel is expensive to mine, transport, and apply in the field. Moreover, once applied, gravel cannot be removed, does not decompose, and might hinder future operations or land uses.

When gravel fragments are large enough they qualify for the term stones, and when stones are large they are called rocks. Many soils in mountainous terrain, in glaciated areas, and—particularly—in desert regions, exhibit an abundance of stone and rock fragments of various sizes, and many outcrop on the soil surface. A recent study on the effect of surface rocks on soil heat and moisture was carried out by Jury and Bellantuoni (1975). They suggested that, in addition to acting as barriers to evaporative loss, surface stones might induce lateral movement of heat and vapor, which could in turn collect water under the stones in amounts sufficient to serve as a source of water for some species of desert plants. The effectiveness of a rock cover can be expected to be most pronounced in relatively dry soil with high soil surface temperatures and where the rock cover is discontinuous (permitting part of the surface to be heated by incoming radiation while the covered regions are partially insulated).

Other mulches which have been applied with varying degrees of success are paper and polyethylene plastic sheeting and sprayable preparations of latex, asphalt, oil, fatty alcohols, and other materials too numerous to mention. The problem, once again, is cost, which at present can be justified only for intensive agriculture with high-income crops such as pineapples and strawberries (the latter are improved in quality by prevention of their contact with the soil).

A novel approach to mulching was recently described by Hillel (1976b). It consists of forming the uppermost soil layer into clods and treating them with water proofing chemicals (e.g., silicones) to make them water repellent. The layer of water-repellent aggregates can then constitute a dry mulch that will serve as an evaporation barrier and as a detention reservoir to prevent runoff and increase infiltration.

Sample Problems

1.   The average daily rate of evaporation from a saturated uniform soil with a high water-table condition is 0.9 cm. Prior studies have shown that the soil's behavior conforms to Eq. (5.9), with the composite coefficient $Aa = 4.88 \text{ cm}^2/\text{sec}$ and the exponential constant $n = 3$. Estimate the threshold depth beyond which the water table must be lowered if evaporation is to be reduced and the water-table depth at which the evaporation rate will fall to 10% of the potential rate. Assume steady-state conditions. Plot the expectable daily evaporation rate as a function of water-table depth.

We begin with Eq. (5.9), which, according to Gardner (1958) estimates the maximal evaporation rate possible from a soil with a high water-table condition:

$$q_{max} = Aa/d^n$$

where d is depth of the water table below the soil surface. Accordingly, the maximal depth of the water table still capable of supplying the surface with the steady flux needed to sustain the potential evaporation rate is

$$d = \left(\frac{Aa}{q_{max}}\right)^{1/n} = \left[\frac{4.88 \text{ cm}^2/\text{sec}}{(0.9 \text{ cm/day})/(86400 \text{ sec/day})}\right]^{1/n}$$

$$\cong 77.7 \text{ cm below the soil surface}$$

Note that any water-table position higher than -77.7 cm could theoretically allow a flux greater than the climatically determined potential rate; hence for a shallow water table the actual evaporation rate is flux controlled at the surface. For deeper water-table conditions, however, it becomes profile controlled.

Now, to calculate the water-table depth $d_{0.1}$ for which q will be 10% of the potential rate, we can write

$$d_{0.1} = \left[\frac{4.88 \text{ cm}^2/\text{sec}}{(0.1 \times 0.9 \text{ cm/day})/(86400 \text{ sec/day})}\right]^{1/3} \cong 167.3 \text{ cm}$$

To plot the q_{max}–d curve, we can obtain the following data:

From $d = 0$ to $d = 77.7$ cm, $q_{max} = 0.9$ cm/day.
At $d = 80$ cm, $q_{max} = 4.88/80^3 = 0.82$ cm/day.
At $d = 100$ cm, $q_{max} = 4.88/100^3 = 0.42$ cm/day.
At $d = 150$ cm, $q_{max} = 4.88/150^3 = 0.12$ cm/day.
At $d = 200$ cm, $q_{max} = 4.88/200^3 = 0.05$ cm/day.
At $d = 250$ cm, $q_{max} = 4.88/250^3 = 0.027$ cm/day.

Again, the actual plotting of these data is left to the enterprising student.

2. Calculate the 10 day time course of evaporation and evaporation rate during the drying process of an infinitely deep, initially saturated, uniform column of soil subjected to an infinitely high evaporativity. Assume the initial volume wetness θ_i to be 48% and the air-dry value θ_f to be 3%, and assume the weighted mean diffusivity to be 100 cm²/day.

The problem posed can be described approximately by the theory of Gardner (1959), Eq. (5.15):

$$E = 2(\theta_i - \theta_f)\sqrt{Dt/\pi}$$

where E is cumulative evaporation. We now solve successively for days 1, 2, 3, etc., to obtain the time course of cumulative evaporation. Thus

after day 1, $E = 2(0.48 - 0.03)\sqrt{100 \times 1/3.1416} = 5.08$ cm;

after day 2, $E = 2 \times 0.45\sqrt{100 \times 2/3.1416} = 7.18$ cm;

after day 3, $E = 2 \times 0.45 \sqrt{100 \times 3/3.1416} = 8.80$ cm;

after day 4, $E = 2 \times 0.45 \sqrt{100 \times 4/3.1416} = 10.16$ cm;

after day 5, $E = 2 \times 0.45 \sqrt{100 \times 5/3.1416} = 11.36$ cm;

after day 6, $E = 2 \times 0.45 \sqrt{100 \times 6/3.1416} = 12.44$ cm;

after day 7, $E = 2 \times 0.45 \sqrt{100 \times 7/3.1416} = 13.44$ cm;

after day 8, $E = 2 \times 0.45 \sqrt{100 \times 8/3.1416} = 14.37$ cm;

after day 9, $E = 2 \times 0.45 \sqrt{100 \times 9/3.1416} = 15.24$ cm;

after day 10, $E = 2 \times 0.45 \sqrt{100 \times 10/3.1416} = 16.06$ cm.

To estimate the mean evaporation rate during each day, we can use E (5.16) to calculate the midday rate:

$$e = dE/dt = (\theta_i - \theta_f)\sqrt{\bar{D}/\pi t}$$

Midday 1: $e = 0.45 \sqrt{100/\pi \times 0.5} = 3.59$ cm/day.

Midday 2: $e = 0.45 \sqrt{100/\pi \times 1.5} = 2.07$ cm/day.

Midday 3: $e = 0.45 \sqrt{100/\pi \times 2.5} = 1.60$ cm/day.

Midday 4: $e = 0.45 \sqrt{100/\pi \times 3.5} = 1.36$ cm/day.

Midday 5: $e = 0.45 \sqrt{100/\pi \times 4.5} = 1.20$ cm/day.

Midday 6: $e = 0.45 \sqrt{100/\pi \times 5.5} = 1.08$ cm/day.

Midday 7: $e = 0.45 \sqrt{100/\pi \times 6.5} = 1.00$ cm/day.

Midday 8: $e = 0.45 \sqrt{100/\pi \times 7.5} = 0.93$ cm/day.

Midday 9: $e = 0.45 \sqrt{100/\pi \times 8.5} = 0.87$ cm/day.

Midday 10: $e = 0.45 \sqrt{100/\pi \times 9.5} = 0.82$ cm/day.

Note that the sum of the daily rates thus estimated does not equal the cumulative evaporation for the 10 day period, since the midday rates calculated for the first three days (when the rate is descending steeply as a curvilinear function of time) underestimate the effective mean rates. Note also that the representation of a continuously drying system in terms of a single, constant "weighted mean diffusivity" is itself an oversimplification. Any more realistic treatment, however, necessarily involves much more complex mathematics.

3. Consider the drying of a uniform soil wetted to a depth of 100 cm at which an impervious horizon exists, allowing no drainage. The initial volume wetness is 24%, and the initial diffusivity is 400 cm^2/day. Assume a diffusivity function of the type $D = a \exp b\theta$, where $a = 1.2$ and $b = 20$. Estimate the evaporation rate as it varies during the first 10 days of the drying process, under an evaporativity of 1.2 cm/day.

The case described can be represented by the finite-column model of Gardner and Hillel, Eq. (5.17):

$$e = -dW/dt = D(\bar{\theta})W\pi^2/4L^2$$

where e is evaporation rate (cm/day), W is total profile water content (cm), t is time (days), D is diffusivity (cm^2/day), which is a function of mean wetness $\bar{\theta}$, and L is length of the wetted profile (cm). The mean wetness $\bar{\theta}$ is related to total profile water content W by

$$\bar{\theta} = W/L \qquad \text{or} \qquad W = \bar{\theta}L$$

Substituting the appropriate values into Eq. (5.17), we calculate the first day's evaporation rate:

$$e_1 = [(400 \text{ cm}^2/\text{day}) \times (0.24 \times 100 \text{ cm}) \times 9.87]/4 \times 100^2 \text{ cm}^2$$
$$= 2.37 \text{ cm/day}$$

This is much greater than the evaporativity, which is "only" 1.2 cm/day. Hence we assume that the evaporation rate is at first controlled by, and equal to, the evaporativity. So we deduct 1.2 cm from the water content of the profile and proceed to calculate the second day's evaporation rate. Our W is now 24 cm $-$ 1.2 cm $=$ 22.8 cm, and our new $\theta = 22.8/100 = 0.228$. Hence, our new $D = 1.2 \ e^{20 \times 0.228} = 238 \text{ cm}^2/\text{day}$. The second day's evaporation rate is therefore

$$e_2 = [(238 \text{ cm}^2/\text{day}) \times 22.8 \text{ cm} \times 9.87]/40,000 \text{ cm}^2 = 1.34 \text{ cm}$$

This is still higher than the potential evaporation rate (the evaporativity), hence we once again deduct 1.2 cm from the water content and proceed to calculate the third day's evaporation rate, updating our variables as follows: $W = 22.8 \text{ cm} - 1.2 \text{ cm} = 21.6 \text{ cm}$, $\theta = 21.6/100 = 0.216$, and $D = 1.2e^{20 \times 0.216} = 90 \text{ cm}^2/\text{day}$.

$$e_3 = [(90 \text{ cm}^2/\text{day}) \times 21.6 \text{ cm} \times 9.87]/40,000 \text{ cm}^2 = 0.48 \text{ cm}$$

This evaporation rate is less than the evaporativity, so we can conclude that the constant-rate, flux-controlled phase of the process has ended and we have entered the falling-rate, profile-controlled phase. Toward the fourth day's events we calculate that $W = 21.6 - 0.48 = 21.12$ cm, $\theta = 21.12/100 = 0.2112$, and $D = 1.2\, e^{20 \times 0.2112} = 82$ cm^2/day:

$$e_4 = [(82 \text{ cm}^2/\text{day}) \times 21.12 \text{ cm} \times 9.87]/40{,}000 \text{ cm}^2 = 0.43 \text{ cm}$$

Again, we update: $W = 21.12 - 0.43 = 20.69$, $\theta = 0.207$, and $D = 1.2\, e^{20 \times 0.207} = 75$ cm^2/day.

$$e_5 = [(75 \text{ cm}^2/\text{day}) \times 20.69 \text{ cm} \times 9.87]/40{,}000 \text{ cm}^2 = 0.38 \text{ cm}$$

Now $W = 20.69 - 0.38 = 20.31$, $\theta = 0.2031$, and $D = 1.2\, e^{20 \times 2031} = 70$ cm^2/day.

$$e_6 = [(70 \text{ cm}^2/\text{day}) \times 20.31 \text{ cm} \times 9.87]/40{,}000 \text{ cm}^2 = 0.35 \text{ cm}$$

Now $W = 20.31 - 0.35 = 19.96$, $\theta = 0.1996$, and $D = 1.2 e^{20 \times 0.1996} = 65$ cm^2/day.

$$e_7 = [(65 \text{ cm}^2/\text{day}) \times 19.96 \text{ cm} \times 9.87]/40{,}000 \text{ cm}^2 = 0.32 \text{ cm}$$

Now $W = 19.96 - 0.32 = 19.64$ cm, $\theta = 0.1964$, and $D = 1.2 e^{20 \times 0.1964} = 61$ cm^2/day.

$$e_8 = [(61 \text{ cm}^2/\text{day}) \times 19.64 \text{ cm} \times 9.87]/40{,}000 \text{ cm}^2 = 0.30 \text{ cm}$$

Now $W = 19.64 - 0.30 = 19.34$ cm, $\theta = 0.1934$, and $D = 1.2 e^{20 \times 0.1934} = 57$ cm^2/day.

$$e_9 = [(57 \text{ cm}^2/\text{day}) \times 19.34 \text{ cm} \times 9.87]/40{,}000 \text{ cm}^2 = 0.27 \text{ cm}$$

Now $W = 19.34 - 0.27 = 19.07$ cm, $\theta = 0.1907$, and $D = 1.2 e^{20 \times 0.1907} = 54$ cm^2/day.

$$e_{10} = [(54 \text{ cm}^2/\text{day}) \times 19.07 \text{ cm} \times 9.87]/40{,}000 \text{ cm}^2 = 0.25 \text{ cm}$$

The total 10 day cumulative evaporation is thus estimated to be

$$1.2 + 1.2 + 0.48 + 0.43 + 0.38 + 0.35 + 0.32 + 0.30 + 0.27 + 0.25$$
$$= 5.18 \text{ cm}.$$

This amount of evaporation is, interestingly, only one-third of the amount we calculated in Problem 2 for the same period from an infinitely deep, initially saturated column of soil subject to infinite evaporativity.

6 Uptake of Soil Moisture by Plants

A. Introduction

Nature, despite its celebrated laws of conservation, can in some ways be exceedingly wasteful, or so it appears at least from our own partisan viewpoint. One of the most glaring examples is the way it requires plants to draw quantities of water from the soil far in excess of their essential metabolic needs. In dry climates, plants growing in the field may consume hundreds of tons of water for each ton of vegetative growth. That is to say, the plants must inevitably transmit to an unquenchably thirsty atmosphere most (often well over 90%) of the water they extract from the soil. The loss of water vapor by plants, a process called *transpiration*, is not in itself an essential physiological function, nor a direct result of the living processes of the plants. In fact, plants can thrive in an atmosphere saturated or nearly saturated with vapor and hence requiring very little transpiration. Rather than by plant growth per se, transpiration is caused by the vapor pressure gradient between the normally water-saturated leaves and the often quite dry atmosphere. In other words, it is exacted of the plants by the *evaporative demand* of the climate in which they live.

In a sense, the plant in the field can be compared to the wick in an old-fashioned lamp. Such a wick, its bottom dipped into a reservoir of fuel while its top is subject to the burning fire which consumes the fuel, must constantly transmit the liquid from bottom to top under the influence of physical forces imposed upon the passive wick by the conditions prevailing at its two ends. Similarly, the plant has its roots in the soil-water reservoir while its leaves are subject to the radiation of the sun and the sweeping action of the wind which require it to transpire unceasingly.

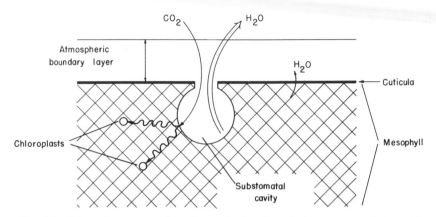

Fig. 6.1. Schematic representation of transpiration through the stomate and the cuticle, and of the diffusion of CO_2 into the stomate and through the mesophyll to the chloroplasts. (After Rose, 1966.)

This analogy, to be sure, is a gross oversimplification, since plants are not all that passive and in fact are able at times to limit the rate of transpiration by shutting the stomates of their leaves (Fig. 6.1). However, for this limitation of transpiration most plants pay, sooner or later, in reduced growth potential, since the same stomates which transpire water also serve as foci for the uptake of the carbon dioxide needed in photosynthesis. Furthermore, reduced transpiration often results in warming of the plants and consequently in increased respiration and hence in further reduction of net photosynthesis.

To grow successfully, a plant must achieve a water economy such that the demand made upon it is balanced by the supply available to it. The problem is that the evaporative demand of the atmosphere is almost continuous, whereas rainfall occurs only occasionally and irregularly. To survive during dry spells between rains, the plant must rely upon the diminishing reserves of water contained in the pores of the soil, which itself loses water by direct evaporation and internal drainage.[1]

How efficient is the soil as a water reservoir for plants? How do plants draw water from the soil, and to what limit can soil water continue to sustain plant growth? How is the actual rate of transpiration determined by the interaction of plant, soil, and meteorological factors? These and related questions are the topics of this chapter.

[1] Perhaps we can identify with the plight of arid zone plants more fully if we imagine ourselves to be living under a government which taxes away 99% of our income while requiring us to keep our reserves in a bank which is embezzled daily.

B. The Soil–Plant–Atmosphere Continuum

Current approaches to the problem of soil-water extraction and utilization by plants are based on recognition that the field with all its parts— soil, plant, and atmosphere taken together—forms a physically integrated, dynamic system in which various flow processes occur interdependently like links in a chain. This unified system has been called the SPAC (soil– plant–atmosphere continuum) by J. R. Philip (1966b). The universal principle which operates consistently throughout the system is that water flow always takes place spontaneously from regions of higher to regions of lower potential energy. Former generations of soil physicists, plant physiologists, and meteorologists, each group working separately in what it considered to be its separate and exclusive domain, tended to obscure this principle by expressing the energy state of water in different terms (e.g., the "tension" of soil water, the "diffusion pressure deficit" of plant water, and the "vapor pressure" or "relative humidity" of atmospheric water) and hence failed to communicate readily across their self-imposed interdisciplinary boundaries.

As we have now come to understand, the various terms used to characterize the state of water in different parts of the soil–plant–atmosphere system are merely alternative expressions of the *energy level*, or *potential* of water. Moreover, the very occurrence of differences, or gradients, of this potential between locations in the system constitutes the force inducing flow within and between the soil, the plant, and the atmosphere. This principle applies even though different components of the overall potential gradient are effective in varying degrees in different parts of the soil–plant– atmosphere system.[2]

In order to describe the interlinked processes of water transport throughout the SPAC, we must evaluate the pertinent components of the energy potential of water and their effective gradients as they vary in space and time. As an approximation, the flow rate through each segment of the system can be assumed to be proportional directly to the operating potential gradient, and inversely to the segment's resistance. The flow path includes liquid water movement in the soil toward the roots, liquid and perhaps vapor movement across the root-to-soil contact zone, absorption into the roots and across their membranes to the vascular tubes of the xylem, transfer through the xylem up the stem to the leaves, evaporation in the intercellular spaces within the leaves, vapor diffusion through the substomatal cavities

---

[2] For example, osmotic potential differences have little effect on liquid water movement in the soil but strongly affect flow from soil to plant. In the gaseous domain, vapor diffusion is proportional to the vapor pressure gradient rather than to the potential gradient as such. Vapor pressure and potential are exponentially, rather than linearly, related.

and out the stomatal perforations to the quiescent boundary air layer in contact with the leaf surface, and through it, finally, to the turbulent atmosphere which carries away the water thus extracted from the soil.

Since we have dealt extensively in this text with the principles governing flow in the soil domain but scarcely at all with the plant domain, it is time for us to digress from our primary concern with the soil per se and attempt to describe, albeit briefly, some pertinent aspects of plant–water relations.

C. Basic Aspects of Plant–Water Relations

Close observation of how higher plants are built reveals much about how they function in their terrestrial environment (Epstein, 1973). To begin with, it is intriguing to compare the shape of such plants to the shape of familiar higher animals. The characteristic feature of animal bodies, in contrast with plants, is their minimal area of external surface exposure. Apart from a few protruding organs needed for mobility and sensory perception, animals are rather compact and bulky in appearance. Not so is the structure of plants, whose vital functions require them to maximize rather than minimize surface exposure both above and below ground. The aerial canopies of plants frequently exceed the area of covered ground by several fold. Such a large surface helps the plants to intercept and collect sunlight and carbon dioxide, two resources which are diffuse rather than concentrated.

Even more striking is the shape of roots, which proliferate and ramify throughout a large volume of soil while exposing an enormous surface area: a single annual plant can develop a root system with a total length of several hundred kilometers and with a total surface area of several hundred square meters. (Estimates of total length and surface area of roots are ten times larger if root hairs are taken into account.) The need for such exposure becomes apparent if we consider the primary function of roots, which is to continuously gather water and nutrients from a medium that often provides only a meager supply of water per unit volume and that generally contains soluble nutrients only in very dilute concentrations. And while the atmosphere is a well-stirred and thoroughly mixed fluid, the soil solution is a sluggish and unstirred fluid which moves toward the roots at a grudgingly slow pace, so that the roots have no choice but to move toward it. Indeed, roots forage constantly through as large a soil volume as they can, in a constant quest for more water and nutrients. Their movement and growth, involving proliferation in the soil region where they are present and extension into ever new regions, are affected by a host of factors additional to moisture and nutrients, e.g., temperature, aeration, mechanical resistance, the possible presence of toxic substances, and the primary roots' own *geotropism* (i.e., their preference for a vertically downward direction of growth).

Green plants are the earth's only true *autotrophs*, able to create new living matter from purely inorganic raw materials. We refer specifically to the synthesis of basic sugars (subsequently elaborated into more complex compounds) by the combination of atmospheric carbon dioxide and soil-derived water, accompanied by the conversion of solar radiation into chemical energy in the process of *photosynthesis*, usually described by the following deceptively simple formulas:

$$6CO_2 + 6H_2O + (\text{sunlight energy}) \rightarrow C_6H_{12}O_6 + 6O_2$$

$$\underset{\text{glucose}}{nC_6H_{12}O_6} \rightleftharpoons \underset{\text{starch}}{(C_6H_{10}O_5)_n} + nH_2O \tag{6.1}$$

We, along with the entire animal kingdom, owe our lives to this process, which not only produces our food but also releases into the atmosphere the elemental oxygen which we need for our respiration. Plants also respire, and the process of respiration represents a reversal of photosynthesis in the sense that some photosynthetic products are reoxidized to yield the original constituents (water, carbon dioxide, and energy) as separate entities. Thus

$$C_6H_{12}O_6 + 6O_2 \rightarrow 6CO_2 + 6H_2O + (\text{thermal energy}) \tag{6.2}$$

In examining these formulas, we note immediately the central role of water as a major metabolic agent in the life of the plant, i.e., as a source of hydrogen atoms for the reduction of carbon dioxide in photosynthesis and as a product of respiration. Water is also the solvent and hence conveyor of transportable ions and compounds into, within, and out of the plant. It is, in fact, a major structural component of plants, often constituting 90% or more of their total "fresh" mass. Much of this water occurs in cell vacuoles under positive pressure, which keeps the cells turgid and gives rigidity to the plant as a whole.

Although all plants are absolutely dependent on water, different types of plants differ in adaptation to environments with varying degrees of water availability or abundance. *Hydrophytes*, or *aquatic plants*, inhabit water-saturated domains. Plants adapted to drawing water from shallow water tables are called *phreatophytes*. Plants which grow in aerated soils, generally in semihumid to semiarid climates, are called *mesophytes*. Most crop plants belong in this category. Mesophytes control their water economy by developing extensive root systems and optimizing the ratio of roots to shoots (the former supply water and nutrients, the latter photosynthesize and transpire) and by regulating the aperture of their stomates. On the dry end of the scale are *xerophytes*, which are adapted to growing in desert environments. Such plants generally exhibit special features (called *xeromorphic*) designed to minimize water loss (e.g., thickened epidermis and a waxy cuticle, recessed stomates and reduced leaf area, and specialized water storage or *succulent* tissues).

Only a very small fraction (generally less than 1%) of the water absorbed by plants is used in photosynthesis, while most (often over 98%) is lost as vapor, in a process known as *transpiration*. This process is made inevitable by the exposure to the atmosphere of a large area of moist cell surfaces, necessary to facilitate absorption of carbon dioxide and oxygen, and hence transpiration has been described as a "necessary evil" (Sutcliffe, 1968). Mesophytes are extremely sensitive, and vulnerable, to lack of sufficient water to replace the amount lost in transpiration. Water deficits impair plant growth and, if extended in duration, can be fatal. It was once believed that transpiration is beneficial in that it induces a greater uptake of nutrients from the soil. However, it now appears that the two processes of water uptake and nutrient uptake are largely independent. The absorption of nutrients into roots and their transmission up the shoots are not merely a passive consequence of the transpiration-induced water flow but are effected by metabolically active processes (Epstein, 1977). A more likely beneficial effect of transpiration is prevention of excessive heating of leaves by solar radiation. Indeed, when water deficits occur and transpiration is reduced by stomatal closure, plant-canopy temperature rises measurably.

D. Water Relation of Plant Cells and Tissues

Water plays a crucial role in the process of plant growth, which involves cell division and cell expansion. The latter process occurs as each pair of divided cells imbibes water and as the resulting internal pressure, called *turgor*, stretches the new cells' elastic walls which, as they stretch, thicken through the deposition of newly synthesized material.

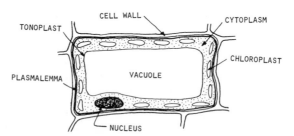

Fig. 6.2. Schematic representation of the internal structure of a mature mesophyll (leaf) cell, showing the solute-confining membranes.

Both the cell wall, constituted mainly of cellulose and other polysaccharides, and the protoplasmic material within, called *cytoplasm*, are generally permeable to water molecules diffusing or flowing into and out of the cell. Since the cell wall contains numerous relatively large interstices, it does not restrict solute movement. However, inside the cell wall (see Fig. 6.2) and surrounding the cytoplasm is a lipoprotein "cell membrane" called the *plasmalemma*, which is selectively permeable and thus helps regulate what enters and leaves a plant cell.[3] Most solutes are prevented from diffusing freely into and out of the cell's internal solution phase, generally contained in a central "cavity" called a *vacuole*, which in mature cells often occupies the greater part of the cell's volume and is itself surrounded by a "vacuolar membrane" called the *tonoplast*. Although some solutes are selectively conveyed through the membranes by a complex process called *active transport*, the cell by and large acts as a small osmometer, drawing water by osmosis from the surrounding aqueous phase into the generally more concentrated solution in the vacuole. The *osmotic potential* of cell water, due to the sugars and salts dissolved in it, is negative with respect to that of pure water (which is taken as the zero reference), and may range, in terms of pressure units, from -5 to -50 bar.

The expansive pressure caused by osmosis is countered by the elastic cell walls, which are tensed as they are stretched and hence tend to contract inwards. This contractile tendency can be expressed in terms of a positive pressure potential. The overall potential of water inside the cell ϕ depends on the osmotic potential of the cell solution ϕ_o and on the effective potential resulting from the pressure exerted by the walls ϕ_p called the *turgor pressure*. If these were the only components of the cell's water potential, we could write

$$\phi = \phi_o + \phi_p \tag{6.3}$$

which is equivalent to the classical expression

$$\text{DPD} = \text{OP} - \text{TP} \tag{6.4}$$

in which DPD is the *diffusion pressure deficit*, OP the *osmotic pressure* of the cell solution, or "sap," and TP is the *turgor pressure*. Note that in this expression the signs of the potential terms are reversed: the negative os-

---

[3] According to Sutcliffe (1968), a biological membrane is believed to consist basically of a bimolecular layer of oriented phospholipid molecules coated on each surface with a layer of protein (Fig. 6.4). It is thought that water penetrates the membrane through minute pores which are too small to allow dissolved substances to pass through easily. This is possible despite the association of water molecules with one another, because they readily become oriented into files of sufficiently small dimensions. The resistance of membranes to diffusion of solutes is thus many times greater than it is to water.

motic potential is expressed as a positive "osmotic pressure," the inward-directed and hence positive turgor pressure is given a negative sign, and the overall potential is taken as a positive DPD (though its negative value relative to the reference state of pure and free water is implied in the word "deficit").

The energy state of cell water is also affected by colloidal hydration, adsorption, and capillary phenomena, which, taken together, can be regarded as an affinity of the tissue's "matrix" for water. It is this affinity which causes the swelling of gelatine and cellulose in water and the initial imbibition of water by seeds. By way of analogy with the matric potential of soil water, this component of plant-water potential (long disregarded by "classical" plant physiologists) has also been called *matric potential* ϕ_m. Thus, the more complete expression for the overall plant-water potential ϕ is

$$\phi = \phi_o + \phi_p + \phi_m \qquad (6.5)$$

This expression is simplistic in that it takes the various components of the overall potential to be independent of one another and hence simply additive. In actual fact, the three components are mutually dependent so that a change in one often necessarily entails a change of the others. For instance, an increase in turgor pressure is generally a consequence of the stretching of cell walls due to the osmotic absorption of water, which, incidentally, reduces osmotic pressure through dilution of the cell sap. Any increase in the degree of hydration of a cell naturally also affects the matric potential.

If we immerse a cell in pure water at atmospheric pressure, the cell will imbibe water until its affinity for water (due to both matric and osmotic effects) is counterbalanced by the cell wall's turgor pressure. The cell will then be fully hydrated and "turgid." (Using the classical and now anti-quated terminology, we would say that the diffusion pressure deficit is zero when the opposing osmotic and turgor pressures are equal in magnitude.) If a plant is "thirsty" (or, according to a currently prevalent expression, if it is subject to "moisture stress," or "distress"), its cells will not be fully turgid and its *relative hydration*; i.e., its water content relative to that when it is fully turgid (at a water potential of "zero") will be less than unity. The functional relationship between the relative hydration of plant cells and their water potential is as important an aspect of plant physiology as the relationship between soil wetness and soil-water potential is in soil physics.

If, instead of immersing a cell in pure water we do so in a solution more highly concentrated than the cell's own solution, outward osmosis will take place and the cell will dehydrate, perhaps to the point of *plasmolysis*, at which the effective turgor pressure falls to zero or might even become negative. At this point the cytoplasm might actually separate from the walls and collapse inwards.

An interesting illustration of the importance of turgor pressure changes is the *stomatal control mechanism*. As mentioned, the opening and closing of stomates, or stomata (plural of stoma, from the Greek word for mouth), is a principal factor controlling the transpiration rate. The aperture or closure of stomates is affected by light intensity and atmospheric CO_2 concentration (i.e., stomates close in the dark and at high CO_2 concentrations). Stomatal aperture is also affected by temperature and the plant's own internal rhythms. Most of all, however, stomates are affected by tissue hydration, and, more specifically, by the turgidity of the paired *guard cells* which constitute what is in effect a highly sensitive valve. Increasing hydration and turgidity causes these cells to bulge outwards, crescentlike, thus dilating the space between them. Low turgidity causes them to collapse against each other, thus shutting off the diffusive outlet for water vapor (also serving as the inlet for carbon dioxide).[4] A transverse section and surface views of a leaf with open and shut stomates are shown in Fig. 6.3.

Methods of measuring plant water potential were described by Barrs (1968).

In the absence of transpiration, plants absorb water from the soil by means of the osmotic mechanism and plant water generally exhibits a positive hydrostatic pressure. Evidence of this positive pressure is found in the tendency of some plants to exude water during the night (a phenomenon called *guttation*) and in the tendency of excised stems or roots to ooze out water (an occurrence often attributed to "root pressure"). However, when transpiration takes place from the leaves the pressure of plant water decreases and generally falls below atmospheric pressure. The resulting tension, or suction, induces upward mass flow from roots to leaves through the tubelike capillary vessels of the xylem. It is remarkable that these capillary vessels can maintain the cohesive continuity of liquid water columns even under a tension of many bars, as needed to draw water from relatively dry soils and transport it to the tops of trees scores of meters high. The rate at which water is extracted from the soil and transmitted to the transpiring leaves obviously depends not only on the magnitude of the potential or pressure gradients but also on the hydraulic properties of the conveyance system, i.e., on the conductance (or its reciprocal, the hydraulic resistance) of the stems, and on the conductance as well as absorptive properties of the roots. It is the latter which we shall consider in the following sections of this chapter.

---

[4] In dicotyledons, the ventral walls of guard cells (those lining the stomatal pore) are typically thicker than the dorsal walls. At high turgor pressure, the more flexible dorsal walls become more convex, thus drawing the central parts of the ventral walls apart to open the pore. At low turgor pressure, the pore closes.

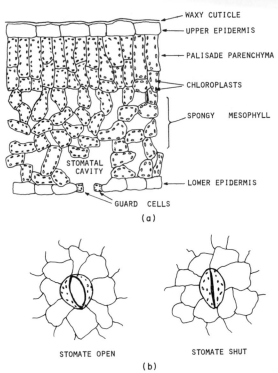

Fig. 6.3.   Illustration of a leaf section with open and closed stomate: (a) transverse view,
(b) surface view.

E. Structure and Function of Roots

The anatomy of roots is designed, as it were, to perform a number of
essential functions, including absorption and conveyance of water and
nutrients from the soil and anchorage of the plant's superstructure. A
longitudinal section of a root reveals several sections, as shown in Fig. 6.4a.
The tip of the root is shielded by a *root cap*, which is often of mucilaginous
consistency and may thus help to lubricate the path of the advancing root.
Following the root cap is a zone of rapid cell division called the *apical
meristem*. Next is a region of *cell elongation*, where the initially isodiametric
cells elongate in the direction of the root axis, thus pushing the root tip
forward and causing the root to extend farther into the soil. Behind the region of cell elongation is that of *cell differentiation*, where different groups
of cells develop specific characteristics and assume specialized functions.
The innermost cells of the root become *vascular* tissue, while the outermost

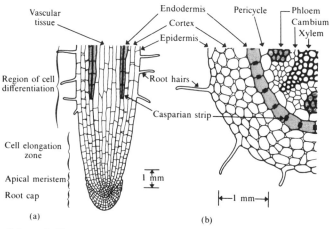

Fig. 6.4. Schematic illustration of the structure of a growing root: (a) longitudinal section through the root cap and the zones of cell division, elongation, and differentiation; (b) cross section, showing the concentric arrangement of the epidermis, cortex, endodermis, pericycle, and vascular bundles. (After Nobel, 1974.)

cells exhibit thin radial protrusions known as *root hairs*, which have the effect of greatly increasing the surface area over which water and nutrients can be absorbed. Farther back, the older section of the root becomes somewhat hardened and less absorbent, and, as the root gradually converges with other roots, it becomes thicker and more stemlike.

A cross-sectional view of a typical root (some distance away from the root tip), illustrated in Fig. 6.4b, shows a central vascular core, called the *stele*, surrounded by three concentric, sleevelike layers of cells: the endodermis, the cortex, and the epidermis.

The vascular system of the roots connects with that in the stems and leaves. It consists of (1) *xylem* vessels, which convey water and mineral nutrients from the soil, along with organic compounds metabolized in the roots, to the upper parts of the plant, and (2) *phloem* vessels, which serve to translocate the organic compounds which are synthesized in the leaves (such as carbohydrates) and distribute these throughout the plant. The two types of vessels are depicted in Fig. 6.5. The xylem vessels are hollow cells, practically devoid of protoplasm, arranged end to end and joined by *perforation plates* to form continuous tubes. The aqueous solution originating in the absorbing sections of the roots, generally having a concentration of the order of $10^{-2} M$ (Epstein, 1977; Nobel, 1970), flows up through the root system, then through the stem and petiole and leaf veins into the leaves, with motion occurring in the direction of decreasing hydrostatic pressure. In contrast with the xylem, the conducting cells in the phloem, called *sieve*

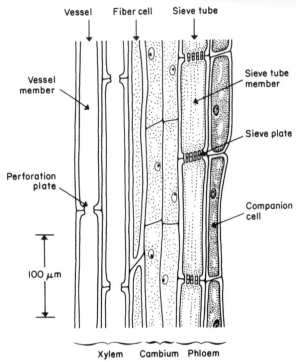

Fig. 6.5. Longitudinal section through xylem and phloem vessels, showing their internal structures. (After Noble, 1974.)

cells, contain cytoplasm and remain metabolically active. The tissue between the xylem and the phloem is the *cambium*, whose cells divide and differentiate to form xylem and phloem. Between the cambium and the endodermis is a layer of cells called the *pericycle*, whose cells can divide and lead to the formation of lateral or branch roots.

The *endodermis* is a layer of cells whose radial walls are impregnated with a waxy, waterproof material forming an impervious shield termed the *casparian strip*. To cross this barrier, water and solutes must either flow along the cell walls or enter into and flow across the cytoplasm. Outside the endodermis are a number of layers of cells known as the *cortex*, which is a rather "loose" tissue with numerous intercellular air spaces through which gases such as O_2 and CO_2 can diffuse. Finally, surrounding the entire root and serving as its "skin" is the *epidermis*, where the root hairs originate.

F. Hydraulic Properties of Roots

Having described some of the anatomical features of roots, we now wish to consider their hydraulic properties. To pose the question in concrete terms: How do roots function as a pumping (and plumbing) system in the soil? The first task is to determine the pathways for flow from the soil into and through the various root segments or tissues described in the preceding section. From this it might be possible to assess the potential gradient and transmission properties of the various pathways and their segments.

Let us define what we mean by "transmission properties." In the soil, we speak of *hydraulic conductivity*, which is defined as the flux per unit hydraulic gradient, the latter being the potential difference across a unit length of the flow path. In the root, however, the geometry is too complex to permit measurement of exact values of gradient and conductivity (or its reciprocal, the *resistivity*) on a microscopic scale. Hence we speak of *hydraulic resistance* (being in theory the ratio of the flow path length to its conductivity), or its own reciprocal, the *conductance*. If we assume that water moves through the root in response to a definable difference in the water potential value and, furthermore, that the flow rate through any segment of root is proportional to the magnitude of the potential drop across that segment, we can formulate the process in terms analogous to those of Ohm's law for the flow of electricity through a resistor:

$$I = \Delta V / R_e \qquad \text{electrical current} = \frac{\text{difference in electrical potential (voltage)}}{\text{electrical resistance}}$$

Similarly,

$$q = \Delta \phi / R_w \qquad \text{water flux} = \frac{\text{difference in effective water potential}}{\text{hydraulic resistance}}$$

If the flux is expressed in terms of volume flowing through a unit area per unit time, say cm^3/cm^2 sec, then the resistance takes on the units of bar sec/cm. If the potential is expressed in head units rather than pressure units, resistance assumes the units of time, and resistivity (resistance per unit length) the units of time per length, thus being the reciprocal of conductivity which is usually given in terms of length per time.

Because the flow of water through a transpiring plant is generally continuous, the water potential varies continuously from roots to leaves. If flux can be assumed to be constant, then flow path segments having greater hydraulic resistance will be sites of greater water potential drop. The sum of all the segmental resistances throughout the flow path will determine the total potential difference necessary to transport any given flux of water from the soil–root interface to the leaf–air interface.

We have already indicated that plant-water potential has several interacting components. Osmotic potential differences result from solute concentration differences. Generally, though not invariably, soil water has a lower total solute concentration than plant water. Hence the osmotic gradient typically acts to drive water from the soil into the root. Gravitational potential is determined by elevation, and the downward gravitational gradient must normally be overcome for the upward transport of water to occur from roots to leaves. The dominant component of water potential in plants is generally the hydrostatic pressure component. During daylight hours, when plants are transpiring, evaporation from the leaves causes the hydrostatic pressure there to become "negative," and the resulting sub-pressure, called tension or suction, "pulls" water from the column below. A continuous pressure gradient is set up from roots through xylem to leaves, which powers the so-called transpiration stream. As mentioned earlier, this stream is largely determined by the evaporative demand (i.e., the potential evapotranspiration) as long as there is sufficient water available to the roots in the soil and as long as the conductance of plant tissues is sufficiently high.

There are several alternative or concurrent paths which water can take in traveling from the soil–root interface through the epidermis, cortex, endodermis, and pericycle to the xylem. Although we can conjecture that flow will occur preferentially along the path of least resistance, we are as yet uncertain which of the possible paths are the major ones under different circumstances. The following are some of the pathways which have been proposed.

1. *Cell-to-cell pathway:* Water travels from the vacuole of one cell to that of another through the vacuolar membrane (the tonoplast), the plasma, the cell membrane (plasmalemma), the cell wall, the intercellular fluid, then through the wall of the next cell and its membranes and plasma into the second vacuole, etc. In this pathway, the major sites of resistance are probably the cell membrane and, perhaps to lesser extent, the cell plasma and vacuolar membrane.

2. *Cell wall pathway:* Water moves through the porous cell walls, traversing the intercellular fluid as it passes from the periphery of one cell to that of the next. Cell walls are generally much more permeable, and hence offer less resistance, than the living cell material (the cytoplasm and its associated membranes). The cell wall path is continuous from the epidermis through the cortex to the endodermis, where, however, water movement in the cell walls is blocked by the casparian strip. Here, apparently, water must move into and traverse at least some cells before it can move into the cell walls on the other side of the endodermis. The major sites of resistance in this path are, therefore, the cell membranes and cytoplasm of the endo-dermal cells (Molz, 1975, 1976).

3. *Intercellular pathway:* Water traverses the tightly aligned epidermis via the cell walls, after which it moves through the cortex via the inter-cellular spaces. At the endodermis, however, it must enter the cells to cross

the casparian strip, following which it can take the cell wall pathway for the remainder of the way into the xylem. The site of major resistance is again the cell membranes and cytoplasm of the endodermal cells. Although wide and therefore seemingly permeable, the intercellular spaces of the cortex can offer considerable resistance to liquid water movement if they contain much air.

4. *Symplasm pathway:* The cytoplasms of cells are often connected by fine strands called *plasmodesmata*, which extend through openings in the cell walls (Ray, 1972). It has been conjectured (Newman, 1974) that the plasmodesmata constitute tubes through which water can flow without having to cross the cell membranes. However, flow through these strands may be limited by their narrowness and small number, as well as by the protoplasmic material itself and the possible presence of membranes within it.

A summary of the various conjectured pathways is given in Fig. 6.6. All the pathways described except, perhaps, the symplasm pathway, require water to traverse the endodermal cell membrane, which is, incidentally, a differentially permeable membrane allowing the selective uptake of nutrients into the xylem, while causing the exclusion of other solutes. In penetrating older secondary root tissues, water must pass through the cork, the phloem, and the cambium before it can enter the xylem. The possible flow pathways are similar to those in primary tissue except that there is no cortex or endodermis to pass through, and some cells have undergone *suberization*, which is the deposition of a waxy material known as suberin. According to Newman (1974), however, suberization of root cells does not necessarily result in reduction of hydraulic conductance, since the deposition is often discontinuous and penetrated by plasmodesmata. Older suberized roots have been observed to contribute significantly to water supply in plants subject to moisture stress.

Thus far we have dealt with the *radial absorption* of water through the peripheral tissues of the root to the xylem at the center. There remains the problem of assessing *axial conduction* through the xylem itself, which consists of hollow vessel and tracheid cells. Since the xylem vessels are essentially similar to continuous capillary tubes (albeit with variable diameters), their hydraulic conductance or resistance can be calculated from knowledge

Fig. 6.6. Hypothetical representation of alternative (or concurrent) pathways for the radial flow of water through the epidermis, cortex, and endodermis toward the root xylem.

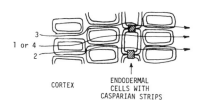

CORTEX   ENDODERMAL CELLS WITH CASPARIAN STRIPS

of their radii and lengths on the basis of Poiseuille's equation (e.g., Emerson, 1954; Wind, 1955; Cowan and Milthorpe, 1968). However, far from being smooth and straight cylindrical tubes, the conducting vessels of the xylem constitute a complex network of intricate passages with numerous bends, interconnections, dead ends, constrictions, air bubbles, etc.

Molz (1975) analyzed the problem of assessing the hydraulic conductivity of root tissue from the specific conductivities and cross-sectional dimensions of the various pathways through which flow can take place. He offered the following equation:

$$K_t = \frac{K_w a_w + k_{mc} a_v \, \Delta x/z}{a_w + a_v} \tag{6.6}$$

wherein K_t is the tissue conductivity, K_w the cell wall conductivity, a_w the cross-sectional area of the cell wall, k_{mc} the permeability of the cell membrane and cytoplasm, a_v the cross-sectional area of vacuolar pathway, and Δx the length of the cell. Using estimated values of K_w and k_{mc}, Molz calculated a value of about 10^{-8} cm^2/sec bar for the composite hydraulic conductivity of root tissue.

Experimental evidence presently indicates that root conductivity is not a constant value but can change with environmental conditions (Newman, 1974). Short of killing the root, any decrease in metabolism seems to decrease its conductivity. (Dead roots, however, have a much greater conductivity than live roots because there is no cell membrane.) Some changes in environmental conditions which lower plant metabolism and are seen to decrease conductivity would be (1) presence of toxic substances, (2) lower temperatures, (3) oxygen stress, and (4) excess CO_2 in soil. The fact that metabolic activity affects conductivity seems to indicate that plants may not be just passive absorbers of water, as is generally thought, but may exert some active control over water transport. The nature of this control is not clearly understood.

Changes in water potential within the root also seem to affect its conductivity. Huck and Klepper (1970) have documented diurnal variations in stem diameter and have correlated these with changes in water potential. In calculating the water potential from stem diameter measurements, they did not assume that conductivity is constant. No particular expression for the variation of conductivity with water potential was given but resistance was simply fitted as a parameter until predicted xylem water potential agreed with calibration points.

Newman (1974) stated that according to most experimental evidence as the water potential gradient increases in the plant, conductivity also increases. The reasons for this are again uncertain. The consequences, however, are that at times of high transpiration pull, the plant will conduct water more easily—a definite advantage.

Root conductance is obviously a very complex problem. Much more information is needed about the way roots behave under different conditions, about metabolic activity in plants affecting water uptake, the specific paths of flow, and the physical properties of tissues and membranes at the major sites of resistance.

G. Variation of Water Potential and Flux in the Soil–Plant System

To characterize the soil–plant–atmosphere continuum physically, it is necessary to evaluate the pertinent components of the energy potential of water and the effective potential gradient as it varies along the entire path of water movement. This includes liquid water movement in the soil toward the roots, absorption into the roots, transport in the roots to the stems and

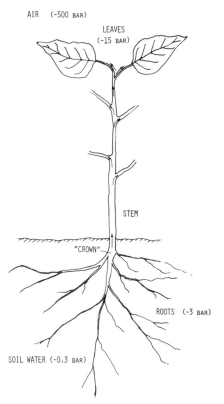

Fig. 6.7. Schematic illustration of the variation of water potential along the transpiration stream.

AIR (-500 BAR)

LEAVES (-15 BAR)

STEM

"CROWN"

ROOTS (-3 BAR)

SOIL WATER (-0.3 BAR)

through the stem to the leaves, evaporation in the intercellular air spaces of the leaves, and vapor diffusion through the substomatal cavities and stomatal openings to the quiescent boundary air layer in contact with the leaf surface, and through it to the turbulent air layer surrounding the plant, whence the vapor is finally transported to the external atmosphere.

The total potential difference between soil moisture and atmospheric humidity can amount to hundreds of bars (as illustrated in Fig. 6.7) and in an arid climate can even exceed 1000 bar. Of this total, the potential drop in the soil toward the roots may vary from a fraction of a bar to several bars. Except where the soil is fairly dry, the potential drop in the roots (from cortex to xylem) is likely to be somewhat greater than that in the soil (Molz, 1976). The potential drop in the xylem from roots to leaves will generally not exceed a few bars. Altogether, therefore, the summed potential drop in the soil and plant will be of the order of, say, 10–30 bar. Thus it is clear that the major portion of the overall potential difference in the SPAC occurs between the leaves and the atmosphere (Philip, 1966b).

Figure 6.8 describes the distribution of water potentials in the SPAC. This figure is not drawn to scale and its purpose is only to illustrate general relationships. Curve 1 is for a low value of water suction in the soil (AB) and hence also at the root surfaces (B). In the mesophyl cells (DE), water

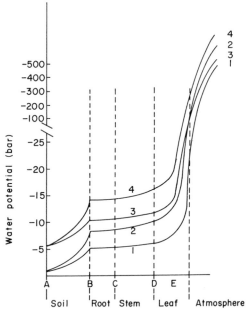

Fig. 6.8. The distribution of potentials in the soil–plant–atmosphere continuum under different conditions of soil moisture and atmospheric evaporativity.

suction (negative potential) is below the critical value at which the leaves may lose their turgidity, hence the plant is able to transport water from the soil to the atmosphere without wilting. E represents the substomatal cavity. In curve 2, soil water suction is equally low but the transpiration rate is higher and water suction in the mesophyl of the leaves approaches the critical wilting values (say, ~ 30 bar). Curve 3 is for the case in which soil-water suction is relatively high but the transpiration rate is low. Curve 4, finally, indicates the extreme condition in which soil-water suction and the transpiration rate are both high, the leaf-water suction exceeds the critical value, and the plant wilts.

An electrical analog of the soil–plant–atmosphere system is shown in Fig. 6.9. This representation is rather simplistic in that, for instance, it does not indicate any appreciable resistance between the root cortex and xylem (i.e., across the endodermis).

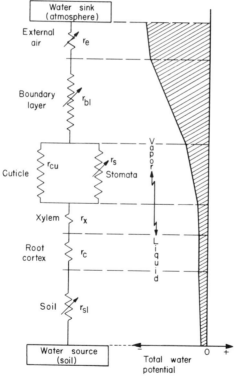

Fig. 6.9. An electrical analog representing resistances to flow in the SPAC. The resistances in the soil, leaf, and atmosphere may vary with variation of soil and meteorological conditions. Flow in the atmosphere is due to the vapor-pressure gradient, and the transfer coefficient is a complex function of such variables as wind, turbulence, etc. (After Rose, 1966.)

Soil-water suction increases as soil wetness decreases. Consequently, the plant-water suction required to extract water from the soil must increase correspondingly. The soil will deliver water to the root as long as the water suction in the latter is maintained greater than in the former. However, as a root extracts water from the soil in contact with it, the suction in the contact zone may increase and tend to equal the root suction. In this case, water uptake cease, unless additional water can move in from the farther reaches of the soil as a result of the soil-water suction gradients which form toward the soil in direct contact with the roots. In order for this additional water to become available to the plant, it must not only be at a suction lower than the root-water suction, but must also move toward and into the root at a rate sufficient to compensate the plant for its own constant loss of water to the atmosphere by the process of transpiration. The dependence of suction on distance from the root during water uptake was calculated by Gardner (1960) for several soils and was found to be quite flat until the initial suction approached 15 bar. This is illustrated in Fig. 6.10.

As long as the plant does not wilt, and as long as the influx of radiation and heat to the canopy results in change of phase only, it is possible to assume steady state flow through the plant. This means that the transpiration rate is equal to the plant transport and that both are equal to the soil-water uptake rate:

$$q = -\Delta\phi_1/R_1 = -\Delta\phi_2/R_2 = -\Delta\phi_3/R_3 = -\Delta\phi_4/R_4 \qquad (6.7)$$

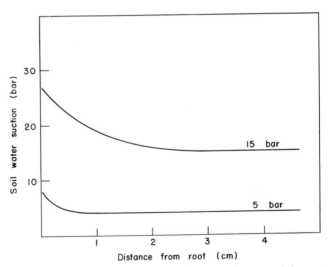

Fig. 6.10. Relation of soil-water suction to distance from the root axis in a sandy soil with an uptake rate of 0.1 cm^3 per centimeter of root per day. The two curves are for two different levels of soil-water suction at a distance from the root. (After Gardner, 1960.)

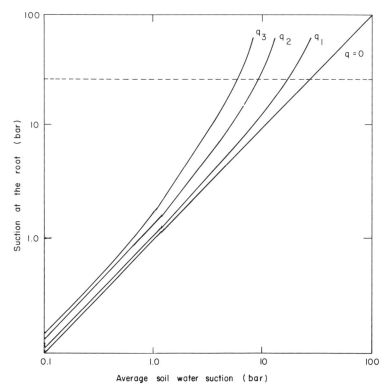

Fig. 6.11. The relation of suction at the root to the average soil-water suction, for different uptake rates $(q_1 < q_2 < q_3)$. The dashed line indicates an approximate value of critical suction at which the plant wilts. The soil-water suction at which wilting occurs is seen to depend upon the uptake and hence the transpiration rate. (After Gardner, 1960.)

where $\Delta\phi_1$ is the potential drop in the soil toward the roots, $\Delta\phi_2$ the potential drop between the soil and root xylem, $\Delta\phi_3$ the potential drop in the plant to the leaves, and $\Delta\phi_4$ the potential drop between the leaves and the atmosphere.[5] As mentioned, the magnitudes of these potential increments are of the order of $\Delta\phi_1 \approx 10$ bar, $\Delta\phi_2 \approx 10$ bar, $\Delta\phi_3 \approx 10$ bar, and $\Delta\phi_4 \approx 500$ bar. It follows that the resistance R_4 between the leaves and the atmosphere might be 50 or more times greater than the resistances of the plant and the soil. When the stomates close, as at noontime during a hot day, the resistance R_4 may become even much greater and result in a decreased rate of transpiration.

[5] Strictly speaking, this representation of resistance may not apply to the transfer of water from the leaves to the atmosphere, as this stage of the process is affected by an influx of external energy (e.g., radiation) rather than by an "internal" potential drop alone.

The suction difference between soil and root ($\Delta\psi$) needed to maintain a steady flow rate depends, as stated, on the conductivity K and the flow rate q, which is the water-extraction rate required of the root by the transpiration process. When soil-water suction is low and conductivity high, $\Delta\psi$ is small and the suction in the root will not differ greatly from the suction in the soil. When soil-water suction increases, and soil conductivity decreases, the suction difference (or gradient) needed to maintain the same flow rate must increase correspondingly. As long as the transpiration rate required of the plant is not too high, and as long as the hydraulic conductivity of the soil is adequate and the density of the roots is sufficient, the plant can extract water from the soil at the rate needed to maintain normal activity. However, the moment the rate of extraction drops below the rate of transpiration (either because of a high evaporative demand by the atmosphere, and/or because of low soil conductivity, and/or because the root system is too sparse), the plant necessarily loses water, and, if it cannot adjust its root-water suction or its root density so as to increase the rate of soil-water uptake, the plant may suffer from loss of turgor and be unable to grow normally. This situation, if it persists, will soon cause the plant to wilt. It follows that, as atmospheric evaporativity increases, and as soil conductivity decreases, the average soil-water suction at which the plant wilts will tend to be lower.

Figure 6.11 shows the hypothetical relation of suction at the root to the average soil-water suction for different rates of water uptake. The curves show that root-water suction need not greatly exceed average soil-water

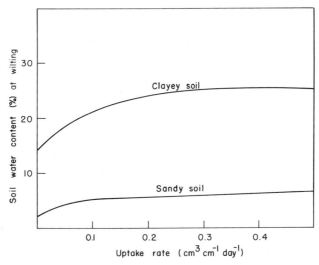

Fig. 6.12.   Relation of soil wetness at wilting to the root uptake flux.

suction as long as the latter does not exceed a few bars. However, as average soil water suction increases beyond 10 bar or so, the suction which must develop at the root to maintain the flow rate constant may be 20 or 30 bar higher, and the difference must increase as the flux increases.

Assuming that wilting occurs at a certain limiting value of plant water suction, it is possible to estimate the dependence of average soil wetness at which wilting occurs upon the transpiration (or uptake) rate (Gardner, 1960). Figure 6.12 shows that the range of soil wetness at which plants might wilt is relatively narrow in a sandy soil and wider in a clayey soil. Thus, both the suction and wetness at wilting are affected by the dyanmics of water uptake.

H. Root Uptake, Soil-Water Movement, and Transpiration

The rate of water uptake from a given volume of soil depends on rooting density (the effective length of roots per unit volume of soil). soil conductivity, and the difference between average soil-water suction and root suction. If the initial soil-water suction is uniform throughout all depths of the rooting zone, but the active roots are not uniformly distributed, the rate of water uptake should be highest where the density of roots is greatest. However, more rapid uptake will result in more rapid depletion of soil moisture, and the rate will not remain constant very long.

Nonuniformity of water uptake from different soil depths has been found in the field, as shown in Fig. 6.13 (after Ogata *et al.*, 1960). In a nonuniform root system, suction gradients can form which may induce water movement from one layer to another in the soil profile itself. In general, the magnitude of this movement is likely to be small relative to the water uptake rate by the plants, but in some cases it can be considerable. As a rough approximation, it is sometimes possible to divide the rooting system into two layers: an *upper layer*, in which root density is greatest and nearly uniform and in which water depletion is similarly uniform, and a *lower layer*, in which the roots are relatively sparse and in which the rate of water depletion is slow as long as the water content of the upper layer is fairly high. The water content of the lower layer is depleted by two sometimes simultaneous processes: uptake by the roots of that layer, and direct upward flow in the soil itself, caused by suction gradients, from the still-moist lower layer to the more-rapidly depleted upper layer.

A mature root system occupies a more or less constant soil volume of fixed depth so that uptake should depend mainly upon the size of this volume, its water content and hydraulic properties, and the density of the roots. On the other hand, in young plants, root extension and advance

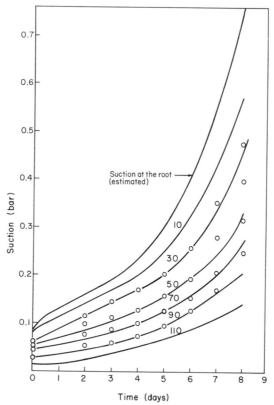

Fig. 6.13. The increase of average soil-water suction with time in different depths in a field of alfalfa. The numbers by the curves represent different depths (cm) within the rooting zone. (After Ogata *et al.*, 1960; Gardner, 1964.)

into deeper and moister layers can play an important part in supplying plant water requirements (Kramer and Coile, 1940; Wolf, 1968).

One possible reason for the differences observed between the responses of pot-grown and of field-grown plants to the soil-water regime is the difference in root distribution with depth. In a pot, root density can be farily uniform, while in the field, it generally varies with depth. Furthermore, the roots present in different layers may exhibit different water uptake and transmission properties. For instance, the roots of the deeper layers may offer greater resistance to water movement within the plant than the roots of the upper layers (e.g., Wind, 1955). The possible contribution of moist sublayers underlying the rooting zone can be especially significant where a high water table is present.

Rose and Stern (1967) presented an analysis of the time rate of water withdrawal from different soil depth zones in relation to soil wetness and hydraulic properties, and to the rate of plant-root uptake. The water conservation equation for a given depth of soil (assuming flow to be vertical only) for a given period of time (from t_1 to t_2) can be written

$$\int_{t_1}^{t_2} (i - v_z - q)\, dt - \int_0^z \int_{t_1}^{t_2} \frac{\partial \theta}{\partial t}\, dz\, dt = \int_0^z \int_{t_1}^{t_2} r_z\, dz\, dt \qquad (6.8)$$

where i is rate of water supply (precipitation or irrigation), q evaporation rate from the soil surface, v_z the vertical flux of water at depth z, θ the volumetric soil wetness, and r_z the rate of decrease of soil wetness due to water uptake by roots.

The average rate of uptake by roots at the depth z is

$$r_z = \int_{t_1}^{t_2} r_z\, dt / (t_2 - t_1) \qquad (6.9)$$

The pattern of soil-water extraction by a root system can be determined by repeated calculations based on the preceding equations for successive small intervals of time and depth. The total (cumulative) water uptake by the roots R_z is given by

$$R_z = \int_0^z r_z\, dz \qquad (6.10)$$

These relationships were used by Rose and Stern to describe the pattern of soil-water extraction by a cotton crop in the field. The results indicated that nearly all water extraction by the crop took place from the top 30 cm during the early stages of growth and that, although the zone of uptake extended downward during later stages, essentially all of the seasonal uptake took place from the top 100 cm.

A similar and detailed field study of water extraction by a root system was carried out by van Bavel *et al.* (1968a, b), using the *instantaneous profile method* of obtaining the hydraulic properties of a complete profile in situ (Watson, 1966; Hillel *et al.*, 1972). (This method requires frequent, independent, and simultaneous measurements of hydraulic head and soil wetness in an internally draining profile, coupled with measurements of evaporation.) The calculated root extraction rates agreed reasonably well with separate measurements of transpiration obtained with lysimeters. Soil-water movement within the root zone of a sorghum crop indicated initially a net downward outflow from the root zone, but this movement later reversed itself to indicate a net upward inflow from the wet subsoil to the root zone above.

Denmead and Shaw (1962) presented experimental confirmation of the effect of dynamic conditions on water uptake and transpiration. They measured the transpiration rates of corn plants grown in containers and placed in the field under different conditions of irrigation and atmospheric evaporativity. Under an evaporativity of 3–4 mm/day, the actual transpiration rate began to fall below the potential rate at an average soil-water suction of about 2 bar. Under more extreme meteorological conditions,

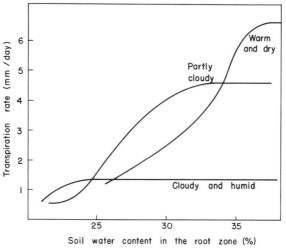

Fig. 6.14. Relation of actual transpiration rate to soil-water content, under different meteorological conditions. (After Denmead and Shaw, 1962.)

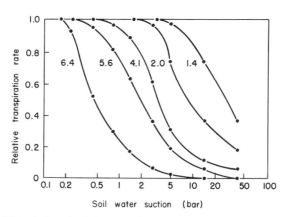

Fig. 6.15. The relation of relative transpiration rate to average soil-water suction, under different meteorological conditions. The numbers represent different rates of potential evapotranspiration. (After Denmead and Shaw, 1962.)

with an evaporativity of 6–7 mm/day, this drop already began at a soil water suction value of 0.3 bar. On the other hand, when the potential evaporativity was very low (1.4 mm/day), no drop in transpiration rate was noticed until average soil-water suction exceeded 12 bar. The volumetric water contents at which the transpiration rates fell varied between 23%, under the lowest evaporativity, and 34%, under the highest evaporativity measured. This is illustrated in Figs. 6.14 and 6.15.

I. Approaches to Modeling Water Uptake by Roots

The roots of a plant function as an elaborate pumping and conveyance system reaching into, probing, and drawing water from various soil depths in a manner which must somehow be "programmed" to maximize the plant's chances for survival and self-perpetuation. Soil moisture content and potential are seldom uniform throughout the root zone, and neither is the density or hydraulic resistance of the roots. How the root system of a plant senses the root zone as a whole and integrates its response so as to utilize soil moisture to best advantage is a phenomenon still imperfectly understood. One classical view (Wadleigh, 1946) was that the root system adjusts its water withdrawal pattern so as to maintain the total soil moisture potential constant throughout the root zone. On the other hand, an often-observed pattern of water withdrawal is such that the top layer is depleted first and the zone of maximal extraction moves gradually into the deeper layers.

Since the soil usually extends in depth considerably below the zone of root activity, it is of interest to establish how the pattern of soil water extraction by roots relates to the pattern of water flow within, through, and below the root zone. Some drainage through the root zone is considered necessary to prevent deleterious accumulation of salts, particularly in the arid zone; yet excessive drainage might involve unnecessary loss of nutrients as well as of water. If a groundwater table is present at a shallow depth, it can contribute to the supply of crop water requirement by upward capillary flow, but it might also infuse the root zone with harmful salts. Considerable upward flow into the root zone might also be possible even in the absence of a water table, depending upon soil moisture characteristics and upon the irrigation regime. In fact, the opposite processes of downward drainage and upward capillary flow can occur in a sequential or alternating pattern at varying rates so that the net outflow or inflow of water for the root zone as a whole must be determined by integrating the flux occurring through the bottom of the root zone continuously over a definite period of time.

In recent years we have witnessed the development of numerous models aimed at quantitative formulation and prediction of the sequential and simultaneous processes involved in the movement of soil moisture and its extraction by plants. These models vary widely in aim, structure, and level of detail. More specifically, available models differ in principle, depending on whether they are based on empirical or mechanistic approaches to the complex soil–plant–water–atmosphere system. Only two decades ago, the literature was replete with a bewildering array of experimental studies not based on any explicit theory, and their seemingly contradictory results could not be reconciled, owing to the absence of a unifying conception. Nowadays it seems that we have the opposite problem, namely, a plethora of theoretical models, which as yet remain largely untested and hence unproven. It has become altogether easier to formulate models than to validate them.

Modelers of soil moisture extraction by roots (e.g., Cowan, 1965; Feddes *et al.*, 1974, 1976; Hansen, 1975; Hillel *et al.*, 1975, 1976; Lambert and Penning de Vries, 1973; Molz and Remson, 1970, 1971; Molz, 1976; Nimah and Hanks, 1973; Rose *et al.*, 1976; Whisler and Millington, 1968; van Bavel and Ahmed, 1976) have differed widely in their quantitative assessment of the relative importance of the root versus soil resistance terms at various stages of the extraction process. Conflicting approaches have either assumed the one term to predominate over the other or vice versa. Theoretical and experimental studies (e.g., Gardner, 1960, 1964; Greacen *et al.*, 1976; Herkelrath, 1975; Newman, 1969, 1974; Reicosky and Ritchie, 1976; So *et al.*, 1976; Taylor and Klepper, 1976; Yang and DeJong, 1971) suggest that both resistance terms can be important: the root resistance term most probably tending to predominate in situations of low soil moisture tension, i.e., high soil hydraulic conductivity, and the soil's hydraulic resistance tending to gain importance as the extraction process continues and causes progressive depletion of soil moisture. In the absence of a consensus, however, the problem remains one of the most vexing ones in our effort to construct a more realistic model of water dynamics in the soil–plant–atmosphere system.

In principle, two alternative approaches can be taken to modeling the uptake of soil water by roots. The first is to consider the convergent radial flow of soil water toward and into a representative individual root, taken to be a line or narrow-tube sink uniform along its length (i.e., of constant and definable thickness and absorptive properties). The root system as a whole can then be described as a set of such individual roots, assumed to be regularly spaced in the soil at definable distances which may vary within the soil profile. This approach, called the *microscopic scale* or *single root* model, usually involves casting the flow equation in cylindrical coordinates and solving it for the distribution of potentials, water contents, and fluxes from the root outward. The solution can be attempted by analytical means (e.g., Philip, 1975; Gardner, 1960; Cowan, 1965), which usually require rather restrictive assumptions, or by numerical means (e.g., Molz and Remson, 1970; Lambert and Penning de Vries, 1973; Hillel *et al.*, 1975), which require the use of a digital computer. Solutions by means of analog and hybrid computers have also been attempted (d'Hollander and Vansteenkiste, 1975). The solutions mentioned have by and large tended to ignore sectional differences in absorptive activity which might result from differences in age or location of roots.

The alternative approach is to regard the root system in its entirety as a diffuse sink which permeates each depth layer of soil uniformly, though not necessarily with a constant strength throughout the root zone. This approach, termed the *macroscopic scale* or *root system* model, disregards the flow patterns toward individual roots and thus avoids the geometric complication involved in analyzing the distribution of fluxes and potential gradients on a microscale. The macroscopic approach was taken by Ogata *et al.* (1960), Gardner (1964), Whisler *et al.* (1968), Molz and Remson (1970, 1971), Nimah and Hanks (1973), Hillel *et al.* (1976), and others. The major shortcoming of the macroscopic approach is that it is based on the gross spatial averaging of the matric and osmotic potentials and takes no account of the increase in suction and in concentration of salts in the soil at the immediate periphery of the absorbing roots.

J. The Single-Root Radial (Microscopic) Model

The geometry of this model is illustrated in Figs. 6.16 and 6.17. It consists of a cylindrical root of radius r_r, surrounded by a homogeneous soil of radius r_s, where r_s is the midpoint distance between adjacent roots. Within the root, another boundary can be defined at $r = r_e$, the radius of the endodermis. As shown by Molz (1975), the composite flow region can thus be divided into a *soil domain*, defined by $r_r < r < r_s$, and a *root tissue domain*, defined by $r_e < r < r_r$. (The possible existence of a distinct, intermediate contact zone between the root and the soil is generally, though perhaps unrealistically, ignored.)

In the soil domain, the radial form of the equation for the transient-state flow of water in the unsaturated soil is given by

$$c \frac{\partial \phi_m}{\partial t} = K \frac{\partial^2 \phi_m}{\partial r^2} + \frac{K}{r} \frac{\partial \phi_m}{\partial r} + \frac{\partial K}{\partial r} \left(\frac{\partial \phi_m}{\partial r} \right)^2 \tag{6.11}$$

where ϕ_m is the matric potential of soil moisture, c the specific soil moisture capacity (the slope of the soil moisture characteristic, $d\theta/d\phi_m$, where θ is volumetric wetness), K the unsaturated soil's hydraulic conductivity (a function of matric potential, or of wetness), and t time.

In the root tissue domain, if solute effects are neglected and local equilibrium is assumed to hold between the water content of each cell and its cell wall (Molz and Ikenberry, 1974), the equation for radial water transport can be written

$$\frac{\partial \phi}{\partial t} = D \frac{\partial^2 \phi}{\partial r^2} + \frac{D}{r} \frac{\partial \phi}{\partial r} \tag{6.12}$$

where ϕ is the tissue water potential and D the tissue's hydraulic diffusivity.

Where soil moisture contains an appreciable concentration of solutes which are by and large excluded by the root, the difference in *total* soil moisture potential between the soil sheathing the root (including the osmotic, as well as the matric, potential) and the root-water potential must be taken into account. A detailed study of water and solute transport from soil to root must consider both the differential hydraulic conductance and the differential osmotic permeability of the various tissues and flow paths. with the osmotic factor being particularly important at the endodermis.

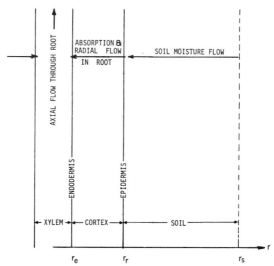

Fig. 6.16. Hypothetical representation of the radial flow region in the soil surrounding the root and in the root surrounding the xylem. (At the xylem, flow becomes axial rather than radial.)

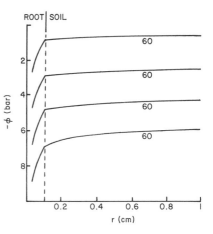

Fig. 6.17. Water potential ψ in the root cortex and soil as a function of radius r. The numbers next to the curves are the initial potential gradients assigned at the endodermis. The radius of the root is 0.1 cm. (After Molz, 1976.)

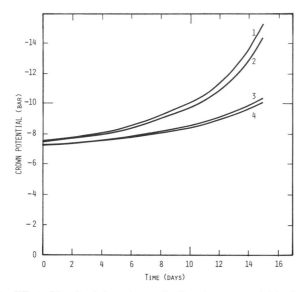

Fig. 6.18. Effect of localized draw-down of soil moisture potential in the vicinity of the root on the development of "crown potential" (plant-water potential at the stem) under a constant transpirational demand of 1 mm/day per 0.1 m of root-zone depth. (1) Draw down of matric and osmotic potentials; (2) draw down of matric potential, uniform osmotic; (3) draw down of osmotic potential, uniform matric; (4) uniform matric and osmotic potentials. (A simulation study by Hillel *et al.*, 1975.)

The soil resistance for any segment of the flow path can be taken as proportional directly to the flow path length and inversely to the hydraulic conductivity prevailing in that segment. The root resistance term is much more difficult to define quantitatively. As pointed out by Leyser and Loch (1972), the overall resistance of the roots can be separated into a *resistance to absorption* and a *resistance to conduction* (xylem resistance). The absorptive resistance is known to depend upon several physiological factors, such as age of the root (which determines the degree of suberization and amount of active root hairs) and water potential, as well as upon such environmental factors as temperature, aeration, specific ions, etc. Finally, root resistance might also depend on the magnitude of the driving force (the potential gradient) and the flow velocity (Slatyer, 1967). In the absence of exact information on these aspects of the problem, we must regard all modeling efforts carried out to date as preliminary. To verify (or refute) some of the predictions of presently available models, we need additional specific experimental data of the type reported by Taylor and Klepper (1976) on the actual process of water extraction by living roots.

Microscopic scale simulations of the moisture extraction process are illustrated in Figs. 6.16–6.21.

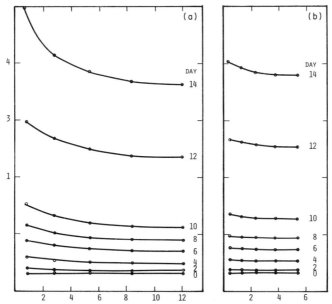

Fig. 6.19. Matric potential as function of distance from root at two-day intervals during extraction by (a) "sparse" root system (1600 roots/m^2) and (b) "dense" root system (8000 roots/m^2). (After Hillel *et al.*, 1975.)

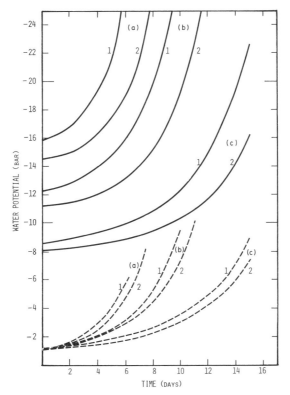

Fig. 6.20. Plant-water potential (solid lines) and total soil moisture potential near root (dashed lines) as function of time during extraction under a transpirational demand of (a) 2, (b) 1.5, and (c) 1 mm/day per 100 mm rooting depth. Numbers 1 and 2 refer to sparse and dense rooting systems, respectively. (After Hillel *et al.*, 1975.)

The microscopic scale simulation by Hillel *et al.* (1975) had as its limited objective the semiquantitative evaluation of the localized draw-downs of both matric and osmotic components of soil moisture potential around the root and hence of possible errors that might result from the macroscopic approach which seems to be the more popular at present. They found that these localized effects can indeed be important, particularly in the case of young plants growing in a soil of low "field capacity" and low unsaturated conductivity values and/or in case the irrigation water or the soil itself is somewhat saline. Indiscriminate application of macroscopic scale models can cause a significant underestimation of the developing plant water stress during the process of soil moisture depletion. The resulting overestimation of the length of time a plant can continue to thrive without replenishment of soil moisture while still avoiding a state of stress can

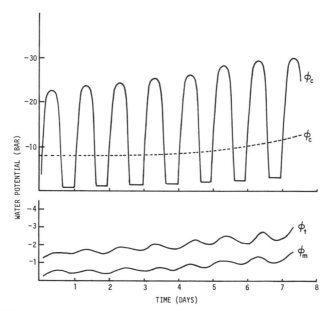

Fig. 6.21. Crown potential ϕ_c, total soil moisture potential ϕ_t, and matric potential ϕ_m of soil contacting the root, under a diurnally fluctuating transpirational demand of 1 mm/day per 100 mm of rooting depth. Dashed line: crown potential under a constant transpirational demand. Rooting density: 1600 roots/m². (After Hillel *et al.*, 1975.)

amount to 30% or even more, particularly in the case of sparse-rooted plants in a coarse-textured soil and where salinity is appreciable.

Field measurements taken with neutron soil moisture meters and tensiometers yield vital data on the gross average of soil wetness and matric potential but cannot indicate the microscopic scale gradients and local values of the moisture potential (osmotic as well as matric) affecting plant roots. Until such time as our measuring techniques are greatly refined, we can only estimate the possible magnitude of these potential differences on the basis of as yet unproven theory.

K. The Root System (Macroscopic) Model

The vertical transient state flow of water in a stable and uniform zone of the soil is described by the following equation:

$$\frac{\partial \theta}{\partial t} = \frac{\partial}{\partial z} \left[K(\theta) \frac{\partial (\phi_m - z)}{\partial z} \right] - S_w \qquad (6.13)$$

in which θ is volume wetness, t time, z depth, $K(\theta)$ hydraulic conductivity (a function of wetness), ϕ_m matric potential head, and S_w a sink term representing the presence of plant roots. The term $\phi_m - z$ is the overall hydraulic head.

The rate of extraction of water from a unit volume of soil can be represented in the following way:

$$S_w = (\phi_{soil} - \phi_{plant})/(R_{soil} + R_{roots}) \tag{6.14}$$

Herein, ϕ_{soil} is the total potential of soil water, which is the sum of the matric ϕ_m, gravitational ($\phi_g = z$), and osmotic ϕ_o potentials, all of which are expressible in head units:

$$\phi_{soil} = \phi_m + \phi_g + \phi_o \tag{6.15}$$

The hydraulic resistance to flow in the soil toward the roots was expressed by Gardner (1964) as inversely proportional to the hydraulic conductivity K (a function of the soil's wetness or matric potential) and to the total length of active roots L in the unit volume of soil:

$$R_{soil} = 1/BKL \tag{6.16}$$

Here B is an empirical constant, which can be taken to represent a specific root activity factor.

The term ϕ_{plant} is the plant-water potential at a point, presumably at the base of the stem, where all roots converge and the plant emerges from the soil with a single water potential which we call the *crown potential* (designated ϕ_c).

The hydraulic resistance of the roots (R_{roots}) can again be taken to be the sum of a resistance to absorption and a resistance to conduction, the latter being a function of the depth of any particular group of roots.

The flow rate $(q_r)_i$ delivered by the roots from any particular layer i in the soil to the crown can be taken as the ratio of the difference in potential between that soil layer and the crown to the total hydraulic resistance encountered:

$$(q_r)_i = [(\phi_s)_i - \phi_c]/[(R_r)_i + (R_s)_i] \tag{6.17}$$

where $(\phi_s)_i$ is the soil moisture potential, $(R_r)_i$ the resistance of the roots, and $(R_s)_i$ the hydraulic resistance of the soil.

The total extraction rate Q from all volume elements or layers of soil, equal to the transpiration rate, is a sum of the contributions of all volume elements within the root zone:

$$Q = \sum_{i=1}^{n} \frac{(\phi_s)_i - \phi_c}{(R_r)_i + (R_s)_i}$$

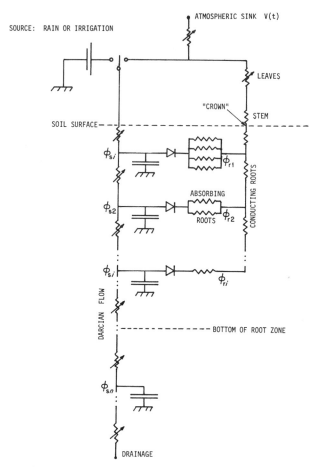

Fig. 6.22. Schematic representation of a root system as a resistance network. Soil layers are shown as capacitors, linked by the variable resistance of unsaturated vertical flow, and discharged by the roots through the variable resistance of the canopy. The roots are represented by a resistance to absorption and a resistance to conduction (the former inversely proportional to rooting density in each layer, and the latter directly proportional to depth). The diodes at each layer indicate one-directional flow into the roots. The atmospheric sink is shown to be of variable potential. The battery at upper left represents a source of water recharging the soil layers during episodes of rainfall. The ϕ_s, ϕ_r values indicate the potential values for water in the soil and roots, respectively, at various levels in the profile. (After Hillel, 1977.)

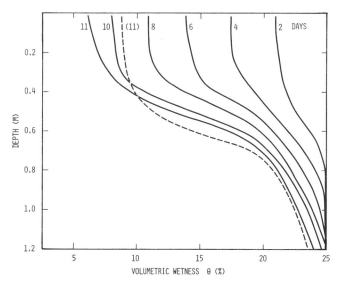

Fig. 6.23. Basic simulation run: Pattern of soil moisture extraction by roots as shown by successive wetness profiles. The numbers alongside the curves indicate days from time zero, at which soil wetness was a uniform 25%. The dashed curve is for a root system 10 times as dense (RTL = 10^5 m) having the same relative distribution in the profile. (After Hillel *et al.*, 1975.)

Hence

$$\phi_c = \sum_{i=1}^{n} \frac{(\phi_s)_i}{(R_r)_i + (R_s)_i} - Q \bigg/ \sum_{i=1}^{n} \frac{1}{(R_r)_i + (R_s)_i} \tag{6.18}$$

where n is the total number of volume elements or layers in the rooting zone. Knowing the value of Q (which, as a first approximation for a freely transpiring plant, is equal to the climatically induced transpirational demand) and the values of $(R_r)_i$ (which depends on the rooting-density distribution in the profile and on the depth of the particular layer) as well as ϕ_s and R_s for each layer, it is possible to obtain the value of ϕ_c at successive times by a process of iteration. This was done by Hillel *et al.* (1975). The electrical analog of their system is shown in Fig. 6.22, and the results of their calculations are illustrated in Fig. 6.23 and 6.24.

Figure 6.23 presents successive soil moisture profiles during the process of extraction from a soil initially at 25% moisture (which is roughly the "field capacity" for the soil simulated) with 99.955% of the roots present in the upper half-meter of the soil profile. An interesting feature of this family of curves is that only the top 30 cm or so of the profile exhibits

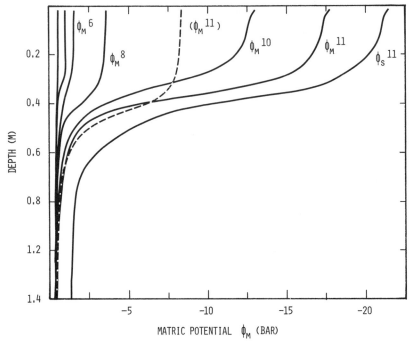

Fig. 6.24. Basic simulation: Successive profiles of matric potential ϕ_m on days 2, 4, 6, 8, 10, and 11 and of total soil-water potential ϕ_s on day 11. The dashed curve is for the dense root system. (After Hillel *et al.*, 1975.)

anything like a uniform value of wetness at any particular time, whereas the zone below is highly nonuniform.

The pattern of moisture extraction is characterized by three discernible phenomena: gradual reduction of wetness within the zone of major root concentration, gradual deepening of the zone of moisture extraction, and gradual steepening of the moisture gradients between the untapped subsoil and the zone of major extraction. On the 11th day, at which our hypothetical plant is presumed to be in a state of stress leading to wilting (the crown potential having fallen below -30 bar), the moisture profile is seen to vary from about 6% in the top 20 cm to about 10% at 40 cm, about 17% at 60 cm, and over 20% below 80 cm.

The highly variable nature of the soil moisture profile is also illustrated in Fig. 6.24, which presents successive matric potential profiles. The concurrently developing profile of total soil moisture potential is shown for the 11th day, at which wilting is presumed to occur. It is seen that the soil moisture potential varies more and more widely within and below the root

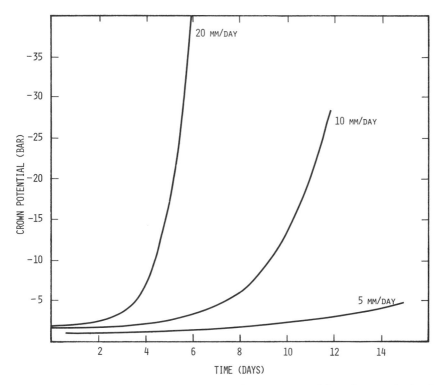

Fig. 6.25. Dependence of the time course of plant-water potential on the transpirational demand (given in millimeters per day). Results of a simulation model by Hillel *et al.* (1975).

zone as the process of water extraction progresses. On the 11th day, the matric suction varies from less than 2 bar at the 50 cm depth to more than 17 in the upper 30 cm. In time, as the topmost layer of soil is progressively depleted by both root extraction and direct evaporation, more and more of the plant's water supply is extracted from the deeper layers even though the roots are very sparse beyond, say, 45 cm. This is apparently a consequence of the considerable flux of water transmitted to the drying root zone from the wetter subsoil layers.

Figure 6.25 illustrates the dependence of plant-water potential on the magnitude of the transpirational demand. As can be seen, an increase of transpirational demand causes an approximately proportional decrease in the length of time the plant can continue to thrive before experiencing any particular level of water stress. Thus, under an extremely high transpirational demand of 20 mm/day, the simulated plant undergoes a sharp decrease of water potential presumably leading to wilting as early as the sixth day, whereas under a moderately high transpiration rate of 10 mm/day it may continue to the 12th day.

An important lesson to be learned from these calculations is the degree of approximation involved in the old but still popular attempt to characterize the water status of the root zone in terms of some fixed quantity of "available" or "extractable" soil moisture. The highly variable profiles of moisture content and potential within and particularly below the root zone make the choice of just where in the profile to measure or how to integrate soil wetness, suction, and salinity over space and time a moot question indeed. The power of the simulation approach is that it can provide an essentially continuous monitoring of the entire system as it varies in response to any number of factors on the basis of cause-and-effect mechanisms. However, the predictions obtained must be validated experimentally before they can be applied with a sufficient degree of confidence.

Attempts at experimental validation of the model described were carried out by Belmans *et al.* (1979) and Feyen *et al.* (1980), who raised close-growing stands of rye grass in tubes packed with sand and loam, with and without the presence of a water table. The pattern of water uptake was monitored by gamma-ray scanning and tensiometry. Correspondence between measured and independently predicted uptake patterns could be improved by assigning a high value of root resistance relative to the soil's hydraulic resistance. Still better fit between predicted and measured patterns of soil moisture extraction could probably be obtained if the conductive component of the root resistance term were taken into account in such a manner as to ascribe progressively greater resistance to the deeper roots, which tend to be thinner and must deliver water through a longer conductive path. A shortcoming of the models described is the assumption of a time-constant value of root absorptivity despite recent suggestions that it may in fact be flux dependent or may otherwise vary with changes in soil or plant water potential or both (Hansen, 1975; Herkelrath, 1975; Seaton *et al.*, 1977).

Future models of soil moisture extraction by roots may need to take into account the possible influence of such "extraneous" factors (in addition to soil and plant hydraulics) as soil aeration, soil mechanical resistance to root penetration, soil temperature variation, distribution of solutes (both nutritive and toxic), presence of soil-borne pathogens or parasites, etc. The number of possible complicating factors is indeed legion. *C'est la vie.*

L. Effect of Root Growth on Soil-Water Uptake

One of the main shortcomings of the macroscopic model, as originally formulated, was the omission of root growth as a process affecting the pattern of moisture extraction and movement in the soil profile and the possible response of the plant in terms of its ability to continue functioning without experiencing excessive water stress. A plant with growing roots can reach continuously into moist regions of the soil rather than depend entirely on the conduction of water over appreciable distances in the soil against a steadily increasing hydraulic resistance, as is the case with a fixed root system. The process of root growth, if rapid enough, can reduce the effect of the localized draw-down of both matric and osmotic potential around each root, as well as increase the effective volume of soil tapped by the root system as a whole.

While we do not as yet have sufficient information on the hydraulics of growing root systems, the limited data already available (e.g., Taylor and Klepper, 1975) suggest that the distribution of roots in the profile can change rather markedly within a period of weeks or even days, particularly in the case of an annual crop. It is therefore of interest to attempt to devise a logical framework for the dynamic simulation of root growth within the context of an overall model of soil-water extraction by variously distributed root systems. Such a model might thus serve as a better criterion for the evaluation of soil moisture availability to different types of plants at different stages of growth.

In principle, and insofar as it relates to water uptake, we can consider overall root growth as consisting of several concurrent processes, including proliferation, extension, senescence, and death. As used in the present context, the term proliferation applies to the localized increase of rooting density (i.e., by branching) within each layer without any increase in the volume of the root zone as a whole. Extension is the additional process by which roots from any layer grow downward so as to invade an underlying layer and increase its rooting density. The process of senescence involves suberization and the gradual reduction of root permeability. With further againg, the older roots eventually become totally inactive and, to all intents and purposes, can be considered dead.

The possible effect of root growth processes on plant–water relations was studied by Hillel and Talpaz (1976), who considered the time course of plant water potential at the "crown" of the roots (where all the roots converge and the stem emerges from the soil with a single value of water potential). Without root extension in depth, this simulation suggests that root proliferation alone can add only about 1 day to the period of time the plant can maintain the potential transpiration rate without experiencing excessive stress (i.e., a crown potential lower than -30 bar). On the other hand, root systems capable of extending themselves deeper into the soil profile can prolong that time span by several days. The predicted soil moisture profiles resulting from extraction by root systems of various growth habits are shown in Fig. 6.26.

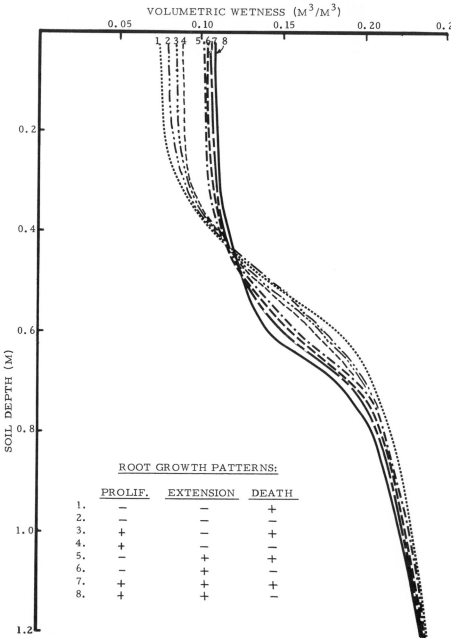

Fig. 6.26.   Soil moisture profiles at the end of a 10-day simulation of moisture extraction by root systems with various growth patterns.

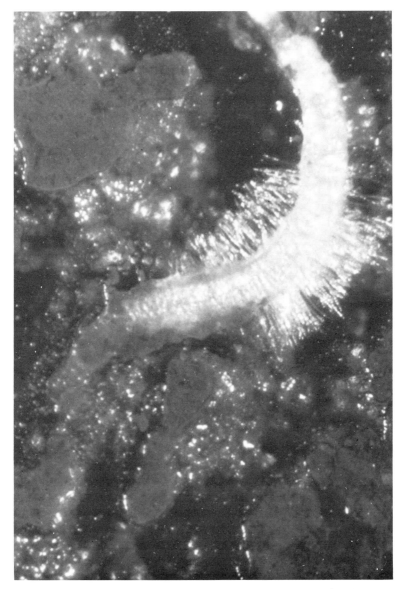

Fig. 6.27.   Root growing in a well-watered, well-aerated soil. (Courtesy of Dr. A. C. Trouse, National Tillage Machinery Laboratory, U. S. Department of Agriculture, Auburn, Alabama.)

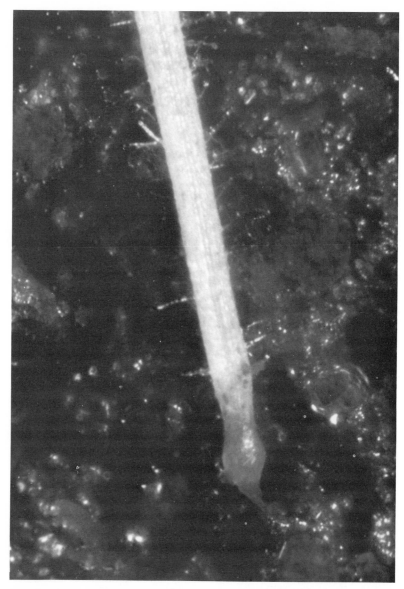

Fig. 6.28. Root subjected to temporary oxygen stress. (Courtesy of Dr. A. C. Trouse, National Tillage Machinery Laboratory, U. S. Department of Agriculture, Auburn, Alabama.)

Fig. 6.29.   Root branching in a tight, cloddy soil. (Courtesy of Dr. A. C. Trouse, National Tillage Machinery Laboratory, U. S. Department of Agriculture, Auburn, Alabama.)

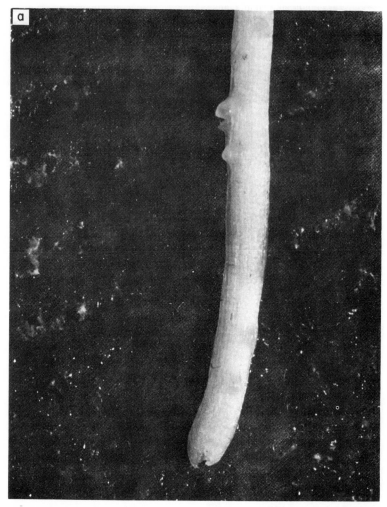

Fig. 6.30. The lateral branching of a maize root: (a) one day after being severed just behind the growing tip, (b) two days after, and (c) six days after. (Courtesy of Dr. A. C. Trouse, National Tillage Machinery Laboratory, U. S. Department of Agriculture, Auburn, Alabama.)

Fig. 6.30. *(Continued)*

The phenomenon of root extension is characteristic of a stand of young plants, such as an annual crop in its early stages of growth. The rate of root system extension probably decreases in older plants and may eventually become nil in the case of mature perennial crops. The vertical extent of root system penetration is often limited by such factors as the lack of aeration or the mechanical impedance of the deeper layers in the soil profile. Growing roots are shown pictorially in Figs. 6.27–6.30.

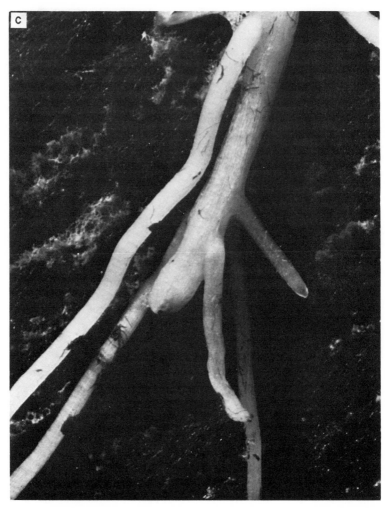

Fig. 6.30.   (*Continued*)

Sample Problem

1.   Consider the pattern of soil moisture extraction by a plant growing in a soil-filled container, initially wet but left unreplenished until the plant wilts. The daily transpiration is 0.5 cm. At first, soil moisture suction is 0.1 bar, plant-water suction (measured at the stem) is 5 bar, and rooting density is 0.2 cm of roots per cubic centimeter of soil. At an intermediate stage, soil moisture suction is 1 bar, plant-water suction is 10 bar, and root-

ing density is 0.5 cm/cm^3. At a still later stage, just prior to wilting, soil moisture suction becomes 10 bar, plant-water suction reaches 20 bar, and the roots have proliferated to 1 cm/cm^3. Assume a soil hydraulic conductivity function $K = a/\psi^3$, where $a = 100$ and ψ is suction head in centimeters. Estimate the relative magnitude of the hydraulic resistances in the soil and root system during the three stages described.

We can estimate the soil resistance term R_{soil} using the expression given by Gardner (1960), Eq. (6.16):

$$R_{\text{soil}} = 1/BKL$$

where K is the soil's hydraulic conductivity (a function of its suction ψ), L is the length of roots per unit volume of soil, and B is an empirical factor which, for want of any specific information, we shall take to be unity. Root system resistance per unit area of soil surface can be estimated from the ratio of the "potential difference" to the "current" (to use a simple Ohm's law analogy):

$$R_{\text{roots}} = \Delta\psi/q$$

where $\Delta\psi$ is the suction difference between the stem and the soil, and q is the transpiration flux.

During the early stage,

$$R_{\text{soil}} = \frac{1}{1 \times (100/100^3) \times 0.2} = 5 \times 10^4 \text{ sec}$$

$$R_{\text{roots}} = \frac{5000 \text{ cm} - 100 \text{ cm}}{0.5 \text{ cm/day} \div 86400 \text{ sec/day}} = \frac{4900 \text{ cm}}{5.8 \times 10^{-6} \text{ cm/sec}}$$

$$= 8.45 \times 10^8 \text{ sec}$$

During the intermediate stage,

$$R_{\text{soil}} = \frac{1}{1 \times (100/1000^3) \times 0.5} = 2 \times 10^7 \text{ sec}$$

$$R_{\text{roots}} = \frac{(10000 \text{ cm} - 1000 \text{ cm})}{5.8 \times 10^{-6} \text{ cm/sec}} = 1.55 \times 10^9 \text{ sec}$$

During the late stage,

$$R_{\text{soil}} = \frac{1}{1 \times (100/10000^3) \times 1} = 10^{10} \text{ sec}$$

$$R_{\text{roots}} = \frac{(20000 - 10000) \text{ cm}}{5.8 \times 10^{-6} \text{ cm/sec}} = 1.72 \times 10^9 \text{ sec}$$

Note: The figures given are entirely hypothetical, meant merely to illustrate the principle that the soil resistance term is generally negligible compared to the root resistance term as long as the soil is relatively wet. As soil moisture is depleted, however, and soil hydraulic conductivity decreases steeply, the soil resistance terms gradually becomes more important. In the formulation given, incidentally, the possible resistance of the contact zone between the soil and the roots was lumped into the root resistance term.

When you can measure what you are speaking about and express it in numbers, you know something about it; but when you cannot measure it, when you cannot express it in numbers, your knowledge is of a meager, unsatisfactory kind.

Lord Kelvin
1824–1907

7 Water Balance and Energy Balance in the Field

A. Introduction

Any attempt to control the quantity and availability of soil moisture to plants must be based on a thorough understanding and a quantitative knowledge of the dynamic balance of water in the soil. The *field-water balance*, like a financial statement of income and expenditures, is an account of all quantities of water added to, subtracted from, and stored within a given volume of soil during a given period of time. The various soil-water flow processes that we have attempted to describe in earlier chapters of this book as separate phenomena (e.g., infiltration, redistribution, drainage, evaporation, water uptake by plants) are in fact strongly interdependent, as they occur sequentially or simultaneously.

The water balance is merely a detailed statement of the *law of conservation of matter*, which states simply that matter can neither be created nor destroyed but can only change from one state or location to another. Since no significant amounts of water are normally decomposed, or composed, in the soil, the water content of a soil profile of finite volume cannot increase without addition from the outside (as by infiltration or capillary rise), nor can it diminish unless transported to the atmosphere by evapotranspiration or to deeper zones by drainage.

The field water balance is intimately connected with the *energy balance*, since it involves processes that require energy. The energy balance is an expression of the classical *law of conservation of energy*, which states that, in a given system, energy can be absorbed from, or released to, the outside, and that along the way it can change form, but it cannot be created or destroyed.

The content of water in the soil affects the way the energy flux reaching the field is partitioned and utilized. Likewise, the energy flux affects the state and movement of water. The water balance and energy balance are inextricably linked, since they are involved in the same processes within the same environment. A physical description of the soil–plant–atmosphere system must be based on an understanding of both balances together. In particular, the evaporation process, which is often the principal consumer of both water and energy in the field, depends, in a combined way, on the simultaneous supply of water and energy.

B. Water Balance of the Root Zone

In its simplest form, the water balance merely states that, in a given volume of soil, the difference between the amount of water added W_{in} and the amount of water withdrawn W_{out} during a certain period is equal to the change in water content ΔW during the same period:

$$\Delta W = W_{in} - W_{out} \tag{7.1}$$

When gains exceed losses, the water-content change is positive; and conversely, when losses exceed gains, ΔW is negative.

To itemize the accretions and depletions from the soil storage reservoir, one must consider the disposition of rain or irrigation reaching a unit area of soil surface during a given period of time. Rain or irrigation water applied to the land may in some cases infiltrate into the soil as fast as it arrives. In other cases, some of the water may pond over the surface. Depending on the slope and microrelief, a portion of this water may exit from the area as surface run-off ("overland flow") while the remainder will be stored temporarily as puddles in surface depressions. Some of the latter evaporates and the rest eventually infiltrates into the soil after cessation of the rain. Of the water infiltrated, some evaporates directly from the soil surface, some is taken up by plants for growth or transpiration, some may drain downward beyond the root zone, whereas the remainder accumulates within the root zone and adds to soil moisture storage. Additional water may reach the defined soil volume by runoff from a higher area, or by upward flow from a water table or from wet layers present at some depth. The pertinent volume or depth of soil for which the water balance is computed is determined arbitrarily. Thus, in principle, a water balance can be computed for a small sample of soil or for an entire watershed. From an agricultural or plant ecological point of view, it is generally most appropriate to consider the water balance of the root zone per unit area of field. The root zone water balance is expressed in integral form thusly:

(change in storage) = (gains) − (losses)

$$(\Delta S + \Delta V) = (P + I + U) - (R + D + E + T)$$

(7.2)

wherein ΔS is change in root zone soil moisture storage, ΔV increment of water incorporated in the plants, P precipitation, I irrigation, U upward capillary flow into the root zone, R runoff, D downward drainage out of the root zone, E direct evaporation from the soil surface, and T transpiration by plants. All quantities are expressed in terms of volume of water per unit area (equivalent depth units) during the period considered.

The time rate of change in soil moisture storage can be written as follows (assuming the rate of change of plant-water content to be relatively unimportant):

$$dS/dt = (p + i + u) - (r + d + e + t_r)$$

(7.3)

Here each of the lowercase letters represents the instantaneous time rate of change of the corresponding integral quantity in the first equation. The change in root zone soil moisture storage can be obtained by integrating the change in soil wetness over depth and time as follows:

$$S = \int_0^z \int_{t_1}^{t_2} \left(\frac{\partial \theta}{\partial t}\right) dz\, dt$$

(7.4)

where θ is the volumetric soil wetness, measurable by sampling or by means of a neutron meter. Note that t is time whereas t_r is transpiration rate in our notation.

The largest composite term in the "losses" part of Eq. (7.2) is generally the evapotranspiration $E + T$. It is convenient at this point to refer to the concept of "potential evapotranspiration" (designated E_{to}), representing the climatic "demand" for water. Potential evapotranspiration from a well-watered field depends primarily on the energy supplied to the surface by solar radiation, which is a climatic characteristic of each location (depending on latitude, season, slope, aspect, cloudiness, etc.) and varies little from year to year. E_{to} depends secondarily on atmospheric advection, which is related to the size and orientation of the field and the nature of its upwind "fetch" or surrounding area. Potential evapotranspiration also depends upon surface roughness and soil thermal properties, characteristics which vary in time (van Bavel and Hillel, 1976). As a first approximation and working hypothesis, however, it is often assumed that E_{to} depends entirely on the external climatic inputs and is independent of the transient properties of the field itself.

Actual evapotranspiration, E_{ta} is generally a fraction of E_{to} depending on the degree and density of plant canopy coverage of the surface, as well as on soil moisture and root distribution. E_{ta} from a well-watered stand of a close growing crop will generally approach E_{to} during the active growing stage, but may fall below it during the early growth stage, prior to full capoy coverage, and again toward the end of the growing season, as the matured plants begin to dry out (Hillel and Guron, 1973). For the entire season, E_{ta} may total 60–80% of E_{to} depending on water supply: the drier the soil moisture regime, the lower the actual evapotranspiration. The relation of yield to ET is still a matter of some controversy.

Another important, indeed essential, item of the field-water balance is the drainage out of the root zone D. A certain amount of drainage is required for aeration and for leaching out excess salts so as to prevent their ‚accumulation in the root zone, a particular hazard of arid zone farming. Where natural drainage is lacking or insufficient, artificial drainage becomes a prerequisite for sustainable agriculture.

The various items entering into the water balance of a hypothetical rooting zone are illustrated in Fig. 7.1. In this representation, only vertical flows are considered within the soil. In a larger sense, any soil layer of interest forms a part of an overall hydrologic cycle, illustrated in Fig. 7.2, in which the flows are multidirectional.

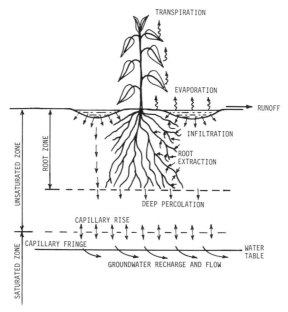

Fig. 7.1.   The water balance of a root zone (schematic).

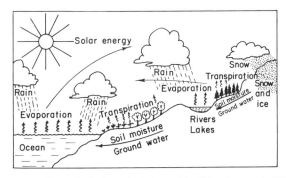

Fig. 7.2. The hydrologic cycle (schematic). (After Bertrand, 1967.)

C. Evaluation of the Water Balance

Simple and readily understandable though the field water balance may seem in principle, it is still rather difficult to measure in practice. A single equation can be solved if it has only one unknown. Often the largest component of the field water balance, and the one most difficult to measure directly, is the evapotranspiration $E + T$, also designated E_t. To obtain E_t from the water balance (Deacon *et al.*, 1958) we must have accurate measurements of all other terms of the equation. It is relatively easy to measure the amount of water added to the field by rain and irrigation $(P + I)$, though it is necessary to consider possible nonuniformities in areal distribution. The amount of run-off generally is (or at least should be) small in agricultural fields, and particularly in irrigated fields, so that it can sometimes be regarded as negligible in comparison with the major components of the water balance.

For a long period, e.g., an entire season, the change in water content of the root zone is likely to be small in relation to the total water balance. In this case, the sum of rain and irrigation is approximately equal to the sum of evapotranspiration E_t and deep percolation D. For shorter periods, the change in soil-water storage ΔS can be relatively large and must be measured. This measurement can be made by sampling periodically, or by use of specialized instruments.[1]

During dry spells, without rain or irrigation, $W_{in} = 0$, so that the sum of D and E_t now equals the reduction in root-zone water storage ΔS:

$$-\Delta S = D + E_t \tag{7.5}$$

---

[1] Of the various methods for measuring the content of water in soil, the neutron meter is the most satisfactory at present since it measures wetness on the volume or depth fraction basis directly and since it samples a large volume and minimizes sampling errors (with repeated measurements made at the same site and depth).

Common practice in irrigation is to measure the total water content of the root zone just prior to an irrigation, and to supply the amount of water necessary to replenish the soil reservoir to some maximal water content, generally taken to be the "field capacity." Some ecologists and irrigationists have tended to assume that the deficit of soil moisture which develops between rains or irrigations is due to evapotranspiration only, thus disregarding the amount of water which may flow through the bottom of the root zone, either downward or upward. This flow is not always negligible and often constitutes a tenth or more of the total water balance (Robins *et al.*, 1954; Nixon and Lawless, 1960; Rose and Stern, 1967a, b).

It should be obvious that measurement of root-zone or subsoil water content by itself cannot tell us the rate and direction of soil-water movement (van Bavel *et al.*, 1968a, b). Even if the water content at a given depth remains constant, we cannot conclude that the water there is immobile, since it might be moving steadily through that depth. Tensiometric measurements can, however, indicate the directions and magnitudes of the hydraulic gradients through the profile (Richards, 1965) and allow us to compute the fluxes from knowledge of the hydraulic conductivity versus suction or wetness for the particular soil. More direct measurements of the deep percolation component of the field water balance may eventually become possible with the development of water flux meters (Cary, 1968). Such devices have not yet proven to be practical, however.

The most direct method for measurement of the field water balance is by use of lysimeters (van Bavel and Myers, 1962; Pruitt and Angus, 1960; King *et al.*, 1956; Pelton, 1961; McIlroy and Angus, 1963; Forsgate *et al.*, 1965; Rose *et al.*, 1966; Harrold, 1966; Black *et al.*, 1968; Hillel *et al.*, 1969). These are generally large containers of soil, set in the field to represent the prevailing soil and climatic conditions and allowing more accurate measurement of physical processes than can be carried out in the open field. From the standpoint of the field water balance, the most efficient lysimeters are those equipped with a weighing device and a drainage system, which together allow continuous measurement of both evapotranspiration and percolation. Lysimeters may not provide a reliable measurement of the field water balance, however, when the soil or above-ground conditions of the lysimeter differ markedly from those of the field itself.

D. Radiation Exchange in the Field

By *radiation* we refer to the emission of energy in the form of electromagnetic waves from all bodies above 0°K. *Solar* (sun) *radiation* received on the earth's surface is the major component of its energy balance. Green plants are able to convert a part of the solar radiation into chemical energy. They do this in the process of photosynthesis, upon which all life on earth ultimately depends. For these reasons, it is appropriate to introduce a discussion of the energy balance with an account of the radiation balance.

Solar radiation reaches the outer surface of the atmosphere at a nearly constant flux of about 2 cal/min cm² perpendicular to the incident radiation.[2] Nearly all of this radiation is of the wavelength range of 0.3–3 μm (3000–30,000 Å), and about half of this radiation consists of visible light (i.e., 0.4–0.7 μm in wavelength). The solar radiation corresponds approximately to the emission spectrum of a blackbody[3] at a temperature of 6000°K. The earth, too, emits radiation, but since its surface temperature is about 300°K, this *terrestrial radiation* is of much lower intensity and greater wavelength than solar radiation[4] (i.e., in the wavelength range of 3–50 μm). Between these two radiation spectra, the sun's and the earth's, there is very little overlap, and it is customary to refer to the first as *short-wave* and to the second as *long-wave radiation* (Sellers, 1965).

In passage through the atmosphere, solar radiation changes both its flux and spectral composition. About one-third of it, on the average, is reflected back to space (this reflection can become as high as 80% when the sky is completely overcast with clouds). In addition, the atmosphere absorbs and scatters a part of the radiation, so that only about half of the original flux density of solar radiation finally reaches the ground.[5] A part of the reflected and scattered radiation also reaches the ground and is called *sky radiation*. The total of direct solar and sky radiations is termed *global radiation*.

Albedo is the reflectivity coefficient of the surface toward short-wave radiation. This coefficient varies according to the color, roughness, and inclination of the surface, and is of the order of 5–10% for water, 10–30% for a vegetated area, 15–40% for a bare soil, and up to 90% for fresh snow.

[2] 1 cal/cm² = 1 langley (Ly). 58 Ly ≈ 1 mm evaporation equivalent (latent heat = 580 cal/gm).

[3] A blackbody is one which absorbs all radiation reaching it without reflection, and emits at maximal efficiency. According to the *Stefan–Boltzmann law*, the total energy emitted by a body J_t, integrated over all wavelengths, is proportional to the fourth power of the absolute temperature T. This law is usually formulated as $J_t = \varepsilon\sigma T^4$ (where σ is a constant, and ε the emissivity coefficient). For a perfect blackbody, $\varepsilon = 1$.

[4] According to *Wien's law*, the wavelength of maximal radiation intensity is inversely proportional to the absolute temperature: $\lambda_m T = 2900$ (where λ_m is the wavelength in microns and T is the temperature on the Kelvin scale). *Planck's law* gives the intensity distribution of energy emitted by a blackbody as a function of wavelength and temperature: $E_\lambda = C_1 / \lambda^5 [\exp(C_2/\lambda T) - 1]$, where E_λ is the energy flux emitted in a particular wavelength range, and C_1, C_2 are constants.

[5] In arid regions, where the cloud cover is sparse, the actual radiation received at the soil surface can exceed 70% of the "external" radiation. In humid regions, this fraction can be 40% or lower.

In addition to these incoming and reflected short-wave radiation fluxes, there is also a long-wave radiation (heat) exchange. The earth's surface emits radiation, and at the same time the atmosphere absorbs and emits long-wave radiation, part of which reaches the surface. The difference between the outgoing and incoming fluxes is called the *net long-wave radiation*. During the day, the net long-wave radiation may be a small fraction of the total radiation balance, but during the night, in the absence of direct solar radiation, the heat exchange between the land surface and the atmosphere dominates the radiation balance.

The overall difference between total incoming and total outgoing radiation (including both the short-wave and long-wave components) is termed *net radiation*, and it expresses the rate of radiant energy absorption by the field.

$$J_n = J_s^{\downarrow} - J_s^{\uparrow} + J_l^{\downarrow} - J_l^{\uparrow} \qquad (7.6)$$

where J_n is the net radiation, $J_s^{\downarrow}$ the incoming flux of short-wave radiation from sun and sky, $J_s^{\uparrow}$ the short-wave radiation reflected by the surface, $J_l^{\downarrow}$ the long-wave radiation from the sky, and $J_l^{\uparrow}$ the long-wave radiation reflected and emitted by the surface. At night, the short-wave fluxes are negligible, and since the long-wave radiation emitted by the surface generally exceeds that received from the sky, the nighttime net radiation flux is negative.

The reflected short-wave radiation is equal to the product of the incoming short-wave flux and the reflectivity coefficient (the albedo α):

$$J_s^{\uparrow} = \alpha J_s^{\downarrow}$$

Therefore,

$$J_n = J_s^{\downarrow}(1 - \alpha) - J_l \qquad (7.7)$$

where J_l is the net flux of long-wave radiation, which is given a negative sign. (Since the surface of the earth is usually warmer than the atmosphere, there is generally a net loss of thermal radiation from the surface.) As a rough average, J_n is typically of the order of 55–70% of $J_s^{\downarrow}$ (Tanner and Lemon, 1962).

E. Total Energy Balance

Having balanced the gains and losses of radiation at the surface to obtain the net radiation, we next consider the transformation of this energy.

Part of the net radiation received by the field is transformed into heat, which warms the soil, plants, and atmosphere. Another part is taken up by the plants in their metabolic processes (e.g., photosynthesis). Finally, a major part is generally absorbed as latent heat in the twin processes of evaporation and transpiration. Thus,

$$J_n = LE + A + S + M \qquad (7.8)$$

where LE is the rate of energy utilization in evapotranspiration (a product of the rate of water evaporation E and the latent heat of vaporization L), A is the energy flux that goes into heating the air (called *sensible heat*), S is the rate at which heat is stored in the soil, water, and vegetation, and M represents other miscellaneous energy terms such as photosynthesis and respiration.

The energy balance is illustrated in Fig. 7.3.

Where the vegetation is short (e.g., grass or field crops), the storage of heat in the vegetation is negligible compared with storage in the soil (Tanner, 1960). (The situation might be different, of course, in the case of the voluminous, and massive, vegetation of a forest.) The heat stored in the soil under short and sparse vegetation may be a fairly large portion of the net radiation at any one time during the day, but the net storage over a 24 hr period is usually small (since the nighttime loss of soil heat negates the daytime gain). For this reason, mean soil temperature generally does not change appreciably from day to day. The daily soil-storage term has been variously reported to be of the order of 5–15% of J_n (Decker, 1959; Tanner and Pelton, 1960). This obviously depends on season. In spring and summer, this term is positive, but it becomes negative in autumn.

In the past, the miscellaneous energy terms [M in Eq. (7.8)] were believed to be a negligible portion of the energy balance. Measurements of carbon

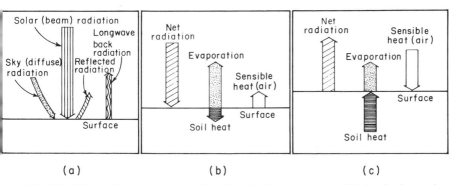

(a)   (b)   (c)

Fig. 7.3. Schematic representation of (a) the radiation balance and (b) the daytime and (c) the nighttime energy balance. [Net radiation = (solar radiation + sky radiation) − (reflected radiation + back radiation).] It is to be remembered that the daytime net radiation during the growing season is much greater than at night. (After Tanner, 1968.)

dioxide exchange over active crops in the natural environment, however, have revealed that photosynthesis may in some cases account for as much as 5% of the daily net radiation where there is a large mass of active vegetation particularly under low-light conditions. In general, though, M is much less than that (Lemon, 1960).

Overall, the amount of energy stored in soil and vegetation and that fixed photochemically account for a rather small portion of the total daily net radiation, with the major portion going into latent and sensible heat. The proportionate allocation between these terms depends on the availability of water for evaporation, but in most agriculturally productive fields the latent heat predominates over the sensible heat term.

F. Transport of Heat and Vapor to the Atmosphere

The transport of sensible heat and water vapor (which carries latent heat) from the field to the atmosphere is affected by the turbulent movement of the air in the atmospheric boundary layer.[6] The sensible heat flux A is proportional to the product of the temperature gradient dT/dz and the turbulent transfer coefficient for heat k_a (cm^2/sec):

$$A = -c_p \rho_a k_a \, dT/dz \qquad (7.9)$$

where c_p is the specific heat capacity of air at constant pressure (cal/cm °C), ρ_a the density of air, T temperature (°C), and z height (cm).

The rate of latent heat transfer by water vapor from the field to atmosphere, LE, is similarly proportional to the product of the vapor pressure gradient and the appropriate turbulent transfer coefficient for vapor.

If we assume that the transfer coefficients for heat and water vapor are equal, then the ratio of the sensible heat transport to the latent heat transport becomes

$$\beta = A/LE \approx \xi_c \, \Delta T/\Delta e \qquad (7.10)$$

where $\Delta T/\Delta e$ is the ratio of the temperature gradient to the vapor pressure gradient in the atmosphere above the field, and ξ_c is the psychometric constant ≈ 0.66 mbar/°C.

The ratio β is called the *Bowen ratio*, and it depends mainly on the temperature and moisture regimes of the field. When the field is wet, the relative humidity gradients between its surface and the atmosphere tend to be

---

[6] "A laminar boundary layer," generally less than 1 mm thick, is recognized in immediate contact with the surface of an evaporating body. Through this layer, transport occurs by diffusion. Beyond this, turbulent transport becomes predominant in the "turbulent boundary layer."

large, whereas the temperature gradients tend to be small. Thus, β is rather small when the energy is consumed mainly in evaporation. When the field is dry, on the other hand, the relative humidity gradients toward the atmosphere are generally small, and the temperature gradients tend to be steep, so that the Bowen ratio becomes large. In a recently irrigated field, β may be smaller than 0.2, while in a dry field in which the plants are under a water stress (with stomatal resistance coming into play), the surface may warm up and a much greater share of the incoming energy will be lost to the atmosphere directly as sensible heat. Under extremely arid conditions, in fact, *LE* may tend to zero and β to infinity. With advection (Section G), sensible heat may be transferred from the air to the field, and the Bowen ratio can become negative.

Whether or not water-vapor transport to the atmosphere from a vegetated field becomes restricted obviously depends not only upon the soil-water content per se, but on a complex interplay of factors in which the characteristics of the plant cover (i.e., density of the canopy, root distribution, and physiological responses to water stress) play an important role.

The assumption that the transfer coefficients for heat and vapor are equal (or at least proportional) is known as the principle of *similarity* (Tanner, 1968). Transfer through the turbulent atmospheric boundary layer takes place primarily by means of *eddies*, which are ephemeral, swirling microcurrents of air, whipped up by the wind. Eddies of varying size, duration, and velocity fluctuate up and down at varying frequency, carrying both heat and vapor. While the instantaneous gradients and vertical fluxes of heat and vapor will generally fluctuate, when a sufficiently long averaging period is allowed (say, 15–60 min), the fluxes exhibit a stable statistical relationship over a uniform field.[7] This is not the case at a low level over an inhomogeneous surface of spotty vegetation and partially exposed soil. Under such conditions, cool, moist packets of air may rise from the vegetated spots while warm, dry air may rise from the dry soil surface, with the latter rising more rapidly owing to buoyancy.[8]

Using the Bowen ratio, the latent and sensible heat fluxes can be written (recalling that $J_n = S + A + LE$, and that $\beta = A/LE$):

$$LE = (J_n - S)/(1 + \beta) \tag{7.11}$$

$$A = \beta(J_n - S)/(1 + \beta) \tag{7.12}$$

[7] It is reasonable to assume that momentum and carbon dioxide, as well as vapor and heat, are carried by the same eddies.

[8] An index of the relative importance of buoyancy (thermal) versus frictional forces in producing turbulence is the Richardson number $R_i = g(dT/dz)/[T(du/dz)^2]$, where dT/dz is the temperature, and g the acceleration of gravity. The air profile tends to be stable when R_i is positive and unstable (buoyant) when R_i is negative (Sellers, 1965).

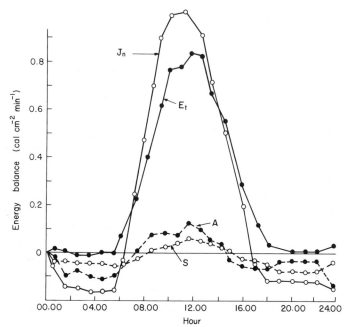

Fig. 7.4. The diurnal variation of net radiation J_n and of energy utilization by evapotranspiration E_t, sensible heating of the atmosphere A, and heating of the soil S. Alfalfa-brome hay on Plainfield sand, 4 September 1957. (After Tanner, 1960.)

Thus, LE can be obtained from micrometeorological measurements in the field (i.e., J_n, S, and β) without necessitating measurements of soil-water fluxes or plant activity.

The diurnal variation of the components of the energy balance is illustrated in Fig. 7.4. The diurnal as well as the annual patterns of the components of the energy balance differ for different conditions of soil, vegetation, and climate (Sellers, 1965).

G. Advection

The equations given for the energy balance apply to *extensive uniform areas* in which all fluxes are vertical or nearly so. On the other hand, any *small field* differing from its surrounding area is subject to lateral effects and can exchange energy in one form or another with neighboring areas. Specifically, winds sweeping over a small field can transport heat into or out of it. This phenomenon, called *advection*, can be especially important in arid regions, where small irrigated fields are often surrounded by an expanse of dry land. Under such conditions, the warm and dry incoming air can transfer sensible heat (which is transformed into latent heat of vaporization) down to the crop (Graham and King, 1961; Halstead and Covey, 1957; Rosenberg, 1974).

The extraction of sensible heat from a warm mass of air flowing *over* the top of a field, and the conversion of this heat to latent heat of evaporation, is called the *oasis effect*. The passage of warm air *through* the vegetative cover has been called the *clothesline effect* (Tanner, 1957). A common sight in arid regions is the poor growth of the plants near the windward edge of a field, where penetration of warm, dry wind contributes energy for evapotranspiration. Where advective heat inflow is large, evapotranspiration from rough and "open" vegetation (e.g., widely spaced row crops or trees) can greatly exceed that from smooth and close vegetation (e.g., mowed grass).

The effects of advection are likely to be small in very large and uniform fields but very considerable in small plots which differ markedly from their surroundings. With advection, latent heat "consumption" can be larger than net radiation. Hence, values of evapotranspiration and of irrigation requirements obtained from small experimental plots are not typically representative of large fields, unless these plots are "guarded" in the upwind direction by an expanse, or *fetch*, of vegetation of similar roughness characteristics and subject to a similar water regime. It should be obvious from the preceding that a small patch of vegetation, particularly if it consists of a spaced stand of shrubs or trees, can at times evaporate water in excess of the evaporation from a free water surface such as a lake, a pond, or a pan.

Advection is not confined to small fields. Large-scale or "macrometeorological" advective effects also occur and were described by Slatyer and McIlroy (1961), who pointed out that even in relatively humid regions, advection caused by the movement of weather systems may temporarily cause latent heat consumption to exceed average net radiation. A case in point is the periodic invasion of semihumid regions along the Mediterranean littoral by searing desert winds, variously called sharkiyeh, sirocco, or khamsin.

H. Potential Evapotranspiration (Combination Formulas)

The concept of *potential evapotranspiration* is an attempt to characterize the micrometeorological environment of a field in terms of an evaporative power, or demand; i.e., in terms of the maximal evaporation rate which the atmosphere is capable of exacting from a field of given surface properties. The concept probably derives from the common observation that when a wet object is exposed and dried gradually in the open air, progressively longer increments of time are generally required to remove equal increments of water. The evaporation rate obviously depends both on the environment and on the state of wetness of the object itself. Intuitively, therefore, one might suppose that there ought to be a definable evaporation rate for the special case in which the object is maintained perpetually in as wet a state as possible, and that this evaporation rate should depend only on the meteorological environment. More specifically, Penman (1956) defined potential evapotranspiration as "the amount of water transpired in unit time by a short green crop, completely shading the ground, of uniform height and never short of water." As such, it is a useful standard of reference for the comparison of different regions and of different measured evapotranspiration values within a given region.

To obtain the highest possible yields of many agricultural crops, irrigation should be provided in an amount sufficient to prevent water from becoming a limiting factor. Knowledge of the potential evapotranspiration can therefore serve as a basis for planning the irrigation regime. In general, the actual evapotranspiration E_{ta} from various crops will not equal the potential value E_{to}, but in the case of a close-growing crop the maintenance of optimal soil moisture conditions for maximal yields will generally result in E_{ta} being nearly equal to, or a nearly constant fraction of, E_{to}, at least during the active-growth phase of the crop season.

Various empirical approaches have been proposed for the estimation of potential evapotranspiration (e.g., Thornthwaite, 1948; Blaney and Criddle, 1950). The method proposed by Penman (1948) is physically based and hence inherently more meaningful. His equation, based on a combination of the energy balance and aerodynamic transport considerations, is a major contribution in the field of agricultural and environmental physics.

The Dalton equation for evaporation from a saturated surface is

$$LE = (e_s - e)f(u) \tag{7.13}$$

where u is the windspeed above the surface, e_s is the vapor pressure (or the absolute humidity) at the temperature of the surface, and e is the vapor pressure of the air above the surface at an elevation sufficient so that e is unaffected by e_s. For the present, we shall disregard the fact that the shape of the function $f(u)$ should depend on the roughness of surface and on the stability (buoyancy) of the air overlying the surface.

If e_a is the saturated vapor pressure of the air, then

$$LE_a = (e_a - e)f(u) \tag{7.14}$$

and

$$E_a/E = 1 - (e_s - e_a)/(e_s - e) \quad (7.15)$$

is obtained by dividing Eq. (7.14) by (7.13) and rearranging of terms. Now, Penman assumed that $S = 0$ (i.e., the soil heat flux is negligible) and he could write Eq. (7.11) as

$$J_n/LE = 1 + \beta \quad (7.16)$$

Since the Bowen ratio may be written

$$\beta = \xi(T_s - T_a)/(e_s - e) \quad (7.17)$$

then

$$\frac{J_n}{LE} = 1 + \xi\left(\frac{e_s - e_a}{e_s - e}\right)\bigg/\left(\frac{e_s - e_a}{T_s - T_a}\right) \quad (7.18)$$

In Eq. (7.18) we may write

$$\frac{e_s - e_a}{T_s - T_a} = \left(\frac{\Delta e}{\Delta T}\right)_{T=T_a} = \Delta$$

where Δ is the slope of the saturated vapor pressure–temperature curve. Now we may rewrite Eq. (7.18)

$$\frac{J_n}{LE} = 1 + \frac{\xi}{\Delta}\left[\frac{e_s - e_a}{e_s - e}\right] \quad (7.19)$$

But since $(e_s - e_a)/(e_s - e) = 1 - E_a/E$ from Eq. (7.15) then by algebraic rearrangement we obtain

$$LE = \frac{(\Delta/\xi)J_n + LE_a}{\Delta/\xi + 1} \quad (7.20)$$

Equation (7.20) is the Penman equation, where $LE_a = 0.35(e_a - e)(0.5 + U_2/100)$ (mm/day); e_a = saturated vapor pressure at mean air temperature (mm Hg); e = mean vapor pressure in air; U_2 = mean wind speed in miles per day at 2m above ground. This equation permits a calculation of the potential evapotranspiration rate from measurements of the net radiation, and of the temperature, vapor pressure, and wind velocity taken at one level above the field.

Actual evapotranspiration from an actively growing crop in the field generally constitutes a fraction, often in the range between 60% and 90%, of the potential evapotranspiration as determined by the Penman equation or by evaporation pans. The Penman formulation avoids the necessity of determining the value of T_s, the surface temperature, just as it disregards the possible fluctuations in the direction and magnitude of the soil heat flux term. Moreover, it makes no provision for surface roughness or air instability (buoyancy) effects. Finally, the Penman theory takes no explicit account of advection. To correct for the differences between potential evapotranspiration from rough surfaces and potential evaporation from smooth water E_0, Penman used the following empirical factors determined in Southern England:

$$E_0(\text{bare soil})/E_0(\text{water}) = 0.9$$

$$E_0(\text{turf})/E_0(\text{water}) = 0.6 \text{ in winter,} \quad \text{ranging to } 0.8 \text{ in summer}$$

It should be emphasized that the representation of potential evapotranspiration purely as an externally imposed "forcing function" is a rather gross approximation. In actual fact, the field participates, as it were, in determining its evapotranspiration rate even when it is well endowed with water, through the effect of its radiant reflectivity, aerodynamic roughness, thermal capacity and conductivity, etc. The often stated principle that all well-watered fields, regardless of their specific characteristics, are subject to, or exhibit, the same potential evapotranspiration is only more or less correct.

The Penman formulation was modified by van Bavel (1966) to allow for short-term variations in soil heat flux and for differences among various surfaces. His method for predicting potential evapotranspiration requires the additional measurements of net radiation and soil heat flux. A roughness height parameter is used to characterize the aerodynamic properties of the surface, i.e., to take account of the fact that, all other things being equal, potential evapotranspiration from a corn field should exceed that from a lawn, which, in turn, should be greater than that from a smooth, bare soil.

Potential evapotranspiration LE_0 is given by

$$LE_0 = \frac{(\Delta/\xi)(J_n - S) + K_v d_a}{(\Delta/\xi) + 1} \tag{7.21}$$

wherein Δ is the slope of the saturation vapor pressure versus temperature curve at mean air temperature, ξ the psychrometric constant, J_n net radiation, S soil heat flux, d_a the vapor pressure deficit at elevation Z_a (namely, $(e_s - e_a)$), and k_v the transfer coefficient for water vapor (a function of wind speed and surface roughness). For the dependence of K_v on mean wind speed U_2, Penman (1948) suggested empirically $k_v = 20(1 + U_2/100) = 20 + U_2/5$, where e_s and e_a are given in millimeters of mercury.

Further improvements of the physically based prediction of evapotranspiration can result from inclusion of air stability or buoyancy effects, in recognition of the fact that vapor transfer is enhanced whenever the thermal structure of the air becomes unstable (Szeicz *et al.*, 1973). The advent of remote-sensing infrared thermometry has made possible continuous monitoring of surface temperature, and hence also allows a better estimation of the vapor pressure at the surface.

The ratio Δ/ξ and the saturation vapor pressure e_s at various temperatures are available in standard tables in many texts on physical meteorology and environmental physics (e.g., Sellers, 1965; Slatyer and McIlroy, 1961; Monteith, 1973). A summary is given in the accompanying table.

T (°C)	10	15	20	25	30	35
Δ/ξ		1.23	1.64	2.14	2.78	3.57	4.53
e_s (mm Hg)	9.20	12.78	17.53	23.75	31.82	42.18
e_s (mbar)	12.27	17.04	23.37	31.67	42.43	56.24

Sample Problems

1. The total incoming global radiation J_s (sun and sky) received by a particular field on a given day is 500 cal/cm², or langleys. The albedo a is 15%. The net outgoing long-wave radiation balance J_l amounts to 10 cal/cm². The sensible heat transfer to the air A is 12 cal/cm², the net heat flow into the soil S is 6 cal/cm², and the metabolic uptake of energy M is 8 cal/cm². Calculate the net radiation, the amount of energy available for latent heat transfer (evapotranspiration), and the day's evapotranspiration in millimeters of water. On the following day, the sensible heat transfer is reversed and evapotranspiration totals 7.5 mm. If everything else remains the same, calculate the amount of advected energy taken up by the field.

To calculate the net radiation J_n we write the radiation balance, Eq. (7.7):

$$J_n = J_s^{\downarrow}(1 - a) - J_l = 500 \text{ cal/cm}^2(1 - 0.15) - 10 \text{ cal/cm}^2 = 415 \text{ cal/cm}^2$$

The latent heat term LE can be calculated from the overall energy balance, Eq. (7.8), when all other terms are known:

$$J_n = LE + A + S + M \quad \text{or} \quad LE = J_n - A - S - M$$

Using the values given, we have

$$LE = 415 - 12 - 6 - 8 = 389 \text{ cal/cm}^2$$

Since roughly 580 cal are required at prevailing temperatures to vaporize

1 gm or 1 cm^3 of water, i.e., $L = 580$ cal, the amount of evaporation is

$$LE/L = 389 \text{ cal/cm}^2/580 \text{ cal/cm}^3 = 0.67 \text{ cm} = 6.7 \text{ mm}$$

On the following day, with a positive influx of sensible heat by advection, evapotranspiration amounts to 7.5 mm, and hence the latent heat term is

$$LE = 0.75 \text{ cm} \times 580 \text{ cal/cm}^3 = 435 \text{ cal/cm}^2$$

The energy balance is therefore

$$\text{income} = \text{disposal}: \quad J_n + A = LE + S + M$$

and the advected energy is

$$A = LE + S + M - J_n = 435 + 6 + 8 - 415 = 34 \text{ cal/cm}^2$$

2. On a given day in early spring the daily net radiation J_n is 350 cal/cm^2, the mean air temperature T_a at standard height (2 m) is 15°C, the mean vapor pressure e_a at that height is 8 mm Hg, and the mean wind speed u_2 is 15 mile/day. On a given day in the late spring the net radiation is 420 cal/cm^2, mean air temperature 20°C, mean vapor pressure 9 mm Hg, and mean wind speed 20 mile/day. Finally, on a given day in summer J_n is 500 cal/cm^2, T_a is 25°C, e_a is 10 mm Hg, and u_2 is 25 mile/day. Estimate the potential evapotranspiration using Eq. (8.21). Assume the net soil heat flux S to be zero in all cases.

Potential evapotranspiration LE_0 is given by Eq. (7.21):

$$LE_0 = \frac{(\Delta/\xi)(J_n - S) + k_v d_a}{(\Delta/\xi) + 1}$$

where d_a is the mean vapor pressure deficit $(e_s - e_a)$ at standard height. Recall that Δ/ξ and e_s at several temperatures are tabulated in Section H, and assume that $k_v = 20 + U_2/5$ (Penman, 1948). Accordingly, for the early spring day we get

$$LE_0 = \frac{1.64 \times (350 - 0) + (20 + 15/5)(12.78 - 8)}{1.64 + 1} = 259 \text{ cal/cm}^2 \text{ day}$$

For the late spring day,

$$LE_0 = \frac{2.14 \times (420 - 0) + (20 + 20/5)(17.53 - 9)}{2.14 + 1} = 351 \text{ cal/cm}^2 \text{ day}$$

and for the summer day,

$$LE_0 = \frac{2.78 \times (500 - 0) + (20 + 25/5)(23.75 - 10)}{2.78 + 1} = 459 \text{ cal/cm}^2 \text{ day}$$

Remembering that approximately 580 cal are required to vaporize 1 gm of water, and assuming a water density of 1 gm/cm^3, we can use 58 cal/cm^2 day as the latent heat flux equivalent to the evaporation of 1 mm of water per day. Hence the values of potential evapotranspiration are estimated to be

$259/58 = 4.5$ mm for the early spring day

$351/58 = 6.1$ mm for the late spring day

$459/58 = 7.9$ mm for the summer day

Whoever could make two ears of corn or two blades of grass to grow upon a spot where only one grew before, would deserve better of mankind, and do more essential service to his country, than the whole race of politicians put together.

Jonathan Swift
1667–1745

8   *Irrigation and Crop Response*

A. Introduction

Water constitutes one of the most important constraints to increasing food production in our hungry world. So tenuous and delicate is the balance between the demand for water by crops and its supply by precipitation that even short term dry spells often reduce production significantly, and prolonged droughts can cause total crop failure and mass starvation. *Irrigation* is the practice of supplying water artificially to permit farming in arid regions and to offset drought in semiarid or semihumid regions. As such, it can play a key role in feeding an expanding population. Even in areas where total rainfall is ample, it may be unevenly distributed during the year so that only with irrigation is multiple cropping possible. In fact, the potential productivity of irrigated land can exceed that of nonirrigated ("rain fed") land four-fold or more, due both to increased yields per season and to the possibility of multiple cropping.

The process of irrigation consists of introducing water into the part of the soil profile which serves as the root zone, for the subsequent use of the crop. A well-controlled irrigation system is one which optimizes the spatial and temporal distribution of water, not necessarily to obtain the highest yields or to use the lowest amount of water possible, but to maximize the benefit-to-cost ratio. Because of the economic considerations involved, which are necessarily specific to each location and set of circumstances, the subject of irrigation management efficiency lies beyond the scope of this

216

necessarily limited book on soil physics. The present chapter is merely a description of some of the physical factors involved in irrigation.

B. Classical Concepts of Soil-Water Availability to Plants

The concept of *soil-water availability*, while never clearly defined in physical terms, has for many years excited controversy among adherents of different schools of thought. Veihmeyer and Hendrickson (1927, 1949, 1950, 1955) claimed that soil water is equally available throughout a definable range of soil wetness, from an upper limit (*field capacity*) to a lower limit (the *permanent wilting point*), both of which are characteristic and constant for any given soil. They postulated that plant functions remain unaffected by any decrease in soil wetness until the permanent wilting point is reached, at which plant activity is curtailed abruptly. This schematized model, though based upon arbitrary limits,[1] enjoyed widespread acceptance for many years, particularly among workers in the field of irrigation management.

Other investigators, however (notably Richards and Wadleigh, 1952), produced evidence indicating that soil-water availability to plants actually decreases with decreasing soil wetness, and that a plant may suffer water stress and reduction of growth considerably before the wilting point is reached. Still others, seeking to compromise between the opposing views, attempted to divide the so-called "available range" of soil wetness into "readily available" and "decreasingly available" ranges, and searched for a "critical point" somewhere between field capacity and wilting point as an additional criterion of soil-water availability.

These different hypotheses, in vogue until quite recently, are represented graphically in Fig. 8.1.

None of these schools was able to base its hypotheses upon a compre-

---

[1] We have already pointed out (Chapter 2) that the field capacity concept (Israelsen and West, 1922; Veihmeyer and Hendrickson, 1927), though useful in some cases, lacks a universal physical basis (Richards, 1960). The wilting point, if defined simply as the value of soil wetness of the root zone at the time plants wilt, is not easy to recognize, since wilting is often a temporary phenomenon, which may occur in midday even when the soil is quite wet. The *permanent wilting percentage* (Hendrickson and Veihmeyer, 1945) is based upon the *wilting coefficient* concept of Briggs and Shantz (1912) and has been defined as the root-zone soil wetness at which the wilted plant can no longer recover turgidity even when it is placed in a saturated atmosphere for 12 hr. This is still an arbitrary criterion, since plant-water potential may not reach equilibrium with the average soil moisture potential in such a short time. In any case, plant response depends as much on the evaporative demand (its variation and peak intensity) as on soil wetness (which itself is a highly variable function of space and time).

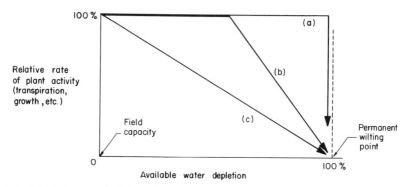

Fig. 8.1. Three classical hypotheses regarding the availability of soil water to plants: (a) equal availability from field capacity to wilting point, (b) equal availability from field capacity to a "critical moisture" beyond which availability decreases, and (c) availability decreases gradually as soil moisture content decreases.

hensive framework that could take into account the array of factors likely to influence the water regime of the soil–plant–atmosphere system as a whole. Rather, they tended to draw generalized conclusions from a limited set of experiments conducted under specific and sometimes poorly defined conditions. Over the years, a great mass of empirical data has been collected, which for a long time no one knew how to explain, correlate, or resolve into a systematic theory based on universal physical principles.

The picture was further confused by failure to distinguish between different types of plant response to soil moisture. While the transpiration rate may be, for a time, relatively independent of soil-water content changes in the root zone, other forms of plant activity may not be. Photosynthesis, vegetative growth, flowering, fruiting, and seed or fiber production may be related quite differently to the content or state of soil water.

It has long been recognized that soil wetness per se is not a satisfactory criterion for availability. Hence, attempts were made to correlate the water status of plants with the energy state of soil water, i.e., with the soil-water potential (variously termed tension, suction, soil moisture stress, etc., indeed a bewildering variety of alternative terms!). The soil-water "constants" were therefore defined in terms of potential values (e.g., $-\frac{1}{10}$ or $-\frac{1}{3}$ bar for field capacity, -15 bar for permanent wilting), which could be applied universally, rather than in terms of soil wetness (Richards and Weaver, 1944; Slater and Williams, 1965). However, even though the use of energy concepts represented a considerable advance over the earlier notions, it still fell short of taking into account the dynamic nature of soil–plant–water relations.

A fundamental experimental difficulty encountered in any attempt at an exact physical description of soil-water uptake by plants is the inherently

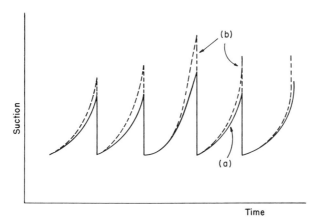

Fig. 8.2. The variation of soil-water suction in the root zone during successive irrigation cycles: (a) the average suction (tensiometric measurement) and (b) the suction of the soil in contact with the root.

complicated nature of the space–time relationships involved in this process. Roots grow in different directions and spacings, and as yet we have no experimental method to measure the microscopic gradients and fluxes of water in their immediate vicinity. The conventional methods for measurement of the content or potential of soil water are based on the sampling or sensing of a relatively large volume and are therefore oblivious to the micro-gradients toward the roots. Water suction of the soil in contact with the roots can be much greater than the average suction, as illustrated in Fig. 8.2.

An additional difficulty in describing the system physically arises from the fact that, up to the present, no satisfactory way has been found to grow plants in a soil of constant water potential. Rather, it is necessary to irrigate the soil periodically, thus refilling its effective reservoir. In a variable soil-moisture regime, plants may be influenced more by the extreme values of the water potential they experience than by the average value (see Fig. 8.2). Furthermore, root distribution is not generally uniform or constant within the root zone. Nor does the water-extraction pattern necessarily correspond to the root-distribution pattern. Hence, the correlation of plant response to soil-water conditions requires integrating each of the two over both space and time. In the process, the actual relationship can become quite obscure.

C. Traditional Principles of Irrigation Management

In practice, the old hypothesis of equal availability of water to crops within a fixed range of soil moisture resulted in a regimen of infrequent irrigation designed to fill the soil reservoir to its "field capacity" followed by a period in which one awaited depletion to nearly the permanent wilting point before replenishing soil moisture by once again making up the "deficit" to field capacity. The traditional irrigation cycle thus consisted of a brief period of infiltration followed by an extended period of soil moisture extraction by the crop. The soil surface was periodically saturated (with a resulting disruption of soil aeration) and then allowed to desiccate excessively, to the detriment of the roots in the surface layer. Practical limitations on the frequency of irrigation by the conventional methods have made it difficult to test alternative regimens based on the continuous maintenance of a more nearly optimal level of soil moisture in the root zone.

The traditional mode of irrigation seemed to make good economic sense because many furrow, flood, or portable sprinkler systems have a fixed cost associated with each application of water. With such systems, it is desirable to minimize the number of irrigations per season by increasing the interval of time between successive irrigations (Rawlins and Raats, 1975). For example, if the cost of portable tubing is a dominant consideration it obviously pays to make maximal use of such tubes by minimizing the amount required per unit area irrigated and by rotating the available tubing from site to site so as to cover the greatest overall area possible before having to return again to the same site for reirrigation. To minimize irrigation frequency in practice has meant maximizing the amount of water stored in the soil by applying heavy irrigations and then maximizing the usage of the stored water by waiting as long as possible before replenishment of the soil "reservoir."

The classical questions involved in irrigation management are *when* to irrigate and *how much* water to apply at each irrigation. To the first question, the traditional reply has been: Irrigate when the available moisture is nearly depleted, as determined by sampling or neutron gauging or tensiometry at some presumably representative depth or depths within the root zone, or by observing indications of incipient moisture stress in the plants. To the second question, the traditional reply was, in substance: Apply sufficient water to bring up the moisture reserve of the soil root zone to field capacity, plus a "leaching fraction" of, say, 10–20% for salinity control. Lately, however, newer concepts of irrigation management have evolved.

D. Newer Concepts of Soil-Water Availability to Plants

In the last two decades, a fundamental change has taken place in our conception of soil–plant–water relationships. With the development of our theoretical understanding of the state and movement of water in the soil, plant, and atmosphere, and with the concurrent development of experimental techniques allowing more exact measurement of the interrelationships of potential, conductivity, water content, and flux, both in the soil and in the plant, the way has been opened for a more holistic approach to the problem. It has become increasingly clear that in a dynamic system such static concepts as "the soil-water constants" (i.e., field capacity, permanent wilting point, critical moisture, capillary water, gravitational water, etc.) are often physically meaningless, as they are based on the supposition that processes in the field bring about static levels of soil-water content or potential. In fact, as we are now very much aware, flow takes place almost incessantly, though in varying fluxes and directions, and static situations are exceedingly rare.

These developments have led to abandonment of the classical concept of available water in its original sense. Clearly, there is no fundamental qualitative difference between the water at one value of soil wetness or potential and another (e.g., across such arbitrary dividing points as $\frac{1}{10}$, $\frac{1}{3}$, or 15 bar of suction), nor is the amount and rate of water uptake by plants an exclusive function of the content or potential of soil water. The amount and rate of water uptake depend on the ability of the roots to absorb water from the soil with which they are in contact, as well as on the ability of the soil to supply and transmit water toward the roots at a rate sufficient to meet transpiration requirements. These, in turn, depend on *properties of the plant* (rooting density, rooting depth, and rate of root extension, as well as the physiological ability of the plant to continue drawing water from the soil at the rate needed to avoid wilting while maintaining its vital functions even while its own water potential decreases); *properties of the soil* (hydraulic conductivity–diffusivity–matric suction–wetness relationships); and also to a considerable extent the *meteorological conditions* (which dictate the rate at which the plant is required to transpire and hence the rate at which it must extract water from the soil in order to maintain its own hydration).

From a physical point of view, evapotranspiration can be viewed as a continuous stream flowing from a periodically replenished *source* of limited capacity and variable potential, namely, the reservoir of soil moisture, to a *sink* of virtually unlimited capacity (though of variable strength or evaporative potential)—the atmosphere. As long as the rate of root uptake of soil moisture balances the rate of canopy loss by transpiration, the stream continues unabated while the plant remains fully hydrated. The moment the uptake rate falls below transpiration, the plant itself must begin to lose moisture. This imbalance cannot continue for any length of time without resulting in loss of turgidity and hence in wilting of the plant.

The concept of *potential transpiration* (Penman, 1949) is an attempt to characterize the evaporative flux extracted from a stand of plants fully covering the ground surface when the supply of soil water is not limiting.[2] Accordingly, it is the meteorological conditions rather than soil or plant conditions which exercise the greatest influence on the transpiration rate as long as the soil is wet enough. However, as soil wetness is diminished, even though not completely depleted, actual transpiration begins to fall below the potential rate either because the soil cannot supply water fast enough and/or because the roots can no longer extract it fast enough to meet the meteorological demand. The point at which this condition is reached depends in a combined way upon the weather, the plants, and the soil. Any attempt to extricate one from the others and attribute phenomena to soil conditions alone is at best futile, and at worst misleading.

E. New Principles of Irrigation Management

Recently, evidence has been accumulating (e.g., Viets, 1966; Rawitz, 1969; Acevedo *et al.*, 1971; Hillel and Guron, 1973) that crops may show a pronounced increase in yield when irrigation is provided in sufficient quantity and frequency to prevent the occurrence of moisture stress at any time during the growing season. Such an effect can be produced by maintaining soil moisture content in the root zone at a high level and soil moisture suction at a low level, while taking care not to wet the soil excessively so as not to impede aeration or leach out nutrients or waste water. The desired conditions were difficult to achieve by the traditional surface and sprinkler irrigation methods prevalent until a decade or so ago, and as a result these new findings have not yet been applied very widely.

The advent of newer irrigation methods (including permanent installations, called "solid set," of low-intensity sprinklers, subirrigation by means of porous tubes, and the technique of drip—or trickle—irrigation) has made it possible to establish and maintain soil moisture conditions at a more nearly optimal level than heretofore. These newer methods may eventually make possible the optimization of soil moisture, salinity, fertility, and aeration simultaneously. Since these new irrigation systems are

---

[2] When the plant canopy completely covers the soil surface, the potential transpiration rate is assumed to be equal to the *potential evapotranspiration* (or total evaporation) rate. As such, it is equivalent to what we have earlier called the *external evaporativity* or the *evaporative demand*. The use of the word "potential" in this context, meaning "maximum possible rate," should not be confused with the potential energy concept and the term "potential" denoting the energy state of water. Unfortunately, the same word is used in the literature for two such widely differing meanings.

capable of delivering water to the soil in small quantities as often as desirable with no appreciable additional cost for the extra number of irrigations, the economic constraints on high-frequency irrigation have been lifted (Rawlins and Raats, 1975).

As the frequency of irrigation increases, the infiltration period becomes a more important part of the irrigation cycle. Changing the irrigation cycle from an extraction-dominated process to an infiltration-dominated process brings into play a different set of relationships governing water flow in the soil. For example, with daily (rather than weekly or monthly) applications of water, the pulses of moisture resulting from intermittent irrigation are damped out within a few centimeters of the surface, so that flow below this point is essentially steady and governed by gravity alone (Gardner, 1964). By adjusting the rate and quantity of application in accordance with such measurable variables as potential evapotranspiration, soil hydraulic conductivity, and the soil solution's salt concentration, it is possible to control the level of suction prevailing within the root zone as well as the throughflow (drainage) rate and thus also the leaching fraction. Control of the crop's soil environment thus passes into the hands of the irrigation manager more completely than ever before. Properly managed, the new systems hold the promise of saving water even while improving growing conditions and increasing yields.

Since a high-frequency irrigation system can be adjusted to supply water at very nearly the exact rate required by the crop, we need no longer depend on the soil's own capacity to store water. The consequences of this fact are far reaching: New lands, until recently considered unsuited for irrigation, can now be brought into production. One outstanding example is the case of coarse sands and gravels, where soil moisture storage is minimal and where surface conveyance and application of water would involve inordinate losses by excess and nonuniform seepage. Such soils can now be irrigated quite readily, even on sloping ground and with hardly any leveling, by means of a drip irrigation system, for instance.

Theoretically, with high-frequency irrigation the irrigator need no longer worry about when available soil moisture is depleted or when plants begin to suffer stress. Such situations are avoided entirely. To the old question, "when to irrigate?" the modern irrigationist answers, "as frequently as practicable; if possible, daily." To the question, "how much water to apply?" he replies, "Enough to meet the evapotranspirational demand and to prevent salinization of the root zone." Potential evaporation can be calculated (e.g., by means of the Penman equation or any of the various locally calibrated formulas), whereas soil salinity can be monitored by sampling the soil or the soil solution or by means of salinity sensors (Oster and Willardson, 1971).

F. Advantages and Limitations of Drip Irrigation

Of particular interest is the method of *drip* (also called *trickle*) irrigation, which has been gaining recognition and is now being introduced into many countries. Although the idea itself is not new, what has finally made it practical is the rather recent development of low-cost plastic tubing and variously designed emitter fittings. System assemblies are now available which are capable of maintaining sufficiently high pressure in the lateral tubes to assure uniform discharge throughout the field, as well as a controlled rate of drop discharge through the narrow-orifice emitters with a minimum of clogging. The application system has been supplemented by ancillary equipment such as filters, timing or metering valves, and fertilizer injectors. Field trials in different locations have resulted in increased yields of both orchard and field crops, perennial as well as annual, particularly in adverse conditions of soil, water, and climate. Drip irrigation has also been found suitable for greenhouses and gardens and lends itself readily to labor saving automation.

The justifiable enthusiasm for the new method, however, carries certain dangers. Hasty adoption of drip irrigation without enough care in adaptation to local crops and conditions can result in disappointment. Drip irrigation offers many potential advantages, yet it is no panacea. Inefficiency, in fact, is just as easy to achieve in the operation of a drip system as it is in the operation of conventional systems. Let us therefore list some of the possible advantages of drip irrigation while mentioning some of its limitations.

One advantage, already mentioned, is the possibility of obtaining favorable moisture conditions even in problematic soils such as coarse sands and clays which are ill suited to conventional ways of irrigation (Goldberg and Shmueli, 1970). Another is the possibility of delivering water uniformly to plants in a field of variable elevation, slope, wind velocity and direction, soil texture, and infiltrability. Still another potential advantage is the ability to maintain the soil at a highly moist yet unsaturated condition, so that soil air remains a continuous phase capable of exchanging gases with the atmosphere. High moisture reduces the soil's mechanical resistance to root penetration and proliferation.

Where salinity is a hazard, as when the irrigation water is somewhat brackish, the continuous supply of water ensures that the osmotic pressure of the soil solution will remain low near the water source. Furthermore, drip irrigation, as it is applied at the soil surface underneath the plant canopy, avoids the hazard of leaf scorch and reduces the incidence of fungal diseases, both of which may occur under sprinkler irrigation. Since drip irrigation wets the soil only in the immediate periphery of each emitter, the greater part of the surface (particularly the interrow areas) remains dry and hence less prone to weed infestation and soil compaction under traffic. In summation, properly managed drip irrigation seems to offer the best opportunity at present to optimize the water, nutrient, and air regimes in the root zone (Rawitz and Hillel, 1974).

The fact that drip irrigation wets only a small fraction of the soil volume can also become a problem. While it is a proven fact that even large trees can grow on less than 10% of what is generally considered to be the normal root zone of a field (provided enough water and nutrients are supplied within this restricted volume), the crop becomes extremely sensitive and vulnerable to even a slight disruption of the irrigation system or schedule. If the system does not operate perfectly and continuously, crop failure can result as the soil-moisture reservoir available to the plants is extremely small.

Apart from the technical problem of maintaining the water delivery system in perfect operation without interruption, there is the initial problem of tailoring the emitter spacing specifically for each crop and stage of development, the optimization of per-emitter discharge rate in relation to soil infiltrability and lateral spread of the water, the optimization of irrigation pulse duration and frequency, and the determination of irrigation quantity variation during the season. The latter can be geared to the daily evaporation as measured by means of a lysimeter or an evaporimeter such as the Class A pan, with due consideration given to the crop's growth stage and canopy development.

Other possible problems associated with drip irrigation are the accumulation of salts at the periphery of the wetted circles surrounding each emitter (an occurrence which can hinder the growth of a subsequent crop) and the excessive through-flow and leaching which can take place directly under the drip emitters. In most cases such problems can, however, be overcome by good management. Clogging of emitters is a frequently encountered problem, either due to the presence of suspended particles or algae or to salts which tend to precipitate. Mechanical clogging can be prevented by proper filtration, whereas chemical and biological clogging can often be reduced by acidification and algacide treatment of the water.

G. Irrigation, Water-Use Efficiency, and Water Conservation

Any concept of efficiency is a measure of the output obtainable from a given input. Irrigation or water-use efficiency can be defined in different ways, however, depending on the nature of the inputs and outputs considered. For example, one can define as an economic criterion of efficiency the financial return in relation to the money invested in the installation and operation of water supply and delivery system. The problem is that costs and prices fluctuate from year to year and vary widely from place to place, and may not be universally comparable. Perhaps a more objective criterion for the relative merits of alternative irrigation systems is an agronomic one, namely, a comparison of the marketable yield per unit of land area or, better yet, per unit amount of water applied. This is merely an illustration of how the measure of efficiency depends on the point of view.

A widely applicable expression of efficiency is the *crop water-use efficiency*, which has been defined (Viets, 1962) as the amount of dry matter produced per unit volume of water taken up by the crop from the soil. As most of the water taken up by plants in the field is transpired (in arid regions, 99% or so!), while generally only a small amount is retained, the plant water-use efficiency is in effect the reciprocal of what has long been known as the transpiration ratio (Briggs and Shantz, 1912), originally defined as the mass of water transpired per unit mass of dry matter produced.

What we shall refer to as *technical efficiency* is what irrigation engineers call irrigation efficiency. It is generally defined as the net amount of water added to the root zone divided by the amount of water taken from some source (Hillel and Rawitz, 1972). As such, this criterion of efficiency can be applied to complex regional projects, or to individual farms, or to specific fields. In each case, the difference between the net amount of water added to the root zone and the amount withdrawn from the source represents the loss incurred in conveyance and distribution.

In practice, many (perhaps most) irrigation projects operate in an inherently inefficient way. In many of the surface irrigation schemes, one or a few farms may be allocated large flows representing the entire discharge of a lateral canal for a specified period of time. Since water is delivered to the consumer only at fixed times and charges may be assessed per delivery regardless of the actual amount used, customers tend to take as much water as they can while they can. This often results in overirrigation, which not only wastes water but also causes project-wide and perhaps even region-wide problems connected with the disposal of return flow, water-logging of soils, leaching of nutrients, and excessive elevation of the water table requiring expensive drainage to rectify. Although it is difficult to arrive at reliable statistics, it has been estimated that the average irrigation efficiency in such schemes is less than 50%. Since it is a proven fact that, with proper management irrigation efficiencies of 80–90% can be achieved in actual practice, there is obviously room, and need, for much improvement.

Particularly difficult to change are management practices which lead to deliberate waste not necessarily because of insurmountable technical problems or lack of knowledge but simply because it appears more convenient or even more economical in the short run to waste water rather than to apply proper management practices of water conservation. Such situations typically occur when the price of irrigation water is lower than the cost of labor or of the automated equipment needed to avoid overirrigation. Very often the price of water does not reflect its true cost but is kept deliberately low by government subsidy, which can be self-defeating (Hillel and Rawitz, 1972).

Where open and unlined distribution ditches are used, uncontrolled seepage and evaporation, as well as transpiration by riparian phreatophytes, are also a major cause of water loss. Even pipeline distribution systems do not always prevent loss. Leaky joints resulting from poor workmanship, corrosion, poorly maintained valves, or mechanical damage by farm machinery may cause large losses. At times the damage is not immediately obvious, as during the failure of a buried pipe.

Surface runoff resulting from the excessive application of water ideally should not occur. Sprinkler irrigation systems should be designed to apply water at rates which never exceed soil infiltrability. In the case of gravity irrigation systems, however, it is often virtually impossible to achieve uniform water distribution over the field without incurring some runoff ("tail water"). Only when provision is made to collect irrigation and rainwater surpluses at the lower end of the field and guide them as controlled return flow can this water be considered anything but a loss.

Evaporative losses associated with water application include any evaporation from open water surfaces of border checks or furrows, evaporation of water droplets during their flight from sprinkler to ground surface, wind drift of droplets away from the target area, and evaporation from wetted crop canopies or from the wet soil immediately after the irrigation, not all of which can be avoided.

In the open field, little can be done to reduce transpiration if the conditions required for high yields are to be maintained. Attempts to use chemical sprays known as "antitranspirants" have generally failed, and attempts to control wind movement above a crop by windbreaks may or may not produce the desired effect economically. It appears at present that the greatest promise for increasing water-use efficiency lies in allowing the crop to transpire freely at the climatic limit by alleviating any water shortages while at the same time avoiding waste and obviating all other environmental constraints to attainment of the fullest possible production potential of the crop. This is particularly important in the case of the new and superior varieties which have been developed in recent years and which can provide high yields only if water stress is prevented and such other factors as soil fertility (availability of nutrients), aeration, salinity, and soil tilth are also optimized. Plant diseases and pests, as well as insufficient fertility of the soil, may depress yields without a proportionate decrease in transpiration and water use. All management practices can thus influence water-use efficiency, and none can be considered in isolation from the others.

H. Transpiration in Relation to Production

Transpiration can be limited either by the supply of water to be evaporated (by precipitation or irrigation) or by the supply of energy needed for vaporization, the latter determined mainly by such climatic factors as radiation, temperature, and humidity. When the supply of water is plentiful, it is the climate, and particularly the net radiation, which determines water use (Penman, 1956). When water is limiting, the assimilation rate, plant growth, and consequently crop yield are all related quantitatively to the water supply, but this relationship may not be a simple one. Interest in the dependence of crop production on water supply has grown in recent years because of the increasing scarcity of water for irrigation. Among the important works in this area are those by de Wit (1958), Jensen (1973). Doorenbos and Pruitt (1975), Shalhevet *et al.* (1976), Stewart *et al.* (1977), Russell (1979), and Hanks (1980).

The classical analysis of the relation between transpiration and yield was published over two decades ago by deWit (1958). He found that in climates with a large percentage of bright sunshine duration (i.e., arid regions) a relation such as

$$y_d = mE_a/E_0 \qquad (8.1)$$

exists between total dry matter yield y_d and the ratio of actual transpiration E_a and free-water evaporation E_0. In climates with a small percentage of bright sunshine (i.e., temperate regions) the relation

$$y_d = nE_a \qquad (8.2)$$

was found (that is to say, dry-matter production is proportional to transpiration and hence water-use efficiency is constant). These relations were found for both container grown and field grown plants, and the values of constants m and n were reported to be characteristic of the crops. Corroborative evidence for deWit's equations was reported by Hanks *et al.* (1969). The assumption of linearity between yield and transpiration has served as the basis for recent efforts toward comprehensive modeling of crop growth as influenced by soil water (e.g., Hanks, 1974).

An empirically based equation to predict yield from known values of evapotranspiration was given by Stewart *et al.* (1977) for dry-matter production:

$$y/y_m = 1 - \beta E_{td} = 1 - \beta + \beta E_t/E_{tm} \qquad (8.3)$$

wherein y is yield, y_m maximum obtainable yield, E_t evapotranspiration, E_{tm} maximum (potential) evapotranspiration, β the slope of relative yield (y/y_m) versus the evapotranspirational deficit ($E_{td} = 1 - E_t/E_{tm}$). To predict yield from this equation, one must know E_t, E_{tm}, y_m, and β. As pointed out by Hanks (1980), the ratio E_t/E_{tm} where y/y_m is zero approximates the portion of E_t that is due to direct evaporation from the soil surface. On the other hand, the portion of E_{tm} that is due to T_m (the maximal, or potential, transpiration) is $1/\beta$. Thus a β value of unity implies no evaporation from the soil and a β of 1.5, for example, implies that one-third of E_{tm} is evaporation and two-thirds T_m. The parameters of Eq. (8.3) must be determined from field measurements.

Determination of the fraction of evapotranspiration due to evaporation from the soil (as apart from transpiration from the plants) can be important, since evaporation does not relate to plant growth and can therefore be considered a loss.[3] This fact has a bearing upon irrigation frequency. According to Hanks (1980), where irrigation is applied frequently, yield can be lower for a given water use than where irrigation is applied less frequently, because the higher irrigation frequency maintains a higher level of soil surface moisture and causes higher evaporation. On the other hand, of course, a low irrigation frequency regime causes soil desiccation and plant-water stress. Therefore, the highest water-use efficiency may result from some intermediate irrigation frequency, even if it produces less than maximum yields.

A micrometeorological approach to yield prediction was taken by Kanemasu *et al.* (1976, 1978), based on the interrelationships among solar radiation, temperature, potential transpiration, and leaf-area index (a measure of the density of the plant canopy cover). Still other models to predict evapotranspiration and yield have been proposed by Nimah and Hanks (1973), Childs and Hanks, (1975), and Childs *et al.* (1977), among others.

Recently, Doorenbos *et al.* (1979) proposed a method for evaluating the yield response of crops to applied water in terms of the following relationship:

$$(1 - y_a/y_m) = F_y(1 - E_{ta}/E_{t0}) \tag{8.4}$$

where y_a is actual yield, y_m is maximum attainable yield when full water requirements are met, F_y is a yield response factor, E_{ta} is actual evapotranspiration, and E_{t0} is maximum (potential) evapotranspiration. Empirically obtained values of F_y have been tabulated for different crops in different climatic regions.

---

[3] Evaporation of soil moisture may not be a total loss, however, since evaporation and transpiration are not mutually independent. A reduction in evaporation may induce an increase in transpiration without commensurate increase in yield.

In many cases, reported relationships between yield and water use pertain to aboveground dry-matter yield. If the yield of interest is grain, fruit, or fiber, its relation to water use by the crop plants can be quite different. Although studies have shown that grain yield bears a constant ratio to dry-matter yield (e.g., Hillel and Guron, 1973), this cannot be taken for granted in every case.

The relationships found between yield and water use under limited water supply may not hold as potential evapotranspiration is attained and water ceases to be a limiting factor in plant growth. Beyond the point where transpiration reaches its climatic limit, the promise of increasing production may lie in identifying and obviating any other possible environmental constraints, such as light distribution in the canopy, carbon dioxide concentration in the air, or nutrients in the soil. Finally, we come up against genetic constraints, which cannot be obviated by any environmental factors. This is the realm where environmental scientists (including soil physicists) must cooperate with plant geneticists in a continuing effort to improve the productive potential of agricultural crops.

Sample Problems

1. In a given season, a field of 4 ha (hectare) was irrigated with a total volume of 48000 m^3 of water. Evapotranspiration amounted to 900 mm. Assuming no significant change of soil moisture storage from beginning of the season to its end, estimate the irrigation efficiency. How much water was drained? What would be the irrigation requirement for the field if an irrigation efficiency of 85% were to be attained? Assume negligible precipitation, run-off, capillary rise, and plant-water storage.

It is convenient to divide the irrigation water volume by the field's area (remembering that 1 ha = 10,000 m^2) to convert the irrigation quantity into depth units:

$$I = 48000 \text{ m}^3/4 \times 10000 \text{ m}^2 = 1.2 \text{ m} = 1200 \text{ mm}$$

"Irrigation efficiency" F_i has been defined as the volume of water used "consumptively" (i.e., in evapotranspiration) divided by the total volume applied to the field. Thus

$$F_i = (E + T)/I = 900/1200 = 0.75 = 75\%$$

We now recall the overall water balance [see Eq. (7.2)]:

$$\Delta S + \Delta V = (P + I + U) - (R + D + E + T)$$

where ΔS is the soil storage increment, ΔV the vegetational storage increment, P precipitation, I irrigation, U upward capillary rise, R run-off, D drainage, E evaporation, and T transpiration. According to the statement of our problem, the terms ΔS, P, U, R, and ΔV are assumed to be negligible. Hence

$$I = D + (E + T)$$

To obtain the amount of drainage, we set

$$D = I - (E + T) = 1200 - 900 = 300 \text{ mm}$$

If irrigation efficiency were 85% and evapotranspiration were unchanged, our irrigation would be

$$I = (E + T)/F_{\text{i}} = 900/0.85 = 1059 \text{ mm} = 1.059 \text{ m}$$

which, over a field of 4 ha, would amount to approximately

$$1.059 \text{ m} \times 40{,}000 \text{ m}^2 = 42{,}360 \text{ m}^3$$

A net savings of $48{,}000 - 42{,}360 = 5460 \text{ m}^3$ would thus be realized.

Note: This calculation of "efficiency" includes evaporation as well as transpiration. In an "open" crop such as a young orchard or row crop, direct evaporation from the exposed soil surface may be a partly avoidable loss. Hence it would stand to reason that evaporation should not be counted as a contribution to efficiency but to inefficiency. To the extent that transpiration by the crop (a largely inevitable consequence of its growth) can be separated from the direct evaporation of soil moisture, an alternative expression of irrigation efficiency can be attempted. However, the two components of evapotranspiration cannot easily be separated, and in any case there usually is a feedback effect such that a reduction of the one would tend to enhance the other, though not necessarily to a commensurate degree.

2. A forage crop was raised under two different irrigation treatments and two fertilization subtreatments. The season's potential evapotranspiration was 1000 mm. The following results were obtained:

(1) In the "wet" treatment, water was applied in several large irrigations totaling 900 mm. The high-fertilization subtreatment produced a drainage of 100 mm and a dry-matter yield of 12 metric ton/ha. The low-fertilization subtreatment drained 200 mm and yielded 7 ton/ha. In neither subtreatment was the end-of-season soil-moisture content different from the start-of-season value.

(2) In the "dry" treatment water was applied in the amount of 500 mm. The high-fertilization resulted in a net depletion of soil moisture amounting to 100 mm and a dry-matter yield of 9 ton/ha. The low-fertilization sub-

treatment depleted soil moisture to the extent of 50 mm and yielded 5.5 ton/ha.

Calculate the evapotranspiration and water-use efficiency values. Assuming direct evaporation from the soil surface to be 10% of evapotranspiration for the "wet" treatment and 15% for the "dry" treatment, calculate the transpiration ratio for each treatment.

To calculate evapotranspiration E_t we subtract the amount of drainage from the amount of irrigation given the "wet" treatment and add the amount of soil water-content depletion to the amount of irrigation given the "dry" treatment:

"Wet" treatment: High fertilization subtreatment:

$$E_t = 900 \text{ mm} - 100 \text{ mm} = 800 \text{ mm}$$

Low fertilization subtreatment:

$$E_t = 900 \text{ mm} - 200 \text{ mm} = 700 \text{ mm}$$

"Dry" treatment: High fertilization subtreatment:

$$E_t = 500 \text{ mm} + 100 \text{ mm} = 600 \text{ mm}$$

Low fertilization subtreatment:

$$E_t = 500 \text{ mm} + 50 \text{ mm} = 550 \text{ mm}$$

To calculate water-use efficiency (WUE), we divide the dry-matter yield by the corresponding per-area mass of water used in evapotranspiration:

"Wet" treatment: High fertilization subtreatment:

$$\text{WUE} = \frac{(12 \text{ ton/ha}) \times (1000 \text{ kg/ton})/(10,000 \text{ m}^2/\text{ha})}{0.8 \text{ m} \times 1 \text{ m}^2 \times 1 \text{ ton/m}^3 \times 1000 \text{ kg/ton}}$$

$$= 1.5 \times 10^{-3} \text{ kg dry matter/kg water}$$

Low fertilization subtreatment:

$$\text{WUE} = (7 \times 1000/10,000)/(0.7 \times 1 \times 1 \times 1000) = 1.0 \times 10^{-3}$$

"Dry" treatment: High fertilization subtreatment:

$$\text{WUE} = (9 \times 1 \times 1 \times 1000)/(0.6 \times 1 \times 1 \times 1000) = 1.5 \times 10^{-3}$$

Low fertilization subtreatment:

$$\text{WUE} = (5.5 \times 1000 \times 10,000)/(0.55 \times 1 \times 1 \times 1000) = 1.0 \times 10^{-3}$$

To calculate the transpiration ratio (TR) we can subtract direct soil-moisture evaporation from evapotranspiration (an arguable procedure, since evaporation and transpiration are not independent) to estimate the amount of transpiration, and then divide the mass of water transpired by the mass of dry matter produced:

"*Wet*" *treatment:* High fertilization subtreatment:

$$TR = \frac{0.8(1 - 0.1)m^3/m^2 \times 1 \text{ ton/m}^3 \times 1000 \text{ kg/ton}}{(12 \text{ ton/ha} \times 1000 \text{ kg/ton})/10{,}000 \text{ m}^2/\text{ha}}$$

$$= 600 \text{ kg water/kg dry matter}$$

Low fertilization subtreatment:

$$TR = [0.7(1 - 0.1) \times 1 \times 1000]/(7 \times 1000/10{,}000) = 900$$

"*Dry*" *treatment:* High fertilization subtreatment:

$$TR = [0.6(1 - 0.15) \times 1 \times 1000]/(9 \times 1000/10{,}000) = 567$$

Low fertilization subtreatment:

$$TR = [0.55(1 - 0.15) \times 1 \times 1000]/(5.5 \times 1000/10{,}000) = 850$$

Just as the rock must first crumble for trees to grow on it, and just as the soil must first be loosened for its fruitfulness to develop, so too can valuable achievement sprout from human society only when it is sufficiently loosened so as to make possible to the individual the free development of his abilities.

Albert Einstein
1879–1955

9   *Tillage and Soil Structure Management*

A. Introduction

Recent years have witnessed a great intensification of land and water use in an effort to increase agricultural production in many parts of the world. These needs are most pronounced in arid regions, where pressure on the limited soil and water resources is the greatest. Numerous technological innovations have been applied in the field in an attempt to increase the efficiency of farming. The urgency of this task, however, has at times led agricultural developers into the pitfall of hastily adopting inappropriate methods which may eventually do more harm than good.

One of the most significant developments is the growing use of increasingly heavy tractors and machinery to plow land more deeply and rapidly and to carry out such operations as planting, spraying, and harvesting with a great saving of labor and time. The blessings of mechanization, however, are not without certain attendant dangers which may not be immediately obvious but which in time may cause the progressive deterioration of the soil. A case in point is the process of soil compaction, which can have a very deleterious effect on growing conditions and which is difficult to correct once it is caused. Equally insidious is the tendency to pulverize the soil excessively, with the consequence of making it extremely vulnerable to erosion by water and wind.

Different problems are encountered in the less developed countries, where farming is practiced in small units and where limitations of power, technology, and economy restrict the large-scale use of machinery from the outset. Under such circumstances, tillage operations may constitute the bottleneck of production because of the necessity for rapid preparation of the land during the crucial planting period. Owing to socioeconomic as well as environmental (e.g., climate and soil) conditions, which are different in the developing countries (many of which are tropical or subtropical) from those prevailing in the (mostly temperate) industrial countries, the machinery and techniques originating from the latter may be inappropriate for the former.

As pointed out by Donahue *et al.* (1971) and by Gill (1979), agricultural management practices vary widely, depending on the soil, climatic regions, and crops. In areas of low rainfall, the tendency is to retain plant residues on the soil surface to reduce evaporation and wind erosion, and the land may be fallowed in some years to accumulate moisture for a subsequent crop. Tools such as chisels, rod weeders, and sweeps loosen the soil without inverting it and causing the loss of soil moisture. In windy areas, attempts are made to reduce wind erosion by maintaining the soil surface in a rough, cloddy condition and by using windbreaks or surface mulches. Sloping lands subject to high-intensity rainfall are contoured, mulched with plant residues, and sometimes double cropped to maintain a plant cover so as to reduce water erosion.

B. Definition and Aims of Tillage

Tillage is usually defined as the mechanical manipulation of the soil aimed at improving soil conditions affecting crop production.

Three primary aims are generally attributed to tillage: control of weeds, incorporation of organic matter into the soil, and improvement of soil structure. An auxiliary function of tillage, still insufficiently well understood, is the conservation of soil moisture, where the processes of rain infiltration, runoff, and evaporation are involved (Hillel *et al.*, 1969).

Two decades ago, the advent of chemical herbicides seemed to reduce the importance of tillage as the primary method for eradication of weeds. However, more lately there have been growing objections to the application of more and more toxic chemicals in agriculture, owing to their residual damage to the larger environment and their increasing cost. Hence there is now once again increasing interest in the weed-controlling aspects of tillage. In the less developed countries, the unavailability as well as the lack of knowledge of appropriate herbicides may limit their use in any case.

The practice of inverting the topsoil in order to bury manures and crop residues has become a less important function of tillage in modern field management, where the use of animal and green manures is rather uncommon. Crop residues can, and in many cases should, be left over the surface as a *stubble mulch* to protect against evaporation and erosion. On the other hand, the use of agricultural land for disposal of waste products may once again reawaken interest in the soil-mixing aspect of tillage.

We come finally to the essential task of *soil structure management*, which has a bearing on planting, germination, water and air exchange, as well as erosion by rain and wind. Here we find that tillage practices suitable in one location may become harmful in another. Arid-zone soils with low organic matter contents and unstable aggregates are particularly vulnerable to compaction, crusting, and erosion. The precise effects of tillage on soil structure must be defined and optimized in each case if tillage is to be transformed from a hit or miss art to a scientifically based, dependable, and sustainable means of production. A basic discussion of soil compaction is given in by Hillel (1980, Chapter 14).

Tillage operations are especially consumptive of energy. The amount of earth-work involved in repeatedly loosening, pulverizing, inverting, and then recompacting the topsoil is indeed very considerable. In a typical small field of 1 h, the topsoil to a depth of only 30 cm weighs no less than 4000 tons. In an extensive farm of 1000 ha the mass of soil thus manipulated may exceed 4,000,000 tons.

In nonmechanized farming, a single tillage operation with a primitive plow requires a 40 km walk on rough land for each hectare!

The consumption of energy, as well as the wear and tear of tractors and implements, increase steeply as the depth of tillage increases. Thus, the cost of deep plowing (to a depth of 45–50 cm) is roughly double the cost of moderately deep plowing (to about 35 cm), quadruple the cost of normal plowing (about 25 cm deep) and tenfold greater than the cost of shallow plowing (approximately 15 cm deep). With the rising costs of fuel, the absolute and relative costs of tillage are almost certain to rise so that certain practices now common may become prohibitive, especially in the context of small-scale farming in the developing countries.

C. Traditional and Modern Approaches to Tillage

The traditional and still widely accepted practice of tillage is based on a series of primary cultivations (aimed at breaking the soil mass into a loose system of clods of mixed sizes) followed by secondary cultivations (aimed at further pulverization, repacking, and smoothing of the soil surface). These practices, performed uniformly over the entire field, often involve a whole series of successive operations, each of which is necessary to correct or supplement the previous operation. In the process, energy is often wasted and natural soil structure may be destroyed (Voorhees and Hendrick, 1977).

The more modern approach to soil structure management conceives of a field typically planted to row crops as consisting of at least two distinctly different zones:

(1) A *planting zone*, where conditions are to be optimal for sowing and conducive to rapid and complete germination and seedling establishment.

(2) A *management zone* in the interrow areas, where soil structure is to be coarse and open, allowing maximal intake of water and air, and minimal erosion and weed infestation.

These two zones differ in function as well as in mode of preparation and management.

Ideally, the planting row should be finely pulverized and loose enough to allow germination, emergence, and early root growth. On the other hand, it should be dense enough to allow intimate contact between seed and soil so as to promote rapid supply of water and nutrients. Sprayable mulches have been developed to enhance moisture and temperature control in the seedbed, but such treatments are as yet too expensive for common practice.

The interrow management zone, which generally consists of the greater part of the surface area in the field, normally remains bare and exposed for prolonged periods. Its condition often determines the water and air economy of the growing crop, as well as the water and soil conservation of the field as a whole.

A third zone or function must be considered. Modern farming almost inevitably involves passage over the field by tractors and implements. Random traffic often causes compaction all over the field, an effect which can be minimized if travel is confined to special tracks and travel over the seedling and water-management zones is avoided (Cooper *et al.*, 1969; Dumas *et al.*, 1975; Trouse, 1978).

The three zones described can be established for a perennial cropping program. Once established, these zones can be maintained consistently over a period of years, in row-cropped fields as well as in orchards and nurseries.

Recent trends in tillage research have been aimed at minimizing tillage operations and travel (both to reduce costs and to avoid soil compaction) while tailoring each operation to its specific zone and objective (Phillips and Young, 1973). This approach, in numerous variations, underlies the methods variously termed "zero tillage," "minimum tillage," "plow–plant," "wheel-track planting," "precision tillage," "permanent bed," etc. Space will not permit a detailed elucidation of each of these methods, some of which overlap in several respects. Suffice it to say in the present context that methods developed in one location may not be suitable for another location, where conditions can differ greatly. Hence there are no universal prescriptions for what constitutes efficient tillage. Some but not all soils have suitable tilth quite naturally and require little if any tillage to serve as favorable media for crop growth. Others, however, exhibit pans or barriers which inhibit root penetration and hence can be improved by appropriate tillage (Gill, 1974).

D. Problems of Tillage Research

The study of soil structure management has two aspects: defining the optimal soil physical state for any given purpose, and determining the most feasible means to achieve such an optimal state. The former task is difficult enough; the latter is *very* difficult. Any attempt to define soil–crop–tillage interactions in a fundamental way involves a complex array of factors relating to the mechanics of implement design and mode of operation, as well as to the dynamics of soil deformation and failure and of soil aggregation. The problem is most often approached empirically from two partial and alternative points of view, the engineering and the agronomic. The engineering approach aims at improving operational efficiency, while the agronomic approach aims directly at improving crop yields. Needed, and often lacking, is a unified approach. Part of the difficulty encountered so far has been the lack of measurable criteria of universal significance.

Various methods of tillage evaluation have been suggested by Nichols (1929), Nichols and Reed (1934), Browning (1950), Lyles and Woodruff (1961), Söhne (1956), Byers and Webber (1957), Payne (1956), Fountaine and Payne (1952), Pereira and Jones (1954), Hawkins (1959), Larson and Gill (1973), and many others. In most cases, pertinent soil physical properties are measured before and after mechanically defined tillage operations, and correlations are sought between soil and tillage parameters. To date, the most comprehensive review and analysis of tillage research is contained in the book by Gill and VandenBerg (1967).

Because of the many factors (mechanical, soil physical, climatic, and agronomic) and the complex interactions encompassed, tillage investigations must necessarily be long-term undertakings. Short-term experiments so often attempted under constraints of time, means, and personnel seldom yield conclusive results even for the location in which they are carried out. Much less are they likely to yield results of universal applicability. Though the tillage methods compared may differ widely in cost, performance, and measurable effect on the soil, final crop yields may indicate no consistent or significant differences. This is probably due in large part to faulty selections of measurable criteria, inexact measurements, and soil variability, as well as to the fact that crop response to tillage often tends to be masked by numerous unpredictable, yet decisive, variables acting in the field (e.g., limitation or excess of water due to vagaries of climate, fertility, pests, diseases). It often seems that the greatest chance for increasing the benefit-to-cost ratio in the field lies not so much in the expectation of obtaining greater yields from new and perhaps more elaborate tillage methods, but more in the possibility of maintaining the same high yields while simplifying and economizing tillage operations. Traditional field practice based on excessive manipulation of the soil (as in repeatedly and alternately loosening and recompacting it, as well as in preparing a seedbed over an entire field when such treatment is only required in the narrow planting rows) is now giving way to management systems based on minimal and precision methods of tillage (Reicosky *et al.*, 1977).

E. Physical Aspects of Machine–Soil Interactions

The ways in which moving tillage tools can affect the soil are exceedingly complex and include compaction, shearing, lifting, sliding, throwing, inversion, and mixing of the soil mass. Of the physical properties which determine the nature of soil reaction to a tillage tool, soil moisture is the most important, since it can cause the soil to vary from a hard condition to a plastic one. Sandy soils are easily manipulated, whereas dry clays may be extremely difficult to till—possessing bricklike mechanical strength. The required strength of tillage machinery, the rate of wear, and the power requirements are direct functions of soil physical conditions during operation (Gill, 1979; Hillel *et al.*, 1969).

According to Gill (1979), the key to the development of a scientific approach to tillage is the establishment of a *soil–machine mechanics* capable of describing and predicting the action of a tillage tool on the soil (Bernacki *et al.*, 1967; Gill and VandenBerg, 1967; Zelenin *et al.*, 1975). Given an initial soil condition and the characteristics and mode of operation of a given tool, what will be the final soil condition and to what extent will it approximate the desired state? To answer this, we must first be able to define the application of forces to the soil and the soil's reaction to variously applied forces. Since the duration of the tillage operation itself is very short but its effects are expected to be long lasting, we are interested both in the immediate response of the soil and in its subsequent resistance to alteration by natural processes (e.g., wetting, drying, freezing, and warming) and by biological factors (plants and animals).

Once a realistic soil–machine mechanics is developed, it can serve to predict soil behavior and help in the selection or design of appropriate tillage tools and in the improvement of tillage efficiency. Although the required level of knowledge does not yet exist and soil reaction cannot yet be predicted or controlled in general, a start has been made toward the description of soil–machine interactions for simple tools (Gill, 1979). The simplest approach, suggested by Söhne (1956), is an attempt to define the motion and action of an inclined wedge as a tillage tool. As the wedge moves through the soil, it causes repetitive compression and shearing of soil blocks, as illustrated in Fig. 9.1. The force necessary to maintain this movement at a constant velocity (i.e., the draft force W) can be estimated according to the following formula:

$$W = N_0 \sin \alpha + \mu' N_0 \cos \alpha + kb \qquad (9.1)$$

where N_0 is the normal load on the inclined tool, α is the lift angle of the tool, μ' is the coefficient of soil–metal friction, and k is the cutting resistance of the soil per unit of width b.

Subsequent attempts have been made to include the influence of velocity in the estimation of draft (Luth and Lismer, 1971). An often used equation is that developed by Goryachkin (1968):

$$W(v) = W_0 + \varepsilon v^2 \qquad (9.2)$$

where W is the draft, a function of velocity v, W_0 is the basic draft independent of velocity (presumably, the draft at a very slow speed of operation), and ε is a constant coefficient.

Simple mechanical relationships have been formulated with some limited degree of success for plows, harrows, bulldozers, scrapers, and other machines which operate in the soil. In all instances, one of the major limitations is the

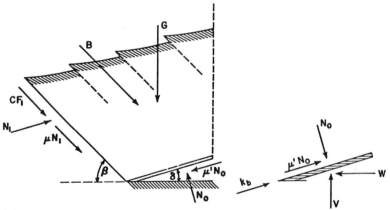

Fig. 9.1. Hypothetical forces and their orientation on a segment of soil reacting to an inclined-plane tillage tool: W is draft force, μ' coefficient of soil–metal friction, μ coefficient of soil–soil friction, N_0 normal load on the inclined tool, N_1 normal load on the soil failure surface, k_b pure cutting resistance of soil per unit width, G weight of soil on the inclined surface, δ lift angle of tool; β angle of soil shear failure surface, B acceleration force of soil, C cohesion of soil, V vertical force applied to soil, and F_1 area of forward shear failure surface. (After Söhne, 1956; NTML Photo 278.)

lack of fundamental criteria or parameters by which to characterize pertinent soil properties and behavior prior to, during, and following the operation. Unlike other industrially processed materials such as steel, the soil varies its properties drastically, as with changing moisture. Hence the search for universally definable "soil constants" has thus far failed to meet with success.

The traditional method by which tillage tools are moved through the soil is to attach them to a tractive unit, which also depends for its own performance upon the soil's dynamic properties. On soft or loose soil, the tractive unit may lose much energy merely to generate the traction necessary to propel itself, as well as the attached tool, forward. Improved operational efficiency, and possibly less damage to soil structure, can perhaps be achieved by the use of dynamic tools that are made to rotate or vibrate, either by means of the tractor's power take-off or hydraulic system, or by means of auxiliary motors (Ulyanov, 1969; Yatsuk, 1971), particularly if the tool–soil interaction is such as to help propel the tillage machine forward.

F. Operation of Tillage Tools

Plows (moldboard and disk types) are used for *primary tillage* to break out, crumble, and invert furrows of soil (Kepner *et al.*, 1972). Subsoilers and chisels, also used for primary tillage, break and loosen the soil without inverting it. Disk harrows, spike harrows, sweeps, drags, cultipackers, and other implements are used to refine coarse soil conditions during *secondary tillage*. Rotary hoes, as well as various harrows, are often used for light tillage, for the control of weeds, and for disrupting soil crusts. A series of special-purpose tools, including hillers, ridgers, subsoilers, and listers, are also used to shape the soil surface zone for planting on the ridge or in the furrow, or for irrigation.

As pointed out by Söhne (1966) the moldboard plow has long been and remains the most important tillage implement for the following reasons:

(1)   It is a classically simple implement, consisting of a share which cuts the soil, a sole and landside which maintain the necessary depth and side stability, and an arched plate (the moldboard) which turns, loosens, and throws the soil sideways.

(2)   The shape of the moldboard can be varied for different soil and climatic conditions, and it can be adapted to different speeds (up to about 8 km/hr at present).

(3)   Finally, the segmented parts of the moldboard plow, particularly the rapidly wearing parts, can be quickly and inexpensively replaced.

The actual operation of the moldboard plow can be described as follows:

A continuous slice of soil is cut by the coulter and share and is forced upward over the curved moldboard. Unless the soil is in a plastic state (i.e., if the soil is friable, as it typically is when plowed) it cannot be deformed and moved over the moldboard without developing shear planes and breaking into fragments or clods of various sizes. Nichols (1929) was the first to describe the pattern by which a furrow slice breaks into primary and secondary shearing planes. As it is lifted, turned, and finally cast off the plow, the furrow slice is accelerated so that when it finally strikes the ground it is generally further shattered. As a result, the soil is deposited as a loose and highly porous assemblage of variously sized clods. The manner in which the furrow slice is deposited and mixed is illustrated in Fig. 9.2. Universal shapes for plow bodies designed to operate at two speeds are illustrated in Fig. 9.3. Note that, because of the greater acceleration imparted to the soil, high-speed plowing generally encounters greater resistance and requires more power than low-speed plowing, as indicated by Eq. (9.1).

Disk plows and disc harrows consist of sharp-edged convex discs which roll obliquely to the direction of movement, thus cutting the soil in a manner similar to that of a moldboard plow. Each disk plow is mounted on its own axis and can be inclined at an angle of up to 20°, whereas all disks of a disk harrow assembly are mounted on the same axis. Vertical force is needed to cause the disks to penetrate into the soil, and at times high loads are necessary. Despite the reduction of friction at the rolling disc, it has been found in practice that drawbar pull (and hence energy consumption) for disk plows is at least as great as for moldboard plows of comparable size. Disk plows are particularly suitable for hard and abrasive soils, but are not widely used in most agricultural areas. Disk harrows, however, have won wide acceptance and are especially useful in the tillage of stubble fields.

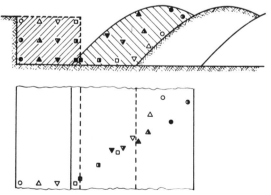

Fig. 9.2. The pattern of deposition and mixing of a furrow slice. The upper part of the figure shows a forward view, and the lower part gives a top view. (After Söhne, 1966.)

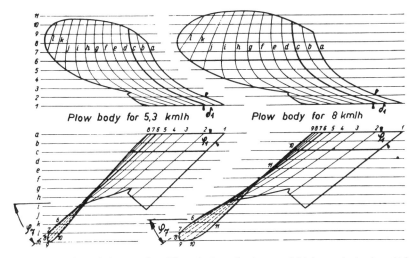

Fig. 9.3. Universal shapes of moldboard plows for low- and high-speed plowing. (After Söhne, 1966.)

Fig. 9.4. Horizontal and vertical rotating tillage tools. (After Söhne, 1966.)

As mentioned in Section E, energy losses are sustained whenever engine power is transmitted to the tillage tool by tractor wheels because of the rolling resistance and slip occurring between wheels and soil. According to Söhne (1966), the efficiency of transmission between tires and soil averages about 60–65% and in unfavorable conditions (soft soil) may fall below 40%. However, attempts to transmit power directly to rotating tillage implements (e.g., using the tractor's power take-off or by means of auxiliary engines) have met with only partial success. Figure 9.4 shows several types of rotating implements, none of which has so far even begun to replace the moldboard plow as a universal implement. Rotary tillers have been found useful on a limited scale in horticultural practice. The thorough pulverization of the soil by such implements can be advantageous, as it allows a seedbed to be prepared in a single operation. As often as not, however, it is disadvantageous, since a rotary-tilled soil can lose its structure if the aggregates are not very stable. Moreover, because of the high cutting velocity and the resulting high acceleration imparted to soil fragments, and because of the much larger cutting surface compared to moldboard plows, rotary tillers require on light-textured soil about 2.5 times and on heavy-textured soils about 3.5 times the power required by a plow for similar width and depth of tillage.

Part II:

EXTENSIONS

10   *The Development and Extension of Penman's Evaporation Formula*

J. L. Monteith

Department of Physiology and Environmental Studies
University of Nottingham School of Agriculture
Sutton Bonington, Loughborough, United Kingdom

A. Introduction

Two conditions must be satisfied before water will evaporate continuously from a wet surface. First, energy stored in the water vapor as latent heat must be supplied by some external source such as the sun. Second, the saturated water vapor in contact with the wet surface must be swept away and replaced by drier air, which becomes saturated in turn. Both these conditions are closely related to the laws of conservation: the conservation of energy requires that the heat used for evaporation be equal to the heat supplied, and the conservation of mass requires that the rate at which vapor is removed from the surface be equal to the rate at which it is removed by the atmosphere, a process which depends on turbulent mixing.

The equation which describes the heat balance of a surface and the corresponding equation of mass balance have one variable in common—the temperature of the evaporating surface. When there is any change in the state of the atmosphere or of the surface, its temperature changes to ensure that the laws of conservation are satisfied. One of the most versatile equations for calculating natural evaporation was first derived by Penman (1948), who solved the heat and mass balance equations simultaneously to eliminate surface temperature—a quantity rarely available from routine measurements and subject to substantial error before precise radiation thermometers became available. In its original form, the equation was used to calculate monthly averages of evaporation from open water surfaces, given the relevant standard climatological records of sunshine, wind speed, temperature, and humidity. Later, the equation was applied to bare soil and to vegetation. Some of the essential features of the equation and its extensions will now be considered.

In nature, the rate of evaporation from an extensive area of open water or wet, bare soil, depends on the vapor pressure e_0 of air in contact with the surface; on the vapor pressure of air e_a at an arbitrary height z, usually within 2 m of the surface to ensure that e_a is related to the state of the surface; and on the mixing of the air by turbulence between the surface and height z, specified by an exchange coefficient, which we shall write simply as $f(u)$ because it is a function of wind speed. It also depends on the value of z and on the roughness of the surface which helps to generate turbulence (Monteith, 1975). The heat used to evaporate water per unit area and per unit time is often written as λE (or LE) and an equation of the form first suggested by John Dalton relates λE to the humidity difference and to windspeed; that is,

$$\lambda E = f(u)(e_0 - e_a) \tag{10.1}$$

The rate C at which a surface loses (or gains) heat by convection can be written in a similar way as

$$C = \gamma f(u)(T_0 - T_a) \tag{10.2}$$

where T_0 and T_a are temperatures at the surface and at height z, respectively. The quantity γ is called the psychrometer constant because it appears in the equation for evaluating vapor pressure from the readings of wet- and dry-bulb thermometers. As it is proportional to the specific heat of air and inversely proportional to the latent heat of vaporization, it has a weak dependence on temperature, but a fixed value of 0.66 mbar/°K is appropriate for most processes at the earth's surface.

With the important proviso that the air in contact with the surface must be saturated at temperature T_0, an identity represented in symbols by $e_0 = e_s(T_0)$, the sum of latent and sensible heat fluxes can now be expressed as a function of one surface variable T_0 and three climatological variables— windspeed, temperature, and vapor pressure at height z.

As the total heat flux can often be estimated independently from the heat balance as H, say, it is possible to find a value of T_0 satisfying the condition

$$\lambda E(T_0) + C(T_0) = H \tag{10.3}$$

When T_0 has been determined from Eq. (10.3), values of λE and C follow immediately from Eqs. (10.1) and (10.2). This procedure is the basis for the formula derived by Penman (1948) to estimate "potential" evaporation, i.e., the maximum rate of water loss in given weather. In the original analysis, T_0 did not appear explicitly and λE had the form

$$\lambda E = \frac{\Delta H + \gamma f(u)[e_s(T_a) - e_a]}{\Delta + \gamma} \tag{10.4}$$

where $e_s(T_a)$ is the saturation vapor pressure at air temperature and Δ is the rate at which it changes with temperature, i.e., $\partial e_s/\partial T$. Strictly, Δ should be evaluated at a temperature between T_a and T_0. In practice, Δ must be evaluated at T_a because T_0 is not known a priori. The resultant error in λE is usually small, partly because the error in Δ appears both in the numerator and the denominator of Eq. (10.4).

In its basic form, the Penman equation describes the evaporation from a water surface. Empirical factors were introduced to describe the rate of "potential" evaporation from short green vegetation with an abundant water supply (Penman, 1948), evaporation from a drying soil, and transpiration from a crop with a restricted supply of water (Penman, 1949). Penman (1953) also derived an expression for the rate of transpiration from a leaf in terms of the diffusion pathway through stomatal pores and through the leaf boundary layer. This treatment was later applied to a crop canopy using a simple resistance analog which will now be described (Monteith, 1965).

B. Extension

An essential condition for the validity of Penman's analysis is that latent and sensible heat exchange must both occur from a surface at the same temperature. It is not essential for the diffusion pathway for water vapor to have the same exchange coefficient as the corresponding pathway for heat: any two functions can replace $f(u)$ in Eqs. (10.1) and (10.2). Moreover, it is not essential for the air in contact with the wet surface to be saturated provided it has a constant relative humidity, independent of the evaporation rate and uniquely related to the free energy of the evaporating water. At this point it is convenient to adopt the resistance nomenclature now widely used in both plant and animal ecology (Monteith, 1975; Campbell, 1977). The resistance to vapor transfer r_V is defined by Eq. (10.1) rewritten in the form

$$\lambda E = \rho c(e_0 - e_a)/\gamma r_V \tag{10.5}$$

where ρc is the volumetric specific heat of air, introduced to make the dimensions of resistance simple—time per unit length. The corresponding resistance for heat transfer is therefore given by

$$C = \rho c(T_0 - T_a)/r_H \tag{10.6}$$

Equations (10.1) and (10.2) can now be regarded as special cases of Eqs. (10.5) and (10.6) with $r_V = r_H = \rho c/\gamma f(u)$. Eliminating surface temperature then gives the evaporation rate in a form similar to Eq. (10.4); that is,

$$\lambda E = \frac{\Delta H + \rho c[e_s(T_a) - e_a]/r_H}{\Delta + \gamma(r_V/r_H)} \tag{10.7}$$

If the water potential of the evaporating surface is ψ so that its equilibrium relative humidity is given by $h = \exp(-\psi/RT)$, Eq. (10.7) becomes

$$\lambda E = \frac{\Delta h H + \rho c[h e_s(T_a) - e_a]/r_H}{\Delta h + \gamma(r_V/r_H)} \tag{10.8}$$

Without deriving this relationship explicitly, Penman (1955) used similar terms to estimate the rate of evaporation from the saline waters of Lake Eyre ($h \simeq 0.9$) after it filled in 1950.

Equation (10.7) has been widely used in two ways: to specify the physiological control of evaporation from a crop in terms of a canopy resistance, and to predict the maximum rate of evaporation from a crop as a function of a minimum value of that resistance. For such analysis, r_H is usually identified with the corresponding resistance for momentum transfer r_M, and r_V is assumed to be the sum of r_M and a canopy resistance r_c in series with r_M.

More exact analysis (Thom, 1975) takes account of the difference between the molecular processes governing momentum transfer at a surface and corresponding processes for heat and mass.

To calculate a rate of evaporation from wet soil or from a crop, the quantity H is usually calculated as the difference between the net radiative exchange R_n of the surface and the rate G at which heat is conducted to the substrate. When vegetative cover is complete, G is often assumed to be a small fixed fraction of R_n, at least for daily totals.

C. Deductions

When appropriate climatological measurements are available, Eq. (10.7) can be used to explore the diurnal and seasonal changes of evaporation for wet soil or from any surface with a fixed value of r_c. Because the diurnal change of saturation deficit $e_s(T_a) - e_a$ is highly correlated with R_n and therefore with H, λE is often nearly proportional to R_n. Another factor maintaining this proportionality is the tendency for both r_c and $e_s(T_a) - e_a$ to be larger in the afternoon than in the morning. Several workers have found that rate of evaporation from vegetation freely supplied with water is given by

$$\lambda E = a\Delta R_n/(\Delta + \gamma) \tag{10.9}$$

where a is an empirical constant with a value of about 1.3 (Davies and Allen, 1973; Stewart and Rouse, 1977). However, there is no theoretical reason why the loss of water from vegetation should follow Eq. (10.9) rather than the exact Eq. (10.7) (Monteith, 1978).

Because Δ and $1/r_H$ appear both in the numerator and in the denominator of Eq. (10.7), evaporation rate is not strongly dependent on either temperature or windspeed. In fact, there is a value of r_H at which E is independent of u. By differentiation of Eq. (10.7), it can be shown that this state is achieved when

$$\lambda E = \Delta R_n/(\Delta + \gamma)$$
$$r_c = r_c^* = \rho c(1 + \gamma/\Delta)[e_s(T_a) - e_a]/\gamma R_n \tag{10.10}$$

When the canopy resistance is smaller than the critical value r_c^*, the evaporation rate increases with wind speed but when $r_c > r_c^*$, evaporation rate decreases as wind speed increases (because the surface temperature changes in such a way that loss of sensible heat increases at the expense of latent heat). Grace (1978) reviewed the experimental evidence for both types of behavior.

Another important implication of Eq. (10.7) is that the control of evaporation by stomatal resistance depends not on the absolute value of r_c but on the ratio r_c/r_a. Provided $r_c = 0$ when leaves are covered with water (e.g., after rain or irrigation) and provided the temperature and vapor pressure of the air are effectively independent of the evaporation rate, the ratio of evaporation E_d from "dry" but transpiring vegetation to the evaporation E_w from the same vegetation when wet is

$$E/E_w = (\Delta + \gamma)/[\Delta + \gamma(1 + r_c/r_H)] \tag{10.11}$$

From field measurements, the minimum value of r_c for many types of vegetation from grassland to forests falls in the range 50–100 sec/m, but r_H has a much wider range because of its dependence on surface roughness. For very short grass, r_c/r_H can be less than unity so that $E \simeq E_w$; i.e., the rate of water loss is almost unaffected by wetting the leaves because it is limited by turbulence. For arable crops r_c and r_H are often nearly equal and E/E_w is about 0.7–0.8. For forests with r_H in the range 1–10 sec, E/E_w may be much less than unity. For example, Rutter (1967) showed that the rate of water loss from the branches of a pine tree increased by a factor of three to five when it was wetted.

The practical significance of Rutter's work has recently been emphasized in a comparison of wooded and unwooded catchments at the headwaters of the Rivers Severn and Wye in Wales. The hydrological balance implied that the annual yield of water from the unwooded area was substantially greater than from the neighboring, mainly wooded, catchment (Clarke and Newsome, 1978), and the difference can be accounted for by much faster evaporation of rain intercepted by conifers (Thom and Oliver, 1977). More work is needed, however, to discover how far the increase of evaporation rate after wetting, as predicted by Eq. (10.11), is offset by a decrease in the saturation deficit of the atmosphere.

D. Plant Physiology and Soil Physics

Several successful attempts have been made to relate the vapor resistance of a canopy r_c, determined from Eq. (10.7) to the corresponding stomatal resistance of individual leaves measured with a diffusion porometer (Black et al., 1970; Gash and Stewart, 1975; Monteith et al., 1965; Szeicz et al., 1973). A canopy with a leaf area index of L can apparently be treated as L layers of unit area index, each layer assumed to have leaves with a uniform stomatal resistance of r_s. Experimental evidence shows that r_c is close to the value expected for L individual resistors in *parallel*; i.e., $r_c = r_s/L$.

Detailed theoretical analysis by Shuttleworth (1976) has shown that the relation $r_c = r_s/L$ is often an acceptable approximation to a much more complex relationship. In general terms, the validity of the parallel resistance model depends on the fact that common values of stomatal resistance in a layer of foliage are usually much larger than the aerodynamic resistance between layers.

The resistance model has therefore helped to reconcile micrometeorological and physiological concepts. Integration with soil-water studies has been less successful for several reasons. In the first place, Eq. (10.7) as it stands is not valid for a drying soil in which the pores from which water is evaporating are usually at a different temperature from the surface at which the exchange of sensible heat with the atmosphere takes place. Failure of the equation for this reason was demonstrated experimentally by Fuchs and Tanner (1967). In the second place, as the relation between stomatal resistance and soil-water availability depends on a long chain of physiological processes and responses, it is impossible to predict with confidence how r_c will change during a growing season as soil water is depleted by root extraction and replenished by rain or irrigation. Irrespective of water

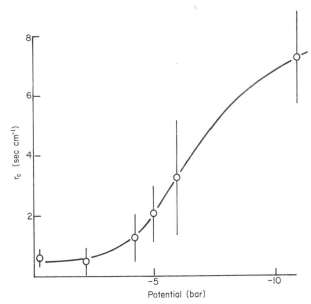

Fig. 10.1. Mean value of r_c for a grass–clover mixture as function of mean soil-water potential between surface and depth of 25 cm. Redrawn from Szeicz *et al.* (1973) with curve fitted by hand.

supply, r_c decreases during early growth (as L increases) and then increases towards maturity as the stomatal guard cells become less active. Moreover, the relation between r_c and the demand for water must depend, inter alia, on the size and vigor of the root system. Seasonal changes in soil-water content are therefore confounded by ontogenetic changes in leaves and roots. Figure 10.1 shows a characteristic example of scatter in the relation between r_c and soil-water potential (Szeicz *et al.*, 1973). Similar evidence was obtained by Russell (1980). For the time being, there are only two ways of calculating the actual rate of evaporation from vegetation as a function of soil-water content or potential: from Eq. (10.7) sometimes referred to as the Penman–Monteith equation, combined with an empirical relationship of the type illustrated by Fig. 10.1; or from much more complex models of the soil–plant–atmosphere system which demand a formidable array of inputs (sometimes available only from the set of experiments on which the model is based). As a matter of personal opinion, not supported by the author of this volume,[1] progress in the immediate future calls for the shrewd interpretation of evidence from carefully designed field experiments rather than the development of more comprehensive simulation models.

[1] The principal author of this volume believes that experimentation and theoretical modeling are mutually complementary.

11   *Freezing Phenomena in Soils*

R. D. Miller

Department of Agronomy
Cornell University
Ithaca, New York

A. Introduction

Most of the land surface in earth's middle latitudes is subject to seasonal freezing. At high latitudes, the regolith may be permanently frozen to great depths (*permafrost*) except for seasonal thawing of an *active layer* at the surface.

To the extent that soil moisture solidifies as polycrystalline ice, the mechanical strength of the soil increases dramatically and its capability for transmitting water declines. Other side effects of the freezing process are slight in gravels and coarse sands but increase in magnitude and importance as particle size decreases. These side effects undergo a change of character, however, as particle size approaches the colloidal range. Freezing of wet silt or coarse-clay fraction, for example, can produce a very large *frost heave* if the load is not large; freezing of very fine clay produces much less heave, but enormous loads can be lifted. If heaving occurs, it will be observed that some of the ice formed is present as seams of pure *segregated ice*. In relatively incompressible soils, these are usually oriented parallel to the isotherms and are called *ice lenses*. In highly compressible soils, one usually sees intersecting veins of ice that form reticulate patterns in three dimensions.

Freezing of moist soil induces movement of water into the zone of freezing which tends to fill with ice. This process of *freezing-induced redistribution* is much too rapid to be explained by vapor transport and may or may not be accompanied by frost heave. If heaving does occur, however, it will be much less than it would have been had the soil been wet before freezing commenced.

Since ice tends to form as a pure phase, solutes tend to concentrate in as-yet unfrozen water. If heaving or freezing-induced redistribution induces upward movement of liquid water into the zone of freezing, a zone of relatively high solute concentration can develop in the vicinity of the freezing front even in well-leached soils of humid zones. Living organisms tend to be strongly desiccated by freezing of the soil environment.

In nature, these processes go forward rather slowly within and below a frozen crust that may have the strength of concrete. They occur during a season of inclement weather. It is not surprising that knowledge of these processes is less than it should be nor that much of what we do know is derived from laboratory tests and experiments.

Side effects of soil freezing are important in a variety of ways. The freezing-induced redistribution process, by itself, means that when unsaturated soil freezes, pores of the frozen zone tend to fill with ice and the infiltration pathway of the hydrologic cycle may be effectively sealed off throughout the winter months until the last vestige of the frozen zone thaws, well below the surface, the following spring. During the period of thawing, soil above the frozen zone remains surfeited with melt water and may suffer a complete loss of strength. This is made worse if there has been heaving; when this soil thaws, excess melt water from segregated ice oozes upward through the surface soil and may be enriched with nutrients dissolved from the fertile surface soil; these find their way into surface waters as "agricultural pollutants." Runoff and erosion hazards are maximal during this period. If soil beneath the frozen zone has been partially dried by redistribution, there may be a very abrupt recovery of strength by the surface soil when, as farmers say, "the frost goes out of the ground" and the partially desiccated subsurface soil absorbs water from the wet surface zone.

When redistribution and heaving take place beneath a road, the onset of thawing leaves the pavement supported only by wet soil and water; wheel traffic breaks through, creating vents, the "potholes" so annoying and so dangerous to motorists. Snow removal in winter enhances heat loss and promotes heaving (see Fig. 11.1).

Fig. 11.1 Frost heave of "susceptible soil" produces ice lenses beneath a paved road. Water in unfrozen ditches enhances heave.

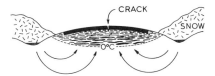

Heaving can develop enormous forces capable of displacing retaining walls, lifting pavements, fenceposts, and utility poles. Heaving can lift foundations, distorting unheated buildings, and is responsible for various other forms of costly mischief. The usual explanation offered is that water expands when it freezes, but this fact has little or nothing to do with the frost heave that people seek to explain in this way. Indeed, geologists and soil scientists are prone to identify expansion of water on freezing as an important agent in physical weathering of porous rocks, but on close examination this explanation of the disruptive action of freezing is rarely tenable (see Fig. 11.2).

It will be some time before we know the real influence of seasonal freezing phenomena on the chemistry of soils or on profile development. Speculations are commonplace, but there is reason to fear that some of these are based on imperfect knowledge of what really goes on when soils freeze.

Soil physicists began serious thought about the physics of ice in soil when Schofield (1935) suggested that the Gibbs free energy of soil moisture could be calculated from a measurement of its freezing point depression. Experimental tests produced results that were good enough to be encouraging but not good enough to be satisfying. Schofield's equation was examined and his assumptions criticized by Edlefsen and Anderson (1943) who explored thermodynamic implications of postulated "adsorption forces" at the surface of soil particles. Their conclusions were indecisive. Disheartened by both experimental vagaries and theoretical imponderables, most soil physicists abandoned research on soil freezing and turned toward more rewarding projects.

In this chapter, we hope to show that people were too easily discouraged. Edlefsen and Anderson overlooked a role for the surface energy (surface tension) of an ice–water interface and chose a less than optimum framework for their analysis of adsorption effects. Had they thought carefully about the experimental findings of Taber (1930) in his studies of frost heaving, or had they noticed an obscure remark about adsorption forces in Beskow's (1935)

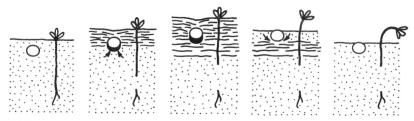

Fig. 11.2. In wet soils, subject to frost heave, buried objects may be displaced upward, relative to surrounding soil, by each freeze/thaw cycle.

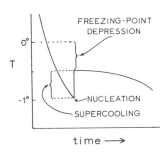

Fig. 11.3. Cooling curve for a moist soil.

monograph on heaving, they might have hit upon the approach to be emphasized in this chapter. This approach is based on elements already familiar to soil physicists in the context of unfrozen soils.

As a porous medium, soil imposes restrictions on the freezing of soil moisture, partly through interactions involving the liquid phase, partly through limitations on the growth of the ice phase. As a result, ice can rarely form inside a soil mass at 0°C. The highest temperatures at which ice could exist in a soil in a given circumstance specifies the *freezing-point depression* of water in the soil. All water in the soil does not freeze at this temperature, however. Lowering the temperature causes more and more of the water to change to ice, a matter that we will consider in terms of a *freezing temperature equation* (Fig. 11.3). There are reports that unfrozen water has been detected in soil materials at temperatures as low as from −40 to −50°C (Anderson and Tice, 1971). If nucleation of freezing is avoided, however, it is possible to *supercool* or *undercool* water in soil. Nucleation cannot initiate freezing until temperature has been lowered at least to the temperature defined by the freezing-point depression.

In this chapter, when we speak of *frozen soil* we refer to soil in which ice and unfrozen water coexist in amounts that depend upon temperature, type of soil, and other variables. We will be primarily concerned with temperatures between 0°C and perhaps −2°C, the range in which most things of interest take place.

B. Terminology: Water

There are strong similarities between the drying-and-wetting of ice-free soil and the freezing-and-thawing of air-free soil. To explore these similarities, we will use a hypothetical apparatus diagrammed in Fig. 11.4.

In this apparatus, there are three distinct locales for liquid water, requiring three distinctive terms. The pure liquid substance will be identified by writing its chemical formula H_2O. The liquid aqueous phase present in soil

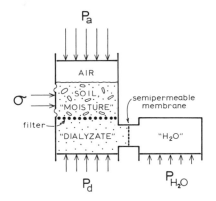

Fig. 11.4. Elements of a scheme for defining equilibrium conditions in a soil–water–air system.

will be identified by the term *soil moisture* or simply *moisture*. A liquid phase in equilibrium with soil moisture but kept free of soil particles by a rigid porous filter, pervious to solutes, will be called *equilbrium dialyzate* or simply *dialyzate*. A tensiometer or piezometer equilibrated with soil moisture will contain equilibrium dialyzate. In principle, H_2O could be equilibrated with soil moisture or its dialyzate if there were such a thing as a truly semipermeable membrane. In that case, the difference between the dialyzate pressure P_d and H_2O pressure P_{H_2O} defines the *osmotic pressure* of the dialyzate Π_d :

$$\Pi_d \equiv P_d - P_{H_2O} \qquad (11.1)$$

Elsewhere in this text the difference between air pressure P_a, acting on soil, and equilibrium dialyzate pressure P_d is called *matric suction* (or with sign reversed, *matric potential*). The difference between P_a and P_{H_2O} is sometimes called *total suction*. In this chapter, we shall avoid using either of these terms. Instead we shall write the quantity $(P_a - P_d)$ or the quantity $(P_a - P_{H_2O})$ as appropriate. We thereby avoid having to invent names for the quantities $(P_i - P_d)$ or $(P_i - P_{H_2O})$ when ice at pressure P_i is present in addition to air or instead of air. In ice-free soils, we are accustomed to the idea that moisture content and hydraulic conductivity are functions of matric suction. These mental associations must be revised if ice is present. The simplest way to avoid confusion is to avoid using the term.

On occasion, we will use the terms *adsorbed moisture* or *film moisture* to refer to that part of the soil moisture that is so close to the surface of a mineral particle that it is involved in direct interactions with the surface. *Capillary moisture* will refer to that part of the soil moisture that is held in spaces that lie outside the effective range of surface-adsorption forces.

The diagram of Fig. 11.4 indicates a variable which will be called *envelope pressure* σ. This is the pressure which will just offset any tendency for a flexible wall enveloping the soil to be distended by action of the soil against it. This quantity will be important in our discussion of the mechanism of frost heave and will be discussed in Section N.

C. The Freezing Temperature Equation

The apparatus of Fig. 11.5 is the same as that of Fig. 11.4 with the addition of a hypothetical osmometer in which an aqueous solution is exposed to ice at the same temperature and pressure as the ice to which soil is exposed on the other side. The only purpose of this addition is to provide the reader with a credible vision of equilibrium between ice and H_2O in which ice pressure P_i exceeds H_2O pressure P_{H_2O}.

Thermodynamic reasoning, as found in standard textbooks, leads to a generalized form of the *Clapeyron equation* which relates P_{H_2O} and P_i to equilibrium temperature T. This equation applies whether the intervening medium is solution in an osmometer or dialyzate and soil. We will write this equation in a form especially convenient for our purposes:

$$P_{H_2O}/\rho_w - P_i/\rho_i = (L/273.15°)T \qquad (11.2)$$

where pressures are *gauge pressures* measured against a standard atmosphere (101.3 kN/m^2) and T is temperature in degrees Celsius; ρ_w is the density of water (1.0 × 10^3 kg/m^3); ρ_i is the density of ice (0.917 × 10^3 kg/m^3); L is the latent heat of fusion (333 kJ/kg). Owing to simplifying physical and mathematical approximations used, this equation is not exact but is nevertheless quite accurate over the range of temperature and pressures of interest in this chapter. Readers will recognize the quantity 273.15° as 0°C expressed in Kelvins.

Recalling the definition of osmotic pressure of the equilibrium dialyzate,

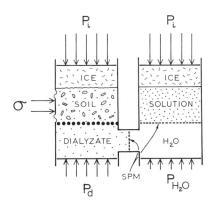

Fig. 11.5. Elements of a scheme for defining equilibrium conditions in a soil–water–ice system.

Eq. (11.1), we can substitute the quantity $(P_d - \prod_d)$ for P_{H_2O} to obtain the still more useful form

$$(P_d - \prod_d)/\rho_w - P_i/\rho_i = (L/273)T \qquad (11.3)$$

where P_d and $\prod_d$ can be measured, in principle, using a tensiometer to measure P_d and analysis of its contents to find $\prod_d$. In this chapter, this equation will be referred to as the *freezing temperature equation.*

Inserting approximate values for the constants and solving for each of the four variables, with gauge pressures in units of kN/m^2, the freezing temperature equation yields:

$$T = 8.2 \times 10^{-4}(P_d - \prod_d) - 8.9 \times 10^{-4}P_i \quad {}^\circ C \qquad (11.4)$$

$$P_i = 0.917(P_d - \prod_d) - 1.12 \times 10^3 T \quad kN/m^2 \qquad (11.5)$$

$$P_d = \prod_d + 1.09P_i - 1.22 \times 10^3 T \quad kN/m^2 \qquad (11.6)$$

$$\prod_d = P_d - 1.09P_i + 1.22 \times 10^3 T \quad kN/m^2 \qquad (11.7)$$

If it could be assumed that ice pressure were always precisely zero (gauge), the freezing temperature equation (11.3) would become the Schofield equation referred to as earlier. In general, this assumption cannot be justified in frozen soil.

D. Frost Heave—A Solution Analog

If heat were extracted at a very slow rate from the system represented schematically in Fig. 11.5, the net result would be the production of additional ice at high pressure and the consumption of H_2O at low pressure. The resultant uplift of the load on the ice would be "frost heave."

On the osmometer side of the apparatus, we can visualize mechanics of the process of ice formation and "heaving" as illustrated in Fig. 11.6. Extraction of heat tends to cool the solution below its equilibrium freezing temperature

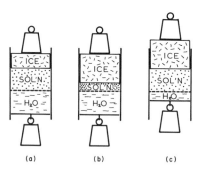

Fig. 11.6. Solution analog of frost heave: (a) osmotic equilibrium; (b) extract heat, form ice; (c) restore osmotic equilibrium.

(a)   (b)   (c)

so that additional ice forms, concentrating the solution and raising its osmotic pressure without increasing solution pressure. As a result, H_2O will be absorbed by the solution to restore osmotic equilibrium. The net result is an increase in the volume of ice (at high pressure) in the upper chamber and the disappearance of H_2O (at low pressure) from the lower chamber. This is a novel heat engine that does work on its surroundings as heat is extracted from the system.

Equation (11.5) gives the upper limit of heaving pressure P_i that could be achieved by an infinitely slow heaving process if ultimate values of P_d, T, and Π_d could be specified. This limit applies equally to soil and to the solution analog.

The solution analog of frost heave demonstrates *coupling* between transports of matter (H_2O) and sensible heat where a phase change is involved. To produce a steady state rate of heave V_H, it would be necessary to reduce the heat content of the system at a steady rate corresponding to the release of latent heat of fusion by H_2O that is converted to ice, i.e., $\rho_i L V_H$. If heat extraction were very slow, it would make no difference in which direction heat flow occurred; it is only necessary to avoid nucleation of freezing in supercooled H_2O in the lower chamber. A natural arrangement would involve steady extraction of heat through the ice at the top and a steady addition of heat through the bottom. If the fluxes of sensible heat were $(q_h)_i$ and $(q_h)_{soln}$ just above and just below the ice/solution interface, these heat fluxes would be "coupled" to the transport of H_2O as required to conserve both thermal energy and matter.

$$(q_h)_i = (q_h)_{soln} + \rho_i L V_H = (q_h)_{soln} + \rho_w L (q_{H_2O})_{soln} \qquad (11.8)$$

where $(q_{H_2O})_{soln}$ is the volumetric flux of H_2O diffusing through the solution. Note that coupling is contingent upon the presence of a mechanism whereby H_2O does not freeze in situ as heat is extracted but must move to a site where phase change can occur, releasing latent heat at that site.

E. Terminology: Soil

It is important to remember that soil scientists commonly use the word *soil* for materials so disparate physically that a heap of dune sand can represent one extreme and a watery suspension of colloidal particles can represent the other. To understand similarities between drying and wetting of ice-free soil and freezing and thawing of air-free soil, it is necessary to think first about each of these soil extremes in order to achieve a vision of what to expect in materials that lie between, or which may be a mixture of the two.

The essential attribute of a particle that exhibits colloidal properties is that at least one of its dimensions is small enough to be comparable to the range over which surface adsorption forces can influence liquid water. These distances amount to some tens of angstroms. This requirement can be met by particles of a montmorillonitic clay—thin sheets with thicknesses of about 9.6 Å. While these particles may be quite broad, say 1000 Å, colloidal behavior can be expected if they do not collect into aggregates of some kind. If the exchange capacity of a montmorillonitic clay mineral is dominated by monovalent ions (e.g., Na^+), colloidal behavior can be expected but if divalent calcium ions are predominant, the particles tend to organize into "tactoids" or "domains" too large to behave as proper colloidal particles. Thus, we rarely encounter a highly colloidal soil system outside the laboratory but to understand tendencies associated with particles that approach true colloidal behavior, we will visualize materials that behave as truly colloidal systems. Our mental model of such a system will involve thin flat sheets as portrayed schematically in Fig. 11.7. We shall identify such a materials as *SLS soil* to signify that the solid particles never touch each other, but are always separated by a space occupied by adsorbed liquid.

We will use the term *SS soil* for materials in which all particles are normally in solid-to-solid contact with adjacent particles, as diagrammed in Fig. 11.7. We expect this to be true of materials in which all dimensions of all particles are large in comparison with the effective range of surface adsorption forces—as in sands, silts, and even in materials fine enough to be included in the clay fraction according to current systems of particle-size classification but not fine enough to behave as colloids.

Most soils encountered in nature will contain some particles that will always be in solid-to-solid contact with some of the adjacent particles. Other particles will be held at a distance from their neighbors by an adsorbed layer of liquid. We will use the term *SSLS soil* to designate such materials.

Before considering freezing of SLS, SS, and SSLS soils, it is necessary to review adsorption and surface tension phenomena.

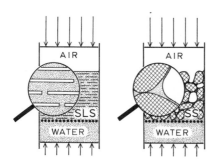

Fig. 11.7. Schematic representations of SLS and SS soils.

F. Adsorption Forces: The Double Layer Model

In his classic paper on the mechanism of frost heaving, Taber (1930) found that he could rationalize his observations of the growth of ice lenses in freezing soil by supposing that ice is always held at a distance, however slight, from neighboring soil particles by a mobile film of absorbed water. He suggested that whatever mechanism explained the hygroscopic character of particle surfaces exposed to air probably explained the presence of this film in freezing soil. Beskow (1935), who confirmed and greatly extended Taber's work, agreed and remarked that perhaps the adsorption mechanism involved a diffuse electrical double layer. Schofield (1946) used the concept of an idealized double layer proposed by Gouy (1910) to compute estimates of the thickness of hygroscopic films as a function of environmental factors, suggesting that the results ought to apply equally well to estimates of swelling properties of clays when the particles were thin flat sheets. Concurrently, Verwey and Overbeek (1948) were developing a more convenient mathematical formulation of the idealized double layer.

A succinct statement of the mathematical model can be found in an article by Bolt and Peech (1953; see Fig. 11.8); others have provided useful curves computed from these equations (e.g., Kemper and Quirk, 1970; Bresler, 1972).

Many experimental studies of clay swelling and ion-exchange phenomena have shown that this model seems to work remarkably well in circumstances where specific-ion effects do not interfere too seriously. The diffuse electrical double layer concept is especially appealing for our purposes and will be used as the basis of discussion. Other models of adsorption phenomena have been proposed, however. In their monumental work on the thermodynamics of soil moisture, Edlefsen and Anderson (1943) postulated an adsorption force field that acts on any substance which comes sufficiently close to a particle surface. Their painstaking analysis of the influence of such a mechanism on freezing led to no useful interpretations and, if anything, discouraged most

Fig. 11.8. The diffuse-electric double-layer for particles with a negative surface change balanced by an excess of cations and a deficit of anions in water close to the surface. (Adapted from Bolt and Peech, 1953.)

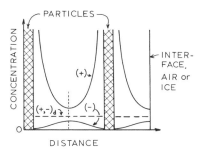

soil physicists from further efforts to understand soil freezing. Some soil scientists favor a model in which adsorbed water acquires, in some degree, an icelike structure, at least on certain clay minerals with lattice dimensions compatible with an icelike structure (cf. Ravina and Low, 1972). We will proceed as if these considerations amount to possible refinements of a more general model that considers only double layer effects.

An electrical double layer comprises a *surface charge* that resides in the surface of a particle and a diffuse *space charge* that resides in the liquid phase. The net space charge must be equal in magnitude but opposite in sign to the surface charge in order to preserve overall electrical neutrality. The space charge is comprised of ions that are dissolved in the liquid phase, but which are not present in either air or ice. If an isolated particle is immersed in water, the space charge will be most concentrated in solution next to the particle surface, tailing off to zero at distances of some scores of angstrom units, depending on the concentration of neutral solution (i.e., dialyzate) found there and depending on the valence of ions comprising the space change, which will be in diffusion equilibrium with ions in the neutral solution. If the space charge required to balance the nearby surface charge is crowded into a narrow space between two particles, or into a thin film between a particle and a film–air (or film–ice) interface, however, the concentration of ions in this space will everywhere exceed what it would be if this restriction were not imposed. Hence we expect that the solution in a very narrow space between particles, or in a thin film, will tend to absorb H_2O (by "osmosis") from a more dilute solution; the soil would swell.

If two particles tend to approach one another for some reason, the space between them, congested with double layer ions, will tend to imbibe H_2O from some less congested region, moving the particles apart again. It follows that a large number of perfectly flat particles would tend to arrange themselves into an array of equally spaced parallel sheets. If we extracted water from this array, the particles would simply move to a closer (but still uniform) spacing in a manner that fits our description of SLS soil. Surface tension effects would tend to minimize the area of an interface that envelops the entire array and would prevent air, or ice, from entering between the individual particles. If the planar particles were thin, with dimensions comparable to those of the double layer, the moisture content of the sample could be quite high.

According to this model, the osmotic pressure, relative humidity, and freezing point depression of moisture at a film–air or film–ice interface (or at a plane midway between parallel plates) can all be regarded as colligative properties of the solution, enriched by ions of the double layer, at a surface of zero electric field.

Finally, according to this model, when SLS soil is subjected to air pressure

P_a in apparatus diagrammed in Fig. 11.4, and equilibrated with dialyzate at pressure P_d, equilibrium will be achieved at that moisture content at which the pressure differential equals the osmotic pressure differential.

$$(P_a - P_d)_{aw} = \prod_s - \prod_d \qquad (11.9)$$

where $\prod_s$ is the osmotic pressure of double layer ions at the plane of symmetry between symmetric particles. The subscript aw signifies that the soil is exposed to air. The significance of Eq. (11.9) is the implication that in SLS soil matric suction (the term on the left-hand side) is due to the osmotic activity of double-layer ions.

If ice were substituted for air above the soil, the condition for equilibrium would be

$$(P_i - P_d)_{iw} = \prod_s - \prod_d \qquad (11.10)$$

where the subscript iw signifies that the soil is exposed to ice.

G. Surface Tension Effects

Readers are already acquainted with the Laplace surface tension equation that relates pressures in two phases that meet at a curved interface. For water at pressure P_w and air at pressure P_a meeting at an interface with mean curvature $(1/r)_{aw}$

$$P_w = P_a + 2\gamma_{aw}(1/r)_{aw} \qquad (11.11)$$

where γ_{aw} is the specific energy (surface tension) of the interface. By convention, curvature is taken to be positive when the center of curvature lies on the water side of the interface.

It has been shown that in very fine glass capillaries at temperatures near $0°C$, an ice–water interface has the same hemispherical appearance as a meniscus formed by an air–water interface in such a capillary (Skapski *et al.*, 1957). One can write the Laplace equation for this interface as well:

$$P_w = P_i + 2\gamma_{iw}(1/r)_{iw} \qquad (11.12)$$

At $20°C$, the handbook value of γ_{aw} for highly purified water is 72.8 mN/m and at $0°C$ the value is slightly higher, 75.6 mN/m. Near $0°C$, γ_{iw} is evidently slightly less than half that of γ_{aw}.

Figure 11.9 portrays particles of SS soil covered by films of adsorbed moisture with capillary spaces partly occupied by air at uniform pressure P_a. The mean curvature of the air–water interface fluctuates from place to place. At point F, for example, $(1/r_f)_{aw}$ is positive; the pressure experienced by water at the interface is greater than the air pressure. At point C, however, the

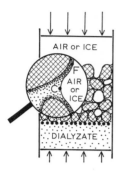

Fig. 11.9. Film moisture (near F) and capillary moisture (near C) in SS soil. Dots suggest distribution of solutes including those that balance surface charge of the particle.

curvature is negative and pressure experienced by water in the interface is lower than the air pressure. In accordance with the double layer model of surface adsorption, this pressure difference in an equilibrium system corresponds to inequalities of osmotic pressures at the respective points. If these are represented by $\prod_f$ and $\prod_c$, respectively, we would have

$$(\prod_f - \prod_c)_{aw} = (P_f - P_c)_{aw} = 2\gamma_{aw}[1/r_f - 1/r_c]_{aw} \qquad (11.13)$$

If moisture at point C is beyond the range of adsorption forces, its pressure and osmotic pressure would be the same as that of dialyzate, i.e., $P_c = P_j$ and $\prod_c = \prod_j$. The higher osmotic pressure at F is attributed to double layer ions crowded into the thin film.

An important consequence of this view of moisture in SS soil is that, at equilibrium, air at *uniform pressure* in SS soil is bounded by liquid whose pressure and ion concentration *fluctuate* from place to place in the interface. All of the liquid phase would be in equilibrium with H_2O held at some fixed pressure P_{H_2O} beyond a semipermeable membrane. Thus, the freezing temperature equation (11.2) could be satisfied if ice at uniform temperature and pressure were substituted for air. At point C, where $P_{H_2O} = (P_d - \prod_d)_{iw}$,

$$(P_d - \prod_d)/\rho_w - P_i/\rho_i = (L/273)T \qquad (11.14)$$

at point F, where $P_{H_2O} = (P_f - \prod_f)$,

$$(P_f - \prod_f)_{iw}/\rho_w - P_i/\rho_i = (L/273)T \qquad (11.15)$$

In summary, it is rational to conclude that at equilibrium in frozen soil at uniform temperature, ice exists at uniform pressure and this pressure can be calculated from the freezing temperature equation if P_d, $\prod_d$, and T are known.

H. Similitude

If we can achieve by a process of freezing-and-thawing a state of moisture content and distribution that is substantially identical to a state of moisture content and distribution achieved by a process of drying-and-wetting, we will say that the two states are *similar*. We should, therefore, compare freezing-and-thawing of air-free soil with drying-and-wetting of ice-free soil. Strictly speaking, the temperatures should always match at similar states. In SLS soils, prior history of changes in moisture content may influence the degree of orientation of a somewhat disorderly array of particles, so that we should expect to achieve similitude only if we have followed similar pathways. The same will be true with respect to SS soils in which prior history determines which "Haines jumps" will have taken place, as described in Chapter 12, Section E.

If hysteresis effects are taken into account, therefore, we expect to reach similar states in SLS soil if

(i) ion populations in dialyzates are identical, and
(ii) ion populations at planes of symmetry between parallel plates are identical.

If we use subscript aw to identify a quantity in a drying–wetting process in the absence of ice and iw for a freezing–thawing process in the absence of air, we conclude that at similar states

$$(\Pi_d)_{aw} = (\Pi_d)_{iw} \tag{11.16}$$

$$(\Pi_s)_{aw} = (\Pi_s)_{iw} \tag{11.17}$$

hence

$$(\Pi_s - \Pi_d)_{aw} = (\Pi_s - \Pi_d)_{iw} \tag{11.18}$$

and

$$(P_a - P_d)_{aw} = (P_i - P_d)_{iw} \tag{11.19}$$

If we use the freezing temperature equation to eliminate either P_i or P_d from the right-hand side, we obtain

$$(P_a - P_d)_{aw} = \left[\frac{\rho_i L}{273} T - \frac{\rho_i}{\rho_w} \Pi_d + \left(\frac{\rho_i}{\rho_w} - 1 \right) P_d \right]_{iw} \tag{11.20}$$

or

$$(P_a - P_d)_{aw} = \left[-\frac{\rho_w L}{273} T + \Pi_d - \left(\frac{\rho_w}{\rho_i} - 1 \right) P_i \right]_{iw} \tag{11.21}$$

Equations (11.16) and (11.17) and the succeeding equations are necessary conditions for similitude for SLS soil but they are not sufficient. We must specify that the ions are distributed in the same way by species.

If hysteresis effects are taken into account with SS soil, we expect to reach similar states if the interfacial curvatures are the same between capillary moisture and air as between capillary moisture and ice, i.e., if

$$(1/r_c)_w = (1/r_e)_{iw} \tag{11.22}$$

Using the respective surface tension equations (11.11), (11.12) we find an alternate statement of this requirement for similitude, namely

$$(P_a - P_d)_{aw} = (\gamma_{aw}/\gamma_{iw})(P_i - P_d) \tag{11.23}$$

Using the freezing temperature equation to eliminate either P_i or P_d from the right-hand side we obtain

$$(P_a - P_d)_{aw} = \frac{\gamma_{aw}}{\gamma_{iw}}\left[\frac{\rho_i L}{273}T - \frac{\rho_i}{\rho_w}\Pi_d + \left(\frac{\rho_i}{\rho_w} - 1\right)P_d\right] \tag{11.24}$$

or

$$(P_a - P_d)_{aw} = \frac{\gamma_{aw}}{\gamma_{iw}}\left[-\frac{\rho_w L}{273}T + \Pi_d - \left(\frac{\rho_w}{\rho_i} - 1\right)P_i\right] \tag{11.25}$$

When we compare equation (11.19) for SLS soil with Eq. (11.23) for SS soil, it becomes obvious that for SSLS soil, there is no way to achieve similitude between a soil subjected to a freezing–thawing process and the same soil subjected to a drying–wetting process. If we use the condition required for similitude of SLS soil, we will overestimate the unfrozen moisture content of the SSLS soil at a given temperature. An underestimate will result if we use the condition for similitude for SS soil.

To the extent that air bubbles are trapped and fail to promptly dissolve during wetting of SS soil, we will underestimate the moisture content of that soil during thawing if we base our estimate on the wetting curve.

Since both ice and air can exist only "outside" SLS soil, both will coexist at the same pressure. Thus it does not matter, in practice, whether freezing takes place in the presence or absence of air so long as we remember that moisture content refers to the ice-free, air-free soil mass.

In SS soil, however, it is crucial that one compare drying and wetting of ice-free soil only with freezing and thawing of air-free soil. If air is present as SS soil is frozen, one will overestimate the residual moisture content. Freezing of unsaturated SS soil is discussed in Section J.

Koopmans and Miller (1966) tested the requirement for similitude for SLS soil, as given by Eq. (11.20) using Na–montmorillonite. The condition for similitude for SS soil as given by Eq. (11.24) was tested using narrow silt fractions. Some results are shown in Fig. 11.10. By pooling data from several silt fractions, over a range of temperatures, they found that the best fit was

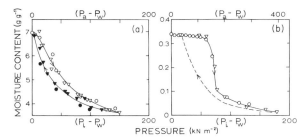

Fig. 11.10. (a) Test of conditions for similitude using Na^+–montmorillonite as an example of SLS soil. (b) Test of conditions for similitude using a 4–8 μm silt fraction as an example of SS soil. Broken line represents thawing data for a replicate sample of SS soil. The symbol ∇ indicates freezing; $\bigcirc$, drying; $\blacktriangledown$, thawing; $\bullet$, wetting. (After Koopmans and Miller, 1966.)

obtained if $\gamma_{aw}/\gamma_{iw} = 2.20$, and this value was used in plotting data for SS soil.

The "working value" of the ratio 2.20 involved air–water interfaces at 20°C. The handbook value of γ_{aw} at 20°C is 73 mN/m. If we use this value, we infer that γ_{iw} near 0°C would be about 33 mN/m. In these experiments, however, it is to be expected that contaminants reduced γ_{aw} some 5% or 10% below its handbook value suggesting that a value of nearer 30 or 31 mN/m would be more realistic. Values suggested by various authors range from about 10 to 50 mN/m but the weight of opinion seems to favor values near 31 mN/m (Everett, 1961; Hesstvedt, 1964).

I. Metastability and Nucleation

When water is exposed to a soluble gas, such as air, the equilibrium concentration of dissolved air present in the water increases as air pressure increases. If air pressure is initially high, so that the concentration of dissolved air is high, an abrupt reduction in pressure leaves the water supersaturated with respect to dissolved air, and it may *effervesce* as air exits from solution, forming bubbles.

The apparatus of Fig. 11.4, used as the basis of our discussion of equilibrium, was characterized as hypothetical for two reasons: (i) No one has invented a membrane that is pervious to H_2O without being at least somewhat pervious to solutes present in soil moisture. (ii) Since dissolved air can diffuse through the filter, then whenever air pressure in the upper chamber is greater than P_d the dialyzate would be metastable, prone to effervesce. The greater the differential pressure $(P_a - P_d)$ we wish to maintain the greater the tendency for effervescence to take place.

One possible mode of "effervescence" would not depend on dissolved air; the air phase would simply displace capillary water from some continuous pathway through the filter and emerge as a steady stream of bubbles at the point where the air phase breaks through the filter into the chamber where dialyzate is being held at a lower pressure. Accordingly, a figure of merit for a filter is its "bubbling pressure," namely the critical pressure differential, $(P_a - P_d)_{cr}$, at which "breakthrough effervescence" can be expected in the lower chamber. This is related to pore sizes in the filter by

$$(P_a - P_d)_{cr} = 2\gamma_{aw}(1/\tilde{r}) \tag{11.26}$$

where $\tilde{r}$ is the "effective" capillary radius of the narrowest pore neck along the optimum pathway for breakthrough of the air phase.

This phenomenon has an analog for the breakthrough of ice. Expressing this in terms of the critical differential pressure $(P_i - P_d)_{cr}$ and the same pathway with the same value of $(1/\tilde{r})$,

$$(P_i - P_d)_{cr} = 2\gamma_{iw}(1/\tilde{r}) \tag{11.27}$$

Hence, if we know the "bubbling pressure" of a filter measured in the air–water mode by $(P_a - P_d)_{cr}$, its "freeze-through" pressure in the ice-water mode should be given by

$$(P_i - P_d)_{cr} = (\gamma_{iw}/\gamma_{aw})(P_a - P_d)_{cr} = 0.45(P_a - P_d)_{cr} \tag{11.28}$$

If $\gamma_{iw} = 31$ mN/m[1], the critical temperature for freeze through is given by

$$T_{cr} = -[5.5 \times 10^{-8}/\tilde{r} + 7.4 \times 10^{-5}P_d + 8.2 \times 10^{-4}\prod_d] \;\; °C \tag{11.29}$$

where $\tilde{r}$ is in meters and pressures are in kilonewtons per square meter.

Example: If $P_d = 0, \prod_d = 0$, and $\tilde{r} = 1$ μm, then $T_{cr} = -0.06°C$.

Returning to the air–water mode, if the pores of the filter are all very small, we may observe that effervescence begins spontaneously before $(P_a - P_d)$ reaches the critical value required for breakthrough. Air at pressure P_a above the filter dissolves in soil moisture, and the dissolved air diffuses through the filter until dialyzate below the filter becomes "saturated" with dissolved air relative to air at pressure P_a. Suppose that an air bubble of radius R_b were inserted into this dialyzate. Air pressure inside the bubble P_b would exceed pressure in the surrounding dialyzate:

$$P_b = P_d + 2\gamma_{aw}(1/R_b) \tag{11.30}$$

The question is how P_b compares with P_a, the air pressure with which the water has been equilibrated. If the bubble is smaller than a critical size, P_b

will be higher than P_a, and the water will be less than saturated with respect to air at this higher pressure. As a result, some of the air in the bubble will go into solution, reducing the bubble size, thereby raising its internal pressure even higher. This will hasten dissolution until the bubble has *evanesced* (vanished). If the bubble is larger than this critical size, however, P_b will be less than P_a; the water will be supersaturated with respect to air at this lower pressure, and the bubble will grow accordingly. As the bubble enlarges, its internal pressure decreases, and its rate of growth increases; effervescence will be underway. The critical bubble size $(R_b)_{cr}$ is that at which air pressure inside the bubble would just equal air pressure in the upper chamber

$$(R_b)_{cr} = 2\gamma_{aw}/(P_a - P_d) \qquad (11.31)$$

Example: If $P_a - P_d = 100 \text{ kN/m}^2$ and if $\gamma_{aw} = 70 \text{ mN/m}$, then $(R_b)_{cr} = 1.4 \ \mu\text{m}$. A bubble smaller than this critical size will evanesce; a bubble larger than the critical size will effervesce.

A very small hydrophobic surface can simulate the role of a "seed bubble" in nucleating effervescence in the water. Materials with low affinity for water (metals, plastics) and a high affinity for hydrophobic contaminants (grease, including fingerprints) are commonplace in conventional tensiometers, and spontaneous effervescence is a commonplace occurrence. The rule of thumb that the useful limit of a tensiometer is "about -70 kN/m^2" should not be interpreted as meaning that effervescence begins at this value. Instead, this limit has more to do with the fact that the release of a small quantity of dissolved air produces a highly visible bubble at rather low pressure. From a volumetric point of view, the rate of effervescence is perceived as intolerable as P_d approaches values near -70 kN/m^2. Spontaneous effervescence can be discouraged by deairing the water at frequent intervals.

Spontaneous effervescence is the rule with pressure-membrane extractors used to obtain soil-moisture release curves. A steady stream of effervescing air emerges from beneath the filter plate even though the air phase has not bubbled through the filter itself.

In the ice–water mode, a similar situation can be expected. Water below the filter will be metastable whenever ice pressure P_i above the filter exceeds the dialyzate pressure P_d below the filter. To seed freezing, an ice "bubble" should exceed a critical size given by the analog of Eq. (11.31).

$$(R_b)_{cr} = 2\gamma_{iw}/(P_i - P_d) \qquad (11.32)$$

If we use Eq. (11.5) to eliminate P_i; we obtain

$$(R_b)_{cr} = 2\gamma_{iw}/[(L/273)T + P_d - (\rho_i/\rho_w)(P_d - \prod_d)] \qquad (11.33)$$

If $\gamma_{iw} = 31$ mN/m, and pressures are in kN/m^2

$$(R_b)_{cr} = -(1.80 \times 10^7 T + 1.34 \times 10^3 P_d + 1.48 \times 10^4 \textstyle\prod_d)^{-1} \quad \text{m} \quad (11.34)$$

Example: If $P_d = 0, \prod_d = 0$, and $T = -1°C$, then $(R_b)_{cr} = 0.06 \ \mu m$.

Ordinary tapwater can usually be supercooled to at least $-2°C$, and often to $-10°C$ before spontaneous nucleation occurs. Extremely pure water in very clean glass capillaries has been supercooled almost to $-40°C$ (Chahal and Miller, 1965). Dilute suspensions of Na–montmorillonite have been supercooled to about $-18°C$ (Chahal, 1961). Nucleation in very pure water at about $-40°C$ is presumably seeded by clusters of water, momentarily organized into an icelike structure. This is called *homogeneous nucleation*. At $-40°C$, the required seed size would be smaller than 10 Å. If foreign bodies induce nucleation, however, this is called *heterogeneous nucleation*. Some crystals with structures compatible with ice (e.g., AgI) are evidently very effective agents for heterogeneous nucleation and are used for seeding freezing of supercooled clouds, setting processes in motion which, it is hoped, will include rainfall.

Supercooling is seldom detected when soils freeze in the field, perhaps because sensors do not happen to be located near the point at which nucleation first occurs, or airborne ice crystals may seed freezing. In the laboratory, however, it is usually necessary to take special measures to avoid unwanted supercooling of several degrees before freezing begins.

J. Three Pore Phases: Freezing-Induced Redistribution

So far, we have considered freezing of air-free soils, avoiding complications that arise when air is also present as a continuous phase in SS or SSLS soils. In nature, most soils are likely to be unsaturated when freezing begins in fall. It is important to have some idea of what happens in these circumstances.

Progressive freezing of unsaturated SS or SSLS soil can produce the extraordinary result illustrated by data shown in Fig. 11.11 (Dirksen and Miller, 1966), the phenomenon called *freezing-induced redistribution*. In that experiment, a column of soil at uniform moisture content was brought to a uniform temperature $+4°C$. One end was abruptly cooled to $-4°C$ and held at that temperature while the other end was held at $+4°C$. No additional water was supplied. After a time, the column was sliced into segments and dried to determine gravimetric moisture content (including ice content). The resultant points were plotted in Fig. 11.11 for the midpoint of each segment and connected by straight line segments. The experiment was repeated with replicate columns sliced after longer freezing periods. In each case, the $0°C$

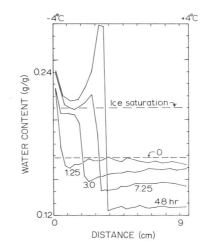

Fig. 11.11. Freezing-induced redistribution in a New Hampshire silt. Water content includes ice. (After Dirksen and Miller, 1965.)

isotherm was somewhere between the two points that bracket the abrupt step in total moisture content seen in each profile. Analysis of the data indicated that the magnitude of the observed freezing-induced redistribution exceeded, by about three orders of magnitude, that which could be explained as vapor transport, a conclusion reached by Beskow (1935) in a similar analysis of similar data. The zone of freezing acts as a strong sink for liquid water, inducing liquid flow into the zone of freezing, with ice accumulation in a zone *behind* the advancing freezing front. As the temperature profile approached a steady state, and the rate of advance of the freezing front slowed to a halt, ice accumulation continued, accounting for the U-shaped profile seen in the ultimate profile (48 hr). It was observed that heaving (under a modest load) occurred only if the total moisture content just behind the freezing front rose above about 90% pore saturation (with ice). Ice lenses appeared at a slight distance *behind* the freezing front.

As ice forms in unsaturated SS soil, we must expect that at any equilibrium state we will have a set of three phase boundaries: air–water, ice–water, and air–ice. To Eq. (11.11) and (11.12) we must add a third surface tension equation:

$$P_i = P_a + 2\gamma_{ai}(1/r)_{ai} \qquad (11.35)$$

where γ_{ai} and $(1/r)_{ai}$ are the surface tension and mean curvature, respectively, of the air–ice interface. Mean curvature $(1/r)_{ai}$ is positive when centered on the ice side of the interface.

According to Antonov's rule,

$$\gamma_{ai} = \gamma_{aw} + \gamma_{iw} \qquad (11.36)$$

This implies acceptance of the idea that the "wetting angle" of water on ice is $0°$. In soil near $0°C$, suggested values are: $\gamma_{aw} = 70\,mN/m$, $\gamma_{iw} = 30\,mN/m$, and $\gamma_{ai} = 100\,mN/m$, in round numbers.

Apparently, the shape of an air–ice interface can change rapidly when the scale is small. Some readers will recall that dendritic snow crystals, formed at very low temperatures, degrade rapidly into the spherical granules that are characteristic of a snowbank at temperatures near $0°C$. Points of contact between small spheres of ice sinter rapidly as if molecules in an air–ice interface are highly mobile near $0°C$, perhaps because of strains associated with the structural discontinuity at the interface. If it is assumed that an air–ice interface adjusts its shape to minimize its area, it is possible to visualize the manner in which air, water, and ice coexist in pores. This view leads to an explanation of redistribution induced by a temperature gradient in the frozen portion of a moist column of soil.

The geometry of an air–water interface in ice-free soil is complex, and difficult to portray in a two-dimensional drawing, even for a system of uniform spheres in a regular array. The problem is much more difficult if ice is also present, with air–ice and ice–water interfaces to contend with. The problem can be simplified by thinking about a "soil" in which the particles are all cylinders of uniform radius R_p, a regular array. Figure 11.12 shows pores between cylindrical particles, viewed in cross section, with some equilibrium configurations for air, water, and ice in those pores. In the examples shown, the radius of the air–water interface R_{aw} is always the same and is one-tenth of the particle radius.[1] It will be assumed that air pressure is always zero (gauge).

We use the four crevices of the pore of Fig. 11.12a to portray configurations at four different temperatures. These configurations have been drawn to scale in a manner that reconciles interfacial curvatures with phase pressures that satisfy, in turn, the freezing-temperature equation in the absence of solutes (Miller, 1973a).

Only water and air (no ice) could exist at temperatures above a critical temperature T^* (12 o'clock in the diagram). At T^*, ice could first appear in the form of a cylinder inscribed between particle surfaces and the air–water interface (3 o'clock); any smaller body would evanesce. Thus T^* is the freezing-point depression of the moisture present. If P_d were held constant and the temperature lowered, the ice body would enlarge and its pressure would increase (6 o'clock) until at a second critical temperature T^{**} a limit would be reached (9 o'clock). Below that temperature no construction can reconcile the geometric requirements for all three interfaces; the three-phase system would be unstable; only water and ice (no air) could coexist at

[1] For a cylindrical interface with principal radii R_{aw} and ∞ the pressure step would be γ_{aw}/R_{aw}.

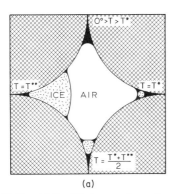 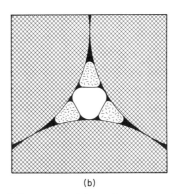

(a)   (b)

Fig. 11.12. (a) Equilibrium configurations of air, water, and ice at four temperatures decreasing clockwise from 12 o'clock in a pore formed by four cylindrical particles with $R_{aw} = 0.1 R_p$. The configuration at T^{**} is unstable and the pore will fill with ice. (b) An unstable configuration for air, water, and ice in a pore formed by three cylindrical particles. The pore will fill with ice when this configuration has been reached if R_{aw} is held constant. (After Miller, 1973a, b.)

equilibrium and one can expect a "jump" in ice content. The stage shown at 6 o'clock was calculated for a temperature midway between T^* and T^{**}. The temperature interval between T^* and T^{**} is small, and the transition from an ice-free pore to an air-free pore is abrupt.

Example: If $P_a = 0, \prod_d = 0$, $R_p = 4$ μm, and $R_p/R_{aw} = 10$, then $P_d = -380$ kN/m^2.

The calculated value of T^* is $-0.29°$C and $T^{**} = -0.37°$C; a temperature change of $-0.08°$C would carry the pore from an ice-free state to an ice-filled state.

The constructions of Fig. 11.12a illustrate the fact that as soil is cooled, holding P_d constant, there is only a narrow temperature range in which air and ice can both be present and in equilibrium within a single pore, even if ice formation is nucleated in only one corner of the pore. If the particles are rearranged, as in Fig. 11.12b, the temperature range for coexistence of air and ice is even narrower. If there is but a single ice body, it will touch the particle on the far side of the pore before reaching the unstable condition portrayed at 9 o'clock in Fig. 11.12a. When contact is made, the pore should immediately fill with ice if P_d is held constant.

If ice is forming in one or both of the other two corners of the pore in Fig. 11.12b, however, the temperature range for coexistence of air and ice would be still narrower. The configuration shown in Fig. 11.12b is the lower limit for coexisting air and ice in that pore when $R_{aw} = 0.1 R_p$. When air–water menisci at the corners of the ice bodies coalesce, a jump to an ice-filled state will occur.

For the simplified "soil" consisting of cylindrical particles packed in the manner of Fig. 11.12b, one can calculate the fraction of the total pore volume that will be occupied by unfrozen water, ice, and air as a function of temperature for various values of R_{aw}. Results are shown in the phase diagram of Fig. 11.13. Above the temperature at which the star is plotted, only air and water (no ice) can exist in the pore. If ice is present in all three crevices at that temperature, the degree of saturation (water plus ice) increases as temperature is reduced until a configuration like that of Fig. 11.12b is reached at the point marked by the solid circle whereupon a jump to saturation occurs. Curves that separate water from ice in the phase diagram of Fig. 11.13 step downward twice. The first step occurs when ice first appears beneath the meniscus in each crevice (as at 3 o'clock in Fig. 11.12a). The second step occurs when water at the corners of the ice bodies freezes as a jump to ice saturation occurs.

We now look again at the ultimate profile of total moisture content achieved by freezing-induced redistribution shown in Fig. 11.11. We see that had we used a temperature coordinate instead of a distance coordinate to plot that profile, and had we suppressed heaving, the ultimate steady-state profile of Fig. 11.11 could have looked very much like the phase diagram of Fig. 11.13.

Finally, we look at the ice-pressure lines at the top of Fig. 11.13. We perceive that when there is a temperature gradient but no gradient of P_d, ice pressure rises with distance behind the freezing front. At some point ice

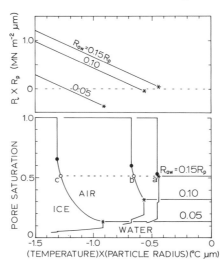

Fig. 11.13. Top: Ice pressures associated with conditions indicated in the phase diagram. Bottom: Phase diagrams for air, water, and ice in the pore diagrammed in Fig. 11.12b for three scaled values of dialyzate pressure. (Adapted from Miller, 1973a, b.)

pressure would be high enough, depending on the load, that heaving could be expected to occur. Ice lenses produced by such a heaving process could only form at a distance behind the freezing front and at a point behind the ice saturation front. This inference agrees with observations of Dirksen and Miller (1966).

The phase diagram of Fig. 11.13 contains other information. Suppose the soil were cooled at constant total moisture content (water plus ice). For the sake of illustration, we will neglect the expansion of water on freezing. The broken line connecting points a, b, and c would then represent progressive freezing of a pore at constant total moisture content (water plus ice). Thus P_d would decrease as temperature fell, from $-\gamma_{aw}/(0.15\ R_p) = -6.7(\gamma_{aw}/R_p)$ at T_a to $-10(\gamma_{aw}/R_p)$ at T_b to $-20(\gamma_{aw}/R_p)$ at T_c, if the degree of pore saturation did not change. Progressive freezing of unsaturated soil would, therefore, induce a hydraulic gradient whereby water would be drawn from unfrozen soil into the zone of freezing, there to fill the pores with ice. Thus we perceive a "capillary sink" mechanism to be an explanation of the freezing-induced redistribution phenomenon.

This capillary model of freezing of moist soil helps us to understand some of the experimental difficulties encountered by those who wished to calculate P_d in moist soils by measuring T^* in accordance with the proposal of Schofield (1935). Not only is P_i not exactly zero, as assumed by Schofield, but one must form a finite amount of ice if freezing is to take place at all. Once freezing commences, however, there will be an immediate and uncontrollable internal redistribution which lowers P_d. It is not surprising that tests of the Schofield equation were confusing and disappointing.

The capillary-sink mechanism just described is in marked contrast to one used in a number of computer exercises intended to simulate freezing-induced redistribution (e.g., Harlan, 1973). These exercises involve a "hydraulic model" and can be reasonably successful in predicting the rate of extraction of water from soil below the frozen zone (Sheppard *et al.*, 1978) but this is, in a sense, fortuitous.

The hydraulic model ignores the capillary-sink phenomenon but achieves approximately the same result, accumulation of ice in a zone behind the freezing front. It differs, however, in presuming that ice is also accumulating *throughout* the frozen zone at *all* times and that if the computed accumulation exceeds the available pore space, the frozen zone is arbitrarily expected to "stretch" wherever the extra space is needed. By contrast, concepts advocated in this chapter regard the ice phase as a continuous rigid body. If heaving occurs, it involves formation of a zone of clear ice (an ice lens) near the freezing front. Heaving involves thickening of this zone. Colder soil and any lenses present there move with uniform velocity equal to the rate at which the basal lense thickens.

The hydraulic model "works" because coupling of flows of water and heat is effectively the same in both models.

If q_w is the volumetric flux of water in the frozen zone, the rate of change of flux with distance at any point, $\partial q_w/\partial x$, is attributed to the rate of ice accumulation with time at that point.

$$\partial q_w/\partial x = -(\rho_i/\rho_w)(\partial\theta_i/\partial t) \qquad (11.37)$$

But if ice is accumulating at a rate $\partial\theta_i/\partial t$, latent heat is being liberated at a rate $\rho_i L(\partial\theta_i/\partial t)$, and this heat is joining the heat flux q_h which must increase with distance. Eliminating $\partial\theta_i/\partial t$ between these two "continuity equations" yields an equation that couples the flows of water and sensible heat:

$$\partial q_h/\partial x = -\rho_w L(\partial q_w/\partial x) \qquad (11.38)$$

In the hydraulic model, it is conveniently (but incorrectly) assumed ice pressure is everywhere zero (gauge). If solutes are neglected, the freezing-temperature equation then reduces to

$$\partial P_d/\partial x = \rho_w L(\partial T/\partial x) \qquad (11.39)$$

where $\partial P_d/\partial x$ is taken to be the driving force for liquid flow and $\partial T/\partial x$ is the driving force in the flow of sensible heat. With reasonable functions to represent a dependence of hydraulic conductivity on moisture content and of thermal conductivity on moisture and ice content, the computer can reconcile these equations with boundary temperatures to yield a time-dependent solution for freezing-induced redistribution. While this approach is comparatively straightforward for those adept in numerical methods it is highly objectionable to those who are concerned with microscopic realities (Bresler and Miller, 1975). The hydraulic model cannot cope with frost heave, expecially in relation to the effect of surface load on rate of heave. Perhaps, with judicious tinkering, the hydraulic model can be modified to yield some useful estimates of heaving behavior but not because it is physically realistic.

K. Regelation

A phenomenon known as *regelation* (refreezing) appears to be an important aspect of the mechanism of frost heave in SS and SSLS soils. The most familiar demonstration of regelation is mechanically induced. It involves a thin wire, with heavy weights at each end, draped over a block of ice with the weights dangling. Observers marvel to see the wire slowly sink through the solid block of ice. Ice beneath the wire is subjected to high pressure, which lowers the freezing temperature below ambient so that ice melts; melt water flows up around the wire and promptly refreezes so that the solid block of ice shows no evidence of the passage of the wire through it. Latent heat of fusion is absorbed as ice melts below the wire, and the temperature there falls. Latent heat is released as the melt water refreezes, raising the temperature above the wire, so that heat carried around the wire as latent heat by moving melt water returns across the wire by thermal conduction. The rate at which the wire will move, therefore, depends on the thermal conductivities of the wire, the water, and adjacent ice together with impedance to flow of melt water around the wire.

A related phenomenon, also mechanically driven, has been demonstrated in a device diagrammed in Fig. 11.14. A thin sheet of ice occupies the space between two rigid wettable filters. Outside the filters are chambers filled with H_2O. The device is immersed in a bath held at a temperature somewhat below $0°C$ so that H_2O in the end chambers is supercooled and ice pressure exceeds H_2O pressure. If H_2O pressure in one end chamber is raised somewhat, it will be observed that H_2O emerges in a steady stream from the other end chamber. Transport is attributed to concurrent formation of new ice and melting of old ice at opposite sides of the central chamber. It has been observed that temperature rises at the inflow face, and falls at the outflow face, in agreement with the idea of ice movement within the chamber as opposed to leakage of fluid along grain boundaries in the ice. This "ice sandwich" device can function as a semipermeable membrane (Miller, 1970). It can also be used as a permeameter for studying the transport of moisture in frozen soil held in the central chamber (Miller *et al.*, 1975). It demonstrates, on a macroscopic scale, how pore ice can move by a regelation process within the pores of SS soil. It demonstrates as well the coupling of transports of matter and heat when phase changes are involved.

If ice contains small pockets of unfrozen salt solution, a temperature gradient will cause these brine pockets to migrate in the direction of rising temperature, an example of *thermally induced regelation*. If the brine pocket remained stationary, the solute content of the brine at the solution–ice interface would have to be greater on the cold side of the pocket than on the warm side in order to satisfy the freezing-temperature equation. Thermal diffusion, however, would tend to equalize brine concentration, with H_2O

Fig. 11.14. The ice sandwich apparatus with supercooled pure water in the end chambers. With frozen soil in the middle chamber, the device serves as a permeameter for frozen soil.

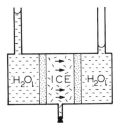

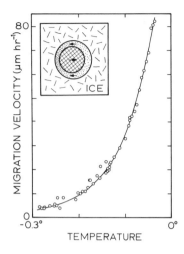

Fig. 11.15. Thermally induced migration of a glass bead (radius 13 μm) in ice with a temperature gradient of 1.13°C/cm. Particle moves from cold to warm ice. (After Römkens and Miller, 1973.)

diffusing toward the cold side and solutes diffusing toward the warm side, so that the freezing-temperature equation cannot be satisfied unless H_2O arriving at the cold interface freezes while a similar amount of ice melts at the warm interface where solutes are arriving. The result perceived would be migration of the pocket from cold to warm. This phenomenon explains why freshly formed sea ice, which normally contains occluded brine pockets, is "purified" by exposure to intense cold at the surface; brine pockets migrate downward and are expelled into relatively warm unfrozen seawater beneath (cf. Hoekstra *et al.*, 1965).

A vision of a soil particle surrounded by ions of a diffuse double layer suggests that such a particle ought to move through ice in a thermally induced regelation process similar to that of the brine pocket. Preliminary tests of this idea indicated that this actually happens (Hoekstra and Miller, 1967), leading to much more detailed studies (Römkens and Miller, 1973). Some data for migration of a silt-sized glass bead through ice are shown in Fig. 11.15. The rapid increase in migration velocity as the particle approached the 0°C isotherm was attributed to thickening of the film as temperature increased, reducing the impedance to moisture movement around the particle.[2]

---

[2] A puzzling aspect of the results was that the migration rate at any temperature did not appear to be directly proportional to the temperature gradient. All gradients were of the order of 1°C/cm. Further study may show this to have been an artifact, perhaps due to a systematic error in interpolated temperatures.

L. Heat Capacity and Thermal Conductivity of Frozen Soil

If the composition of a soil with respect to volume fractions of particles, water, ice, and air (θ_p, θ, θ_i, θ_a) remains virtually unaffected by a small change in temperature, the quantity of sensible heat ΔH that must be added to achieve a change in temperature ΔT is

$$\Delta H = \sum c_j \theta_j \, \Delta T \tag{11.40}$$

where c_j is the volumetric specific heat of component j, occupying volume fraction θ_j in the soil. Table 11.1 lists some values for c_j for typical components (cf. de Vries, 1963). The volumetric heat capacity for air is trivial by comparison with other components. Volumetric heat capacities of quartz (often the dominant mineral), some clay minerals, organic matter, and ice are each about half that of water. The volume fraction of all solids combined is given by $\theta_p = (1 - \theta - \theta_a)$. The volumetric heat capacity of the soil c_v is

$$c_v = \sum c_j \theta_j \simeq (1 + \theta - \theta_a)/2 \quad \text{cal/cm}^3 \, {}^\circ\text{C}$$
$$\simeq 2.1 \, (1 + \theta - \theta_a) \quad \text{MJ/m}^3 \tag{11.41}$$

If ice is present, a small increase in temperature ΔT causes some part of the ice, however small, to melt. If air is present there will be some increase in the mass concentration of water vapor $\Delta \rho_v$. Masses of H_2O involved are $\rho_i \Delta \theta_i$ and $\theta_a \, \Delta \rho_v$, respectively. Added sensible heat, ΔH will be partitioned between stored sensible heat in the amount $\sum c_j \theta_j \, \Delta T$ and sensible heat converted to latent heat in the amount $\theta_a L_v \, \Delta \rho_v - \rho_i L_f \, \Delta \theta_i$ where L_f is latent heat of fusion and L_v is latent heat of vaporization of water. $L_f = 79.7$ cal/gm deg = 333 KJ/Kg; $L_v = 596$ cal/gm = 2.49 MJ/Kg at 0°C. Near 0°C vapor densities are so low ($\rho_v = 5$ g/m³) that the terms involving $\Delta \rho_v$ can be neglected despite the very high value of L_v. The *apparent specific heat* of the soil $\tilde{c}_v$ will be

$$\tilde{c}_v = \Delta H / \Delta T = \sum c_j \theta_j - \rho_i L_f \, \Delta \theta_i \tag{11.42}$$

Thus for soil with high ice content $\tilde{c}_v$ will pass through a maximum in the temperature range in which most of the ice melts.

Elsewhere in this text one will find a summary of efforts by de Vries (1963) and others to develop formulas for estimating thermal conductivities of soils, including corrections for vapor transport induced by a temperature gradient in unsaturated ice-free soils. Approximate values of thermal conductivities of typical soil components are given in Table 11.1. Values for many silicate minerals are about one-third that of quartz.

De Vries (1963) presents a sample calculation for a soil in which pores

Table 11.1

WORKING VALUES OF THERMAL PROPERTIES OF SOIL CONSTITUENTS NEAR 0°C

Constituent	Thermal conductivity		Specific heat	
	(W/m °C)	(mcal/cm sec °C)	(kJ/kg °C)	(cal/gm °C)
Air	0.024	0.057	1.0	0.24
Water	0.56	1.34	4.2	1.00
Ice	2.2	5.3	2.1	0.50
Quartz	8.5	20.0	0.8	0.19
Common silicates	2.9	7.0	0.8	0.20
Organic matter	0.3	0.7	1.9	0.45

were fully occupied by water and geometrical factors were estimated from independent measurements, with $\theta_p = 0.573$, of which 89% by volume was quartz and 11% was "other minerals." The calculated thermal conductivity of the water-saturated sample was 6.0 mcal/cm sec °C. For the same soil when dry, the calculated result was one-tenth as large, 0.60 mcal/cm sec °C. Both results were in remarkable agreement with measured values for that soil. We have repeated this calculation with an assumption that the pore space was fully occupied by ice. We obtained a value about twice that of the water-saturated material, namely 12 mcal/cm sec °C. Thermal conductivities of air, water, and ice are in the ratio 0.044:1.0:3.8. Thermal conductivities calculated for the soil with pore spaces occupied by air, water, and ice were more nearly equal and were in the ratio 0.10:1.0:2.0. Unfortunately, there were no measurements with frozen soil.

Kersten (1949) measured heat capacities and thermal conductivities of comparatively large samples of soil with various degrees of compaction and wetting in both frozen and unfrozen states. His data show that freezing of nearly saturated samples increased their measured thermal conductivities by a factor of 1.35–1.45. This factor decreased, more or less linearly, to about unity as the degree of saturation decreased to about 0.3. He achnowledged, however, that there was evidence of freezing-induced redistribution in the samples; freezing destroyed the initial homogeneity so that the values reported are averages for nonuniform material.

In warm unsaturated soil, vapor pressure is high and vapor transport induced by a temperature gradient can be substantial. Since moving vapor transports latent heat, sensible heat absorbed in a region of net vaporization and transported as latent heat is released as sensible heat in a region of net condensation. In this way, thermal conduction of sensible heat can be augmented by convective transport of latent heat. Thus, in warm, moist soil in a closed chamber, the apparent thermal conductivity of a water-saturated soil may actually be increased if pore water, with high thermal conductivity, is partially replaced by air with low thermal conductivity (de Vries, 1963). This effect diminishes rapidly as temperature is reduced owing to the reduction of vapor pressure and the resultant decrease in vapor transport.

If the cool end of a moist column is cold enough for ice to form, however, the redistribution phenomenon appears. We have described a "capillary sink" mechanism to explain that water is drawn into the freezing zone from unfrozen soil as empty pores fill with ice. It has been shown (Beskow, 1935; Dirksen and Miller, 1965) that freezing-induced redistribution observed experimentally, Fig. 11. 11, can exceed by three orders of magnitude that which could be explained as vapor transport. Thus, in this circumstance, latent heat of fusion is being transported by moving water rather than latent heat of vaporization by moving vapor.

In a zone of ice accumulation, if heaving is not in progress, moving water is being converted to stationary ice, releasing a corresponding quantity of latent heat to join the flow of sensible heat. In this region, neglecting vapor transport, conservation of mass is expressed by

$$\rho_w \, \nabla q_w = -(\rho_i \, \partial \theta_i / \partial t + \rho_w \, \partial \theta / \partial t) \qquad (11.43)$$

Neglecting convective transport of sensible heat as trivial relative to convective transport of latent heat, conservation of thermal energy is expressed by

$$\nabla q_h = -\rho_i L_f \frac{\partial \theta_i}{\partial t} + \frac{\partial}{\partial t} \left(\sum c_j \theta_j T \right) \qquad (11.44)$$

M. Hydraulic Conductivity—SS Soil

In Chapters 8 and 9 of Hillel (1980), the reader will find a discussion of Darcy's law and its mathematical expression,

$$q_w = -K(\theta)_i = -K(\theta) \, \nabla(P_d / \rho_w g + z) \qquad (11.45)$$

where $K(\theta)$ is the coefficient of hydraulic conductivity which depends, as indicated, upon moisture content θ. Darcy's law is valid for SS soils over a wide range of moisture contents. In SLS soil, however, transport is much more complicated owing to interactions between surface ions, ions, and H_2O. We shall not attempt to discuss transport in SLS soil and possible roles of cracks occupied by air or ice. We shall restrict ourselves to flow of water in SS soil and shall limit ourselves to conditions in which net transport via adsorbed films is negligible in comparison with flow through capillary spaces that contain capillary water.

In ice-free SS soil, as $P_a - P_d$ increases, air intrudes into an interconnected network of relatively large pores, leaving the remaining spaces filled with capillary moisture. To some extent, air-filled spaces are bounded by exposed surfaces of adsorbed moisture with very low hydraulic conductivity but most of the conduction is in air-free capillary spaces and bypasses these films. Thus, it should make little difference if ice, a rigid body, were substituted for air, a fluid, in the large spaces. Perhaps, when moisture content becomes very low, the residual conductivity is primarily due to conduction through absorbed films but since the total quantity of adsorbed moisture is very small, the hydraulic conductivity can be so slight that it ceases to be of any importance and the question of drag against a rigid pore phase, as opposed to a fluid pore phase, is no more than an academic curiosity.

If this view is tenable, then the range of water contents in which K varies by several orders of magnitude is one in which the only question is to know what part of the capillary water has been replaced by air, or by ice, as the case may be. Invoking similitude, we infer that in SS soil if

$$[K(\theta)]_{aw} = [K(\theta)]_{iw} \qquad (11.46)$$

then

$$[K(P_a - P_d)]_{aw} = \{K[(\gamma_{aw}/\gamma_{lw})(P_i - P_d)]\}_{iw} \qquad (11.47)$$

As this is written, tests of similitude for hydraulic conductivity in frozen and unfrozen soils have not been made. Unpublished data for frozen soils only (Sahin, T., 1973) obtained with an ice-sandwich permeameter (Miller *et al.* 1975) and data of Williams and Burt (1974) resemble what one would expect the data for the unfrozen soils to look like over a substantial range of values of θ (See Fig. 11.16).

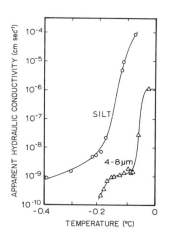

Fig. 11.16. Temperature dependence of apparent hydraulic conductivity of frozen soils as measured in permeameters by Williams and Burt (1974) for a silt and by Sahin (1973) for a 4–8 μm silt fraction. The flat portion may represent transport by ice, moving by regelation, at rates limited by the thermal conductivity of soil constituents.

Permeameter studies of $K(\theta)$ must take into account simultaneous transport in the form of pore ice moving by regelation. At high values of θ, this transport is undoubtedly negligible by comparison with fluid flow. At very low values of θ, the reverse may be true.

N. Stress Partition

Many years ago, Fisher (1928) and Haines at Rothamstead published a running dialogue on computations of the influence of moisture content on the tensile strength of systems of uniform spheres. Results were inconclusive but led Haines (1930) to his famous capillary model of hysteresis and to his method of measuring soil-moisture characteristic data. Their original objective, an understanding of tensile strength of moist soils, is directly related to initiation of ice lenses in soil during frost heaving.

Since the time of Haines and Fisher, however, soil physicists have hardly improved their understanding of tensile strength of moist soils. In a different context, geotechnical engineers have confronted the same questions and we shall adopt the method of Bishop and Blight (1963) to formulate the problem in terms of a stress-partition factor χ.

The scheme can be portrayed in terms of force balance at an inert wall pressed against a soil, Fig. 11.17. The wall stress, or *envelope pressure σ*, is assumed to be balanced by two components. One of these is reaction of the particle matrix. In SS soil, this is visualized as stressing of intergranular contacts and is called *effective stress σ* , because it will influence the resistance of the matrix to shearing deformation or to compression. The remaining component is reaction of the pore contents to the wall stress and has been called the *neutral stress σ_n*. If there is only one pore phase, then, to the extent that particle contacts approach point contacts, neutral stress will approach the liquid pressure, and in SS soil this is taken to be equal to P_d as measured by a tensiometer. At force balance, one can write

$$\sigma = \sigma_e + \sigma_n = \sigma_e + P_d \qquad (11.48)$$

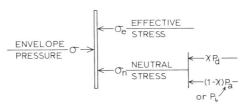

Fig. 11.17. The stress-partition scheme of Bishop and Blight (1963) for unsaturated ice-free SS soil and its analog for air-free frozen SS soil.

This is the *Terzaghi equation* for saturated soils, the fundamental equation of geotechnical engineering.

If there are two continuous pore phases, however, reaction of pore contents must be partitioned between the two. For example, if the pores contain capillary water and air, at different pressures, we can follow the scheme of Bishop and Blight and write

$$\sigma = \sigma_e + \sigma_n = \sigma_e + [\chi P_d + (1 - \chi)P_a] \qquad (11.49)$$

where the stress-partition factor χ is 1.0 in saturated soil; χ decreases to the limit 0.0 as moisture content decreases to zero.

If the second pore phase is ice, the concept of similitude implies that we can write, for air-free frozen SS soil,

$$\sigma = \sigma_e + [\chi P_d + (1 - \chi)P_i] \qquad (11.50)$$

According to this viewpoint, neighboring particles will be pressed against each other whenever effective stress is greater than zero. To initiate an ice lens in soil under some load σ, it would be necessary to reduce effective stress to zero if the particles are to move apart so that an ice lens can form between them. Since P_d will in general be negative, and since χ is less than unity, ice pressure within the pores must *exceed* the load if an ice lens is to be initiated.

To "break" unsaturated ice-free soil in which air pressure is zero (gauge), effective stress must be reduced to zero by a negative envelope pressure σ. This would represent the tensile strength of moist SS soil. Equation (11.49) reduces to $\sigma = \chi P_d$, where P_d is negative and χ ranges from 1.0 at saturation to values approaching zero as moisture content approaches zero. In effect, the dialogue between Haines and Fisher concerned efforts to calculate χ for systems of uniform spheres.

In SLS soil, the concept of effective stress is an artifact since the particles do not touch each other. Geotechnical engineers continue to use the term effective stress, however, to represent the difference between envelope pressure σ and water pressure P_d. In reality, this difference is an expression of the affinity of the matrix for water and corresponds to the "swelling pressure" of the SLS soil.

O. Frost Heave—Contrasts

If soil expands when it freezes, the volume change is called frost heave. Water expands 9.1% when it changes to ice but this fact is purely incidental to the process of heaving as it usually occurs in nature. To provide perspective, we shall describe what can be expected in terms of three examples.

1. COARSE SS SOIL

In nature, freezing of clean gravel or coarse sand does not result in frost heave. The only exception would be gravel or sand that is (i) saturated or nearly saturated with water and (ii) a *closed system*, i.e., one from which water cannot escape as freezing proceeds. In nature, it is very unusual to find both of these requirements satisfied; to become filled with water, the systems must be "open." One natural exception sometimes occurs after a lake in an area of permafrost has been drained. Before drainage, when the lake is unfrozen (in summer), it absorbs much more heat than the adjacent land where tundra, dessicated in the long days of summer, serves as surface insulation. Wetted by fall rains, the tundra is less effective as an insulator in winter so that it serves as a seasonal "check valve" that favors low soil heat content. This valve is absent for a lake. The result is that beneath the lake there is a deep depression in the *permafrost table*, the top of the permanently frozen zone, whereby the lake can be underlain by a deep pocket of water-saturated unfrozen sediment, a *talik*. If the lake is drained, the exposed bottom vegetates, and permafrost formation soon seals off the surface of the talik, which must eventually freeze, as well. As marginal freezing shrinks the talik, expansion of water being converted to ice is accommodated by uplift of the weakest part of the overlying "skin" of frozen lake bottom, characteristically that beneath a shallow residual pond left by incomplete drainage. As this blisterlike uplift occurs, water being injected by continued freezing of the talik also freezes, resulting in an ice-filled hill, mantled with frozen soil. Such a hill may be the most prominent feature of an otherwise flat landscape (Fig. 11.18). When formed in this manner, the feature is called a *closed-system pingo*.[3] Mackay (1978) has studied pingo formation in the field and reports that a bore hole through a growing pingo encountered a basal lens of water at high pressure. As water spurted from the bore hole, the pingo subsided until the hole resealed itself by freezing.

2. SLS SOIL

Freezing of highly colloidal materials is characterized by (i) the appearance of a network of intersecting ice-filled cracks and (ii) consolidation of ice-free soil between these cracks. Growth of the ice phase is mostly or entirely at the expense of water extracted from adjacent soil which shrinks accordingly. Thus the net expansion (heave) corresponds to the volume change on freezing of the water extracted by ice formation. Because the hydraulic conductivity of such soils is very low, it makes little difference whether the system is open

---

[3] A pingo is but one of many exotic features associated with permafrost. Ice-wedge polygons, stone nets, and stone stripes are examples of "patterned ground." "Icings" form when groundwater emerges at the surface and freezes above ground.

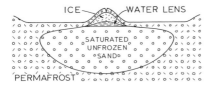

Fig. 11.18. Growth of an ice-filled hill (a closed-system pingo) in an area of permafrost. (After Mackay, 1978.)

or closed; there is little exchange of water with the surroundings. The dynamics of heaving of such materials have not been studied to any extent but it is useful to remark that the cracks are formed by ice growth, contrary to what some have postulated (the formation of cracks which subsequently fill with water and freeze).

3. FINE SS SOIL

Under favorable conditions, freezing of soils comprised of particles in the size range of silts and coarse (noncolloidal) clay can produce phenomenal heaves. Such materials are described as highly *frost-susceptible soils*. Favorable conditions include (i) little or no surface load, (ii) an open system, and (iii) a water table that is maintained at a level close to the freezing front. In the usual circumstance, heaving produces a succession of tabular ice bodies, *ice lenses*, which appear and thicken, one at a time, as a freezing front descends through the soil. Heave corresponds to the aggregate thickness of these lenses. When a new lens appears, its predecessor ceases to grow; heaving proceeds at a continuous smooth rate. The rate of heave decreases rapidly as surface load is increased or as water table is lowered. If the soil is wet, a closed system will produce heaving but at a rate which declines as the supply of water stored in the unfrozen zone is depleted.

The extraordinary heave of highly susceptible soils poses a challenge to soil physicists who seek to understand the mechanism involved. Succeeding sections are a response to this challenge.

P. Primary Frost Heaving

Until Taber (1930) published results of his experiments on frost heaving, it was generally inferred that heaving was due to "buckling" of soil, induced by expansion on freezing of moisture present. An ice lens was presumed to form after the fact; a crack formed by buckling filled with groundwater which, when frozen, became an ice lens. Successive ice lenses were presumed to be evidence of a succession of freezing cycles and it was thought that extensive heaving depended on repeated cycles.

Taber could not reconcile such reasoning with his field observations and

Fig. 11.19. Schematic representation of Taber's demonstration of frost heave as a continuous process accompanying progressive freezing induced by exposure of a loaded column to cold air. The aggregate volume of ice lenses corresponded to the volume of water that disappeared from the basal reservoir.

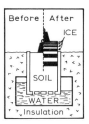

when electric refrigerators became readily available he proved his point, Fig. 11.19. He showed that a soil wetted with benzene or nitrobenzene would heave; these liquids *contract* on freezing. He showed that continuous exposure to continuous cold produced continuous heaving with appearance of a succession of ice lenses, formed one at a time. There was no perceptible change in the rate of heave as one ice lens ceased to thicken and a new one appeared and commenced to grow at a lower level. Tabor showed that with incompressible soil the total heave equaled the volume of ice lenses formed and was equivalent to the volume of water absorbed from a reservoir.

Taber's vision of lens growth is diagrammed in Fig. 11.20a. He reasoned that the ice must be held at a distance from the particles on which it rests if the lens is to thicken. He reasoned that as heat was lost (upward), some of the film supporting the lens would freeze, thickening the lens by reducing film thickness, whereupon the film would restore itself by imbibing moisture from adjacent pores. Extraction of moisture from these pores would induce a hydraulic gradient and upward flow in the unfrozen soil to sustain the process. Beskow (1935) developed these ideas in greater detail and proposed measurements of the "capillary" characteristics of a soil as a way of estimating the probable susceptibility of a soil to frost heave. Taber's mechanism

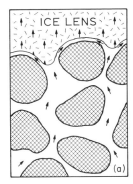 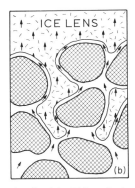

Fig. 11.20. (a) Primary heaving in the mode visualized by Taber, Beskow, and others. (b) Secondary heaving.

is perfectly plausible as far as it goes and is consistent with the surface-tension/double-layer model of moisture just described.

Two decades passed before Gold (1957) introduced surface tension of the ice–water interface as a factor in frost heaving. Penner (1959) brought the idea into clearer focus in relation to the Taber–Beskow model of lens growth. Miller *et al.*, (1960) extended it to include the osmotic model of film behavior and showed, by experiment, that the freezing temperature equation could be verified, at least within limits of geometric approximations for interfacial curvature. Everett (1961) provided a rigorous derivation which avoided geometric approximations in dealing with the role of surface tension, but he did not try to explain film behavior.

Gold (1957) had suggested that if ice grew through pore necks beneath the lens, the lens would become "anchored" to the underlying soil, and heaving would cease. According to this view, the limiting value of the ice pressure, relative to water pressure, would be fixed by pore-neck size. The critical value would be given by

$$(P_i)_{max} = P_d + 2\gamma_{lw}(1/\bar{r}) \tag{11.51}$$

where $\bar{r}$ would be the "effective radius" of pore necks below the lens. If this were true, one could measure $\bar{r}$ by air intrusion tests and Williams (1966) developed apparatus for this purpose. The implication was that if one could correctly anticipate what P_d would be for a given circumstance, one could calculate the maximum heaving pressure $(P_i)_{max}$ that could be developed.

This model has a serious limitation, however. It underestimates observed upper limits of heaving pressures (Loch and Miller, 1975). Neither can it be used to explain how a new lens could form at a distance below an existing lens and at precisely the moment that the upper lens ceases to grow. From our discussion of stress partition, it is evident that in ice-free soil below an ice lens where P_d is negative, effective stress would be high, even greater than that due to the load being lifted by heaving. To form a new lens, particles in contact must separate, and effective stress would have to be zero at that point if a new lens were to appear. It is hard to imagine how this could happen in ice-free soil. Hence, the Taber–Beskow model (known in Europe as the Everett model) of heaving, hereafter referred to as *primary heaving*, evidently cannot account for heaving in which more than one ice lens appears. If heaving is accompanied by the appearance of a sequence of lenses, the usual mode in heaving in nature, this must be attributed to another mode of heaving, *secondary heaving* (Miller, 1978), which occurs when the ice lens has developed a fringe that extends into the underlying pores, Fig. 11.20b.

It appears that primary heaving in the Taber–Beskow mode is usually limited to (i) SLS soil or (ii) ice lenses formed at a textural discontinuities in

SS or in SSLS soils. In both cases, primary heaving would give way to secondary heaving when the critical conditions for pore entry are exceeded. In SLS soil, this would not occur until soil beneath the ice lens had been desiccated (by ice segregation) and consolidated to its shrinkage limit, shifting the material to an SSLS or SS state. If this limit is not reached, however, the "single lens" that characterizes primary heaving may appear in a peculiar way. Three-dimensional consolidation that may accompany lens formation can be induced by a three-dimensional habit of growth. Instead of a single horizontal lens, a reticulate network of segregated ice, resembling mortar in a masonry wall, may develop, isolating blocks of ice-free soil. The segregated ice constitutes, however, a single continuous particle-free ice body. Since the segregated ice forms from moisture extracted from adjoining soil, which consolidates accordingly, total heave will correspond to the volume change of moisture converted to ice.

When a freezing front, moving downward through coarse SS or SSLS soil, encounters a stratum of much finer material, advance of the ice front may be arrested, and an ice lens may form at the top of the fine layer. This may be primary heaving, at least for a time, until ice begins to enter the fine pores, whereupon heaving reverts to the secondary mode.

Perhaps the most striking example of primary heaving at a textural discontinuity occurs at the soil surface. Isolated clumps of columnar ice, "needle ice," emerge like coarse whiskers from the face of the earth, especially during a clear frosty night. Each clump or needle may support a bit of debris which, chilled by radiation losses, nucleated freezing at that spot on the surface of wet ice-free soil.

Q. Secondary Frost Heaving

Gold's (1957) idea that growth of an ice lens must cease if it develops a fringe of ice within the underlying pores was an obvious first thought. If we remember, however, that there must be an unfrozen film between an ice lens and underlying particles to sustain heaving in the primary mode, then we have to concede that a similar film would exist between pore ice below an ice lens and the stationary particles which form those pores. In the case, the "anchoring" effect comes into question. The anchor would not be firm, it could "drag," allowing lens growth to continue with the ice of its frozen fringe moving, within rigid pores, by regelation. On closer examination of this situation, however, we come to a very different point of view, especially in circumstances that normally prevail. The frozen fringe within the pores is not a restraint on heaving; instead, the heaving "engine" lies within frozen soil below the ice lens when such a fringe is present! Primary heaving is then perceived as the limiting case of a more general mechanism already referred to as secondary heaving.

Development of the concept of secondary heaving is incomplete. It appears that all of the equations required for computer simulation of experimental data are known in principle (Miller, 1978). Successful tests however, must await proper evaluation of the stress-partition factor χ discussed in Section N. Numerical solution of the equations appears to be a formidable undertaking. It would be premature, therefore, to present the model in detail in this chapter. The concept does appear to explain how secondary heaving can occur and provides a basis for understanding the spacing and thickness of ice lenses. We shall discuss this matter in a qualitative way.

The reader will recall the movement of isolated particles imbedded in ice by thermally induced regelation. Particles migrated from cold toward warm ice and velocity increased as temperature increased. One would expect, therefore, that if the particles were held stationary, thermally induced regelation would tend to move pore ice in the opposite direction, from warm to cold. Thus, within stationary pores of a frozen soil, ice would tend to move in the direction of decreasing temperature. The tendency would be strongest in the warmest part, i.e., at the freezing front, at least to judge from the behavior of isolated particles. (See Fig. 11.21.)

The tendency for pore ice to move within pores between stationary particles will not coincide with the tendency of an isolated particle to move within stationary ice at the same temperature. The fact that ice does not completely surround each particle will alter matters. Broadly speaking, however, we expect that the tendency for movement will be high where the temperature is high and low where temperature is low, but will be enhanced by asymmetry of envelopment by ice—and hence enhanced by a gradient of ice content. However, to the extent that pore ice moves at all, it must move with uniform velocity since it is a continuous rigid body.

Let us suppose that for a given temperature gradient, load σ, and water table, there is a certain velocity of heave. All ice, throughout the frozen zone,

COLD

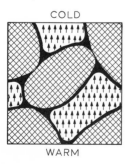

WARM

Fig. 11.21. Thermally induced regelation of ice in stationary pores of a "frozen fringe" in SS soil. The ice is a single rigid body throughout the frozen zone and moves with uniform velocity.

will be moving with this same velocity. Where the soil is cold, the tendency for ice to move (relative to the particles) is small, and if this tendency is small enough, the particles should be carried upward as the ice moves upward. Where the soil is warm, the tendency of the ice to move may exceed the velocity with which it is actually moving. In this region, the particles will tend to move downward despite the upward movement of ice in the pores. If the particles are prevented from moving downward by stationary particles below, however, each particle will exert a net force on its neighbor below. As a result, effective stress would increase, progressively, as we go downward through this region, reaching a maximum as we pass through the freezing front into unfrozen soil.

Somewhere between particles that are pressing downward in the warmer part of the frozen soil and the particles that are being carried upward by moving ice in the colder part of the frozen soil, there will be a zone in which the moving ice contains no particles, and this zone would be perceived as a growing ice lens. If the soil were cooling at all times, but the ice velocity is constant, we should expect the proper location for such a zone to move downward. Thus, the proper location for an ice lens would progress downward as cooling continued and the freezing front descended.

At first thought, we should expect any given particle to be "caught" and carried upward by the moving ice as soon as its temperature falls to that at which the regelation velocity of the individual particle became less than the ice velocity. If this were true, we should expect to find isolated particles distributed through the ice above that level in a more or less continuous way. At the moment each particle detached, the effective stress would be zero just beneath it. If we think about it carefully, however, we realize that the dynamics of the system are such that effective stress first reaches zero at a finite distance *below* the base of an ice lens. In other words, a whole layer of particles will be detached at one time, whereupon no more particles can be detached until a new ice lens appears at a certain distance farther down. Thus we shall see a series of discrete ice lenses interspersed with lens-free bands of soil.

The explanation for this behavior is found in the manner in which temperature and dialyzate pressure control ice pressure and moisture content (in air-free SS soil) and the manner in which moisture content, in turn, affects the hydraulic conductivity and the stress-partition factor χ in the frozen zone.

Let us look first at Fig. 11.22 to see what we should expect from air-free SS soil in which there is a temperature gradient but which is just prevented from heaving by a load, σ. In the example, σ equals the maximum ice pressure P_i; the surface temperature is $-0.2°$. The water table is so close to the $0°C$ isotherm that a tensiometer would measure a dialyzate pressure P_d of 0. Since there is no heave, there is no flow of moisture and P_d will be only that

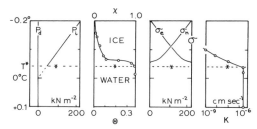

Fig. 11.22. Static profiles of dialyzate pressure, ice pressure, unfrozen water and ice content, the stress-partition factor, effective stress, neutral stress, envelope pressure, and hydraulic conductivity in the vicinity of the freezing front (∗) for a 4–8 μm silt fraction. Envelope pressure σ has been set exactly equal to ice pressure and is large enough to prevent frost heave. Solutes are negligible. (After Miller, 1978.)

due to gravity, which we shall neglect. Solutes are substantially absent from the dialyzate ($\prod_d = 0$). Using the freezing point equations, we can plot the equilibrium ice-pressure profile as a function of temperature. The moisture content profile in the figure is that for the 4–8 μm silt fraction studied by Koopmans and Miller (1966). If we assume that the stress-partition factor is equal, nearly enough, to the degree of saturation with unfrozen moisture, we can then calculate a profile for neutral stress σ_n as shown. This profile coincides with the profile of P_d in the ice-free soil where $\chi = 1.0$, but as χ decreases, the profile of σ_n branches away from the P_d profile and converges on the ice-pressure profile as χ approaches zero. As neutral stress increases, closely approaching the load σ, effective stress σ_e decreases, approaching zero. Some tentative data for hydraulic conductivity of this same soil are also shown, using a logarithmic scale because K decreases very rapidly as moisture content decreases.

In the example of Fig. 11.24, the system can almost heave; the neutral stress is almost high enough to lift the load σ. To initiate heaving, we have only to lower the temperature, thereby increasing ice pressure until σ_h reaches σ and σ_e reaches zero. Once heaving starts, however, we find that surface temperature must be lowered somewhat further to sustain heaving.

Figure 11.23 has been drawn to indicate schematically how things must be when heaving is in progress. Observe that in particle-free ice, the rate of heave and ice flux are one and the same. In the frozen fringe beneath an ice lens, ice flux is reduced by a factor equal to the fraction of space occupied by ice θ_i. The balance must be made up by water moving upward through the frozen fringe in accordance with Darcy's law. Since hydraulic conductivity decreases more or less exponentially in the frozen fringe, P_d must decrease at an ever-increasing rate. Since the temperature gradient is nearly linear in the frozen fringe and since the freezing-temperature equation tells us that the difference between P_i and P_d increases in proportion to the decrement of

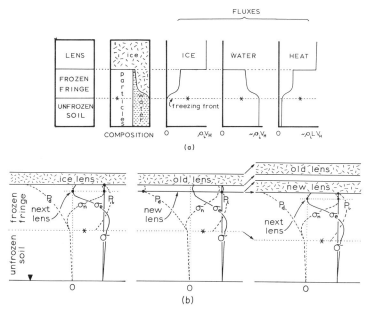

Fig. 11.23. (a) Schematic representation of conditions near the freezing front when secondary heaving is in progress. (b) Profiles of dialyzate pressure, ice pressure, envelope pressure, neutral stress, and effective stress near the freezing front during secondary heaving. At left, profiles represent the moment when a new lens is about to appear. At center, these profiles have shifted moments after a new lens has been initiated. At right, another new lens is about to appear at a lower level at some later time. (After Miller, 1978.)

temperature, the ice-pressure profile must also be strongly curved, passing through a maximum before declining to equal σ at the base of the ice lens where P_i must equal the load borne by the lens. As χ decreases, the profile of neutral stress σ_n again branches away from P_d and converges on the profile of P_i, also passing through a maximum. Where σ_n passes through a max-

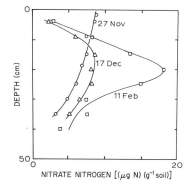

Fig. 11.24. Concentration of solutes near the freezing front began after 27 November and continued through 11 February in a soil near Guelph, Ontario, in the winter of 1975–1976. (B. Kay, unpublished report, 1976.)

imum, σ_e passes through a minimum. As the temperature falls, P_i rises, and σ_n rises. When the maximum value of σ_n reaches σ, the minimum value of σ_e reaches zero, satisfying the condition for initiating a new lens. This happens at a distance *below* the existing ice lens.

The new lens will form at a level at which ice pressure exceeds the load pressure σ. As soon as the new lens is established, however, its pressure cannot exceed σ, for that is the load being supported by the ice of the new lens. Hence, P_i drops as the new lens is established; σ_n drops and σ_e rises above zero, inhibiting the loss of any more particles into the moving ice. At the base of the new lens P_i remains constant and equal to σ, but as temperature continues to fall, P_i rises below the new lens, and in time we repeat the lens-initiation cycle at a lower level.

This vision of heaving provides a qualitative rationalization for the large influence of surface load in reducing the rate of heave. Before an ice lens can form at any given rate of heave, neutral stress must rise to a level equal to the load. If the load is increased, this can only be achieved by a reduction in velocity of heave which, by reducing the gradient of P_d, allows the maximum ice pressure to rise high enough to achieve the required increase in neutral stress. This is not a simple function, however, owing to the involvement of the variable stress-partition factor. This explains, no doubt, why efforts to find empirical relationships between load and rate of heave have succeeded only in carefully regulated circumstances (Penner and Ueda, 1978; Linell and Kaplar, 1959).

It is evident that, according to this view, the real heaving engine, in secondary heaving, is thermally induced regelation in the frozen fringe below the thickening ice lens. The fact that effective stress reaches zero at the plane where a new lens is about to appear signifies that, at this level, thermally induced regelation is just keeping abreast with an ice velocity induced in warmer soil below. Regelation at the level of incipient lens formation is contributing nothing to the work being done. Hence, the appearance of a new lens is only incidental to the workings of the engine and will not significantly perturb heaving rate. This agrees with observations. We now view the ice lens not as the working element of the engine but as a feature of its exhaust system. A series of ice lenses can be compared to a series of discrete drops that form and fall from the tip of a capillary tube in which flow is driven by a pressure gradient in the tube and is opposed by fluid viscosity inside the tube. The drops are highly visible evidence of flow but are purely incidental details of the exhaust system. In secondary heaving, ice lenses are visible evidence that ice is moving with uniform velocity, being driven by thermally induced regelation below the lens itself, and are incidental to the heaving mechanism.

According to this view, primary heaving is simply a limiting case of secondary heaving in which thermally induced regelation (or more properly, thermally induced gelation) involves only the upper hemispheres of particles just beneath the ice lens, a view consistent with the Taber–Beskow vision of heaving.

To see how this vision of secondary heaving stands up to further development and testing, the reader will have to watch the literature. Developments may be slow. In our narrative, we bypassed details such as the influence of expansion of moving moisture that is being converted to moving pore ice in a frozen fringe that is cooling. This can produce expulsion of water at the bottom of the frozen fringe at the same time that heaving is occurring at the top, especially when the freezing front is descending rapidly. Although each equation describing each detail is elementary, their multiple interactions defy analytical solutions. Numerical solutions will be a tedious and intricate matter.

R. Freezing and Solutes

In much of the discussion so far it has been convenient to minimize complications that arise when soil moisture contains a significant amount of soluble material. These complications can be severe. Ice tends to form as a pure phase. Thus, if solute-bearing water is moved upward by freezing-induced redistribution or by heaving, or both, solutes tend to accumulate in water in contact with ice that is forming. While the resultant gradient of concentration will tend to induce diffusion of these solutes back in the direction from which they came, the interactions of rate of heat extraction, water flow, and solute diffusion become complicated indeed. One can visualize that with very rapid rates of heat extraction, solute-enriched layers may be bypassed and trapped behind the advancing freezing front. These trapped brine pockets will, however, tend to migrate through the frozen soil in the direction of the freezing front, and if the front slows and stops, the solutes may arrive there in due course and be ejected into the unfrozen zone. Data in Fig. 11.24 illustrate the development of a solute-enriched zone in the vicinity of a freezing front that has advanced downward into an agricultural soil. Such enrichment may cause precipitation of solutes as freezing proceeds. One can understand that such phenomena can be of importance in soils, especially in relation to irreversible processes such as fixation of potash or phosphate.

These processes have a certain relationship to behavior of solutes in a soil being dried by surface evaporation. Crystallization of solutes within a soil mass will be subject to limitations associated with surface tension of the solution–crystal interface. The mechanics of nucleation and growth of crystals from solution will resemble, in certain ways, nucleation and growth of ice lenses (Miller, 1973b). It is known that large lenticular crystals of pure "segregated" gypsum can form in soil. It is not hard to visualize their growth as a variation of the primary heaving mechanism. Salt crystals that appear as efflorescent crusts on the surface of wet saline soils, seem to be an analog of "needle ice." It is tempting to speculate that the crystalline linings of apparently closed cavities in porous rocks (geodes) may have formed in the manner of needle ice.

S. Closure

Efforts to understand freezing phenomena in soils are justifiable in themselves. At the same time, such efforts tend to broaden the perspective with which we view phenomena that occur in unfrozen soils. We have tried to emphasize the importance of reconciling ideas relating to freezing with ideas about drying. Some of our thoughts may not survive closer scrutiny. We hope that some readers will find inspiration to uncover the real truths, however they may differ from the as yet sparsely documented intuitive relationships that have been recounted here.

This chapter is not a scholarly review of landmark papers. Most papers have been cited to document a detail, while others have been mentioned to provide a point of entry into a body of relevant literature. Conspicuously absent are references drawn from the vast literature on permafrost and soil freezing that has developed in the Soviet Union. The language barrier has not been penetrated by translations of selected articles, and a reader quickly realizes that few translators can cope with the jargon and conventions developed by isolated schools of thought. If readers are dismayed by disarray among soil physicists who write about soil and water in English, they must sympathize with a translator who evidently encounters similar problems among scientists who write in Russian. A valiant effort is the English translation of Tystovich's book, as edited by G. K. Swinzow, published in 1975 under the title *The Mechanics of Frozen Ground* (Tystovich, 1975). Proceedings of the Third International Conference on Permafrost (1978) include some 45 current papers (in Russian, with abstracts in English and French). These two volumes can lead readers into the Russian literature. Those who enter this literature will find that Soviet scientists attribute frost heave to mechanisms resembling the secondary heaving mechanism described in this chapter but evidently differing in important details. Their concepts of freezing-induced redistribution evidently involve diverging viewpoints, some of which resemble the one presented in this chapter, at least in some degree.

Their concepts of adsorption and capillary effects are similar but not as elementary as the concepts used in this chapter. Among readers of this chapter may be those who will succeed in unifying the works of the two largely independent efforts to understand the behavior of freezing soils.

Acknowledgment and a Recent Development

Some suggestions made by Professor B. Kay of the University of Guelph during the preparation of this chapter were gratefully accepted; perhaps others ought to have been.

As this chapter goes to press it can be reported that Dr. Kevin O'Neill, of the U.S. Army Cold Regions Research and Engineering Laboratory has succeeded in obtaining numerical solutions of equations for the model of secondary frost heave described in Section Q. An example of his results will appear in proceedings of the Second International Symposium on Ground Freezing, Trondheim, June 1980.

12   *Similitude and Scaling of Soil-Water Phenomena*

E. E. Miller

Departments of Physics and Soil Science
University of Wisconsin
Madison, Wisconsin

A. Introduction

The analysis of general problems in applied physics should begin with a preliminary process known as scaling. This step simplifies problems by expressing them in the smallest possible number of "reduced" variables. Any solution for a problem that has been worked out in reduced formulation holds true for an infinite variety of actual systems which, although they differ physically, are simply "scale models" of each other.

1. SCALING EXAMPLE: SIMPLE PENDULUM

To clarify the scaling concept, let us take a familiar example from elementary physics: analysis of the simple pendulum. Call the bob mass m, the string length l, and the string angle measured from the vertical α. For $t = 0$, denote by α_0 the initial value of α, and by $\dot{\alpha}_0$ the initial value of its time rate of change $\dot{\alpha}$. The force of gravity component along the string-constrained path is $mg \sin \alpha$, which by Newton's law equals m times the acceleration $l\ddot{\alpha}$ along this path. The m's cancel out, giving for the differential equation of motion

$$l\, d^2\alpha/dt^2 = -g \sin \alpha \qquad (12.1)$$

We can simplify this equation a bit further by "reducing" it to hide from view the two dead-wood parameters which stay fixed for any given pendulum, i.e., l and g. You might say this step consists of "dissolving" all possible passive parameters into the active variables of the differential equation. In this example, since only α and t are active variables the choice is obvious: α is dimensionless already so only t can be used as the "solvent" for getting rid of l and g. Except for a scale factor we shall want the reduced time t_* to behave just like the physical time t; accordingly t must enter with the power $+1$ into the expression for t_*. We shall hold to this plan for every reduced variable in this discussion. We shall also denote each reduced variable by the symbol of its corresponding physical variable, adding an asterisk as a subscript. From the form of Eq. (12.1) it is clear that there is only one possible way to dissolve l and g into t:

$$t_* \equiv \{t/\sqrt{l/g}\} \qquad (12.2)$$

Observe that since Eq. (12.1) is dimensionally homogeneous and α is already dimensionless, t_* will automatically be dimensionless. Note also the curly brackets which we will consistently employ to denote the reduced combination of a physical variable with its appropriate reducing parameters. The denominator $\sqrt{l/g}$ is seen to be a "characteristic" physical time for a physical system having particular values of l and g; it is of course the "period" of the pendulum except for a factor of 2π. The reduced version of the differential equation (12.1) is obviously the simplest possible version, i.e.,

$$d^2\alpha/dt_*^2 = -\sin \alpha \qquad (12.3)$$

with initial conditions $\alpha \to \alpha_0$, $\dot{\alpha} \to \dot{\alpha}_0$ as $t_* \to 0$. It is slightly simpler of course, but it is also more sophisticated.

2. TELESCOPING OF SOLUTIONS

Any solution that we can work out for given initial conditions is immediately applicable to a host of distinct physical systems which, though they must all be alike in α_0 and $\dot{\alpha}_0$, may differ in their values of the physical parameters m, l, and g. This collapsing of an infinity of physical systems into one reduced solution which quantitatively describes them all is referred to as the "telescoping" of many physical solutions into a reduced equivalent.

We could have obtained the reduced time expression (12.2) by a more elementary procedure known as dimensional analysis, which (crudely speaking) consists of simply writing down all the variables and parameters that enter a problem and then trying to clump them together into dimensionless groups. But dimensional analysis yields nothing that cannot be obtained by the preceding method which is known as similitude analysis, i.e., directly reducing the appropriate differential equations themselves, along with the attached boundary conditions of a problem, so that all physical solutions—calculable or not—will also be in reduced form and therefore telescoped in terms of these similitude-reduced variables. In the far more complicated problem of scaling systems of porous media, one cannot get very far at all with dimensional analysis. In that problem, most of the reduced variables which are developed from similitude analysis contain various combinations of a microscopic characteristic length λ representing particle size and a macroscopic characteristic length L representing soil profile depth. Both being lengths, λ and L are completely indistinguishable by dimensional analysis, but they are easily sorted out by similitude analysis. The results are astonishing to most people on first encounter, but of course experiment has proven the pudding or you would not be reading this.

B. Microscopic (Pore Level) Scaling of STVF Porous Media

For an unsaturated soil in which the pore spaces are not too small, the shapes of the air–water interfaces are governed by the laws of surface tension, while the flow patterns within the water-filled spaces obey the laws of viscous flow. The following analysis is designed for media and conditions for which these assumptions constitute useful approximations. We shall use the abbreviation STVF media (for surface tension, viscous flow). For these media the similitude analysis, involving two laws (two differential equations), will be much more complex than that for a pendulum, but the principles are exactly the same.

1. Shapes of Microscopic Interfaces

For any small bit of air–water interface, the surface tension σ pulling all around the rim of the bit generates a resultant normal force component, provided that the surface is bent. This resultant is balanced by the normal force generated by a pressure difference between the two sides of the interface acting upon the surface area of the bit. The bending of the bit is correctly described by the "Gauss mean curvature" of the interface, which we will call $1/\langle r \rangle$. For a simple spherical shape, $1/\langle r \rangle$ reduces to the reciprocal of the sphere radius. (The angular brackets just denote the concept[1] of *mean* curvature for a general surface shape.) Balancing these two normal forces gives a familiar equation for porous media,

$$p = 2\sigma(1/\langle r \rangle) \tag{12.4}$$

where p denotes the water pressure relative to air. (The sign of p is thus negative for unsaturated media, as is the sign of $1/\langle r \rangle$; σ is taken to be positive.) This may not look like a differential equation, but $1/\langle r \rangle$ is the mean of two curvatures, each of which represents a second derivative. We know what to do now. A differential equation has bobbed up, so we have to "dissolve" any system parameters that appear—σ in this case—into the action variables. From a *macro*scopic viewpoint the suction varies from place to place in a soil profile, but on the *micro*scopic scale of a pore, p is essentially constant. In one sense, Eq. (12.4) is a relation between three fixed parameters—suction, surface tension, and interface curvature. Given the suction and the surface tension, Eq. (12.4) evaluates for us the microscopically constant value of $1/\langle r \rangle$. In another (microscopic) sense, Eq. (12.4) is a differential equation for the shape of a surface characterized by the constant curvature $1/\langle r \rangle$ at all points. To solve it we need its boundary conditions, which consist of the terribly complicated shapes of all the solid grains of the porous medium along with a specified contact angle γ for all intersections of the interface with the surfaces of the solid grains. We may know or assume something about the contact angle γ, but there is no way to know the solid shapes—or if we did know, no remotely practical way to solve for the ornate interface shapes obeying $1/\langle r \rangle$ is a specified constant.

2. SIMILAR MEDIA; SIMILAR STATES

Is there anything we can do in this messy situation that will allow us to draw useful scaling conclusions from a similitude analysis? There had better be and there is. We can say, suppose we made a scale model copy of this matrix of solid grains, identical in every detail except for the difference of scale. Could we then think about scale modeling the interface shapes as well and learn something useful that way?

Media which are exactly alike geometrically except for scale have the same relation to each other as do similar triangles. Hence we shall call

---

[1] The Gauss mean curvature is obtained from any two mutually perpendicular planes whose intersection is normal to the interface; the interface must cut each of these two planes in an ordinary curve, and $1/\langle r \rangle$ is then defined as the arithmetic mean of the curvatures of these two ordinary curves (taken to be negative in the normal unsaturated soil condition, i.e., concave on the air side).

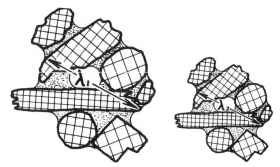

Fig. 12.1. Illustration of two "similar media" in "similar states." Note that the two characteristic lengths λ_1 and λ_2 connect corresponding points in the two media. (From Miller and Miller, 1956.)

them, by analogy, *similar media*. Figure 12.1 is a drawing of a medium (with some reasonable looking air–water interfaces as well) which has been reproduced photographically at two different scales. An arbitrary microscopic characteristic length λ has been included in the drawing, and the two subscripts 1 and 2 have been inked in afterward to distinguish the two. These drawings serve to exemplify two such similar media with their respective scales characterized by λ_1 and λ_2. The inclusion of the interfaces in the original drawing means that the two media are also shown in what we shall call similar states. If we multiply Eq. (12.4), by the parameter λ we can put it in a reduced form that telescopes the behavior of all physically different similar media having the same values of contact angle γ,

$$p_* = 2(1/\langle r \rangle_*)\qquad(12.5)$$

where

$$1/\langle r \rangle_* \equiv \{\lambda(1/\langle r \rangle)\}\qquad\text{and}\qquad p_* \equiv \{(\lambda/\sigma)p\}$$

When we magnified the picture in Fig. 12.1 we also magnified the radius of curvature of the interfaces by the same factor, and of course we did not change γ. This idea is expressed generally by Eq. (12.5), i.e., the concept that two similar media with the same γ's can indeed also be in similar states provided that the reduced pressure $\{(\lambda/\sigma)p\}$ is the same for the two.

Can be. Whether they *are* depends on their histories, too, as we should know from our awareness of hysteresis behavior in porous media.

3. Scaling with Hysteresis; Haines Jumps

What about hysteresis? For STVF media, the entire root of hysteresis lies in the multistable nature of the solutions of Eq. (12.5). The familiar concept of the ink bottle pore is shown in Fig. 12.2. Anywhere between

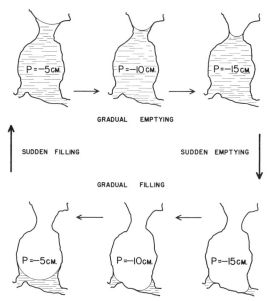

Fig. 12.2. The shape of the gas–liquid interface is determined not only by the differential pressure but also by the previous history of this pressure. (From Miller and Miller, 1956.)

$p = -5$ cm and $p = -15$ cm there are two possible solutions of Eq. (12.5); one a "full" state, the other an "empty" state. At -5 cm the empty state becomes unstable and executes a sudden and irreversible "Haines jump" to the full state at $p = -5$ cm. Conversely at -15 cm the full state Haines-jumps to the empty state. In practice these jumps occur in milliseconds, so the pressure at which they occur is independent of the time rate of approach to that pressure. In other words, the state of the interface (whether full or empty in Fig. 12.2) certainly depends on the history $p(t)$, but this dependence is time-scale invariant. We can stretch or shrink the time scale without affecting the interface shape that corresponds to a given stage of p. Now if we convert the foregoing discussion to reduced geometry (dividing all lengths by λ and all pressures by the characteristic pressure σ/λ), the same conclusions (in terms of p_*) become applicable to all similar media.

Actual media are much more complicated than an assembly of ink bottle pores with Haines jumps all neatly paired off. However, if we imagine a reduced geometrical picture of the solid grains, for each p_* we will have a certain (finite) number of interface shapes possible as solutions for Eq. (12.5). Now strain your imagination a bit to think of all these solutions as a continuous function of p_*, really a four-dimensional model. At discrete values of p_*, one of these multiple solutions becomes unstable and will Haines-jump over to some other solution. If we start with one solution, then change p_* with time in some sequential plan $p_*(t)$, this model will serve as a conceptual road map. Following the sequence $p_*(t)$ we can find out which of the many possible solutions will be in effect at any desired stage of the history $p_*(t)$. Since we have made the argument in terms of a reduced model, the end result telescopes all physically different but similar media into the same final reduced solution for interface shape at any stage expressed in terms of p_*. (Because of time-scale invariance, we need not worry about reducing the time; its scale does not matter in this argument. Time serves only to keep the changes of p_* in the same *order* of succession.)

C. Macroscopic Scaling (Averaged over Many Pores)

1. Reduced Wetness Characteristic $\theta_H(p_*)$

Let us now use the reduced microscopic geometry of the interfaces to draw a simple conclusion at the macroscopic level for similar media. Think of the foregoing reduced model of a given medium—all lengths having been divided by λ—and of the reduced interface shapes determined after a given history of $p_*(t)$. From these microscopic shapes we can integrate over the water-filled volumes to find the fraction of volume that is occupied by water. This water-volume fraction is the familiar θ. Being a (volume/volume) fraction derivable from our reduced model, it is obviously both reduced and dimensionless—as was α in our pendulum example.

Over what size of region should we try to define θ in this way? Obviously it must be bigger than a pore or it will not vary smoothly with position (or time), but the region should nevertheless be small enough that p_* will not vary significantly within its borders. This is the usual tightrope act of defining macroscopic properties, of whatever nature, from discrete microscopic entities, whether pores or atoms; we need not dwell on it. Thus θ can sensibly be associated with a point at the center of such a midsize region and, macroscopically, θ will then vary over system-size distances, along with p_*. At one point in space, dependence of θ upon p_* is of course a history-dependent relationship, time-scale invariant in character. It is not a function but a functional, a time-scale invariant functional of p_*. This is too many words, so we will abbreviate this as a *hysteresis function* of p_* using a capital subscript H: $\theta_H(p_*)$.

Since we saw earlier that the development of detailed interface geometries corresponding to a given $p_*(t)$ history represented all STVF similar media at once, so also must $\theta_H(p_*)$, which is integrated directly from the reduced model, telescope the $\theta_H(p)$ characteristics of all similar media.

Expressed in terms of p_*, all similar media must exhibit identical static moisture characteristics, $\theta_H(p_*)$.

2. STATISTICALLY SIMILAR MEDIA VERSUS EXACTLY SIMILAR MEDIA

This would be an important conclusion except for one blemish. Where is one going to find two media that are *exactly similar* in microscopic geometry? Fortunately, if our only interest is in macroscopic conclusions, we can relax this impossible condition. The way to see this is to think for a moment about, not *similar*, but *identical* media. Take a barrel of soil, mix it; remove a sample; do an experiment to measure some desired aspect of $\theta_H(p)$—say, the main-branch hysteresis loop. Now take out a second sample and repeat. Within experimental limitations one cannot see any difference; the static characteristics $\theta_H(p)$ are always the same for identical samples taken out of this barrel. But the *micro*scopic geometries of any two such "identical" samples differ completely. It is for average or macroscopic aspects that we describe the two samples as identical. Fair enough. Why not just extrapolate this easygoing statistical viewpoint from *identical* to *similar* media? One sample from the barrel, after being magnified in exact detail by some factor, can be regarded, statistically, as "similar" to another sample freshly taken out of the barrel. It is this averaged or statistical concept that we will henceforth mean when we speak of similar media. (In certain microscopic arguments one may at times still want to use a modifier: *exactly* similar media.) Now our conclusion that $\theta_H(p_*)$ is the same for all similar media has practical consequences of obvious practical interest.

3. WATER FLOW PATTERNS WITHIN WET PORES

We are ready now to confront the dynamic or flow aspect of the problem. As previously stated, we shall postulate viscous flow within the water-filled microvolumes within the soil. The classical flow equation of Navier–Stokes can be truncated by omission of the kinetic energy and momentum dependent terms, leaving only the viscous drag term, a superb approximation for slow flow through tiny channels. With primes to indicate the microscopic point of view, this equation becomes

$$\mathbf{f} - \nabla p'_w \equiv \mathbf{f}'_T = \eta \, \nabla^2 \mathbf{v}' \tag{12.6}$$

where $\mathbf{f}'_T$ denotes the total force per unit volume on a small bit of water moving with velocity $\mathbf{v}$, and η is viscosity. The body force (gravity, in practice) acting on the bit is denoted $\mathbf{f}$ (being constant it needs no prime), and the water pressure is denoted p'_w. At this point we shall assume that for most practical purposes the air pressure p_a is independent of position within a soil. Therefore for differential operations such as $\nabla p'_w$ we can replace p'_w with $p' \equiv p'_w - p_a$.

To achieve similitude of flow patterns, we must obviously consider appropriate requirements on the boundary conditions. These are set geometrically by the solid grain walls and the interfaces, and physically by some condition on $\mathbf{v}'$ as one approaches the microboundaries of the water-filled spaces. Clearly there is no possibility for similitude of flow patterns in two systems unless the complicated boundaries of the water-filled spaces are geometrically similar, i.e., unless the two systems are composed of (exactly) similar media in (exactly) similar states. Physically the usual boundary condition for viscous flow in liquids is that $\mathbf{v}' \to 0$ as solid–liquid boundaries are approached, but for the air–liquid interfaces one might not expect the same condition. Rather one might think that the surface would be completely free to move along the interfaces. However, the speed of small air bubbles rising through water can be explained only with the nonslip assumption, i.e., that $\mathbf{v}' \to 0$ on the bubble surface. On the grounds of this and other evidence, we shall adopt the simple assumption that the same $\mathbf{v}' \to 0$ condition holds at water–air boundaries as at water–solid boundaries.

4. Darcy's Law from Flow-Pattern Similitude Made Macroscopic

With this boundary condition Eq. (12.6) is linear[2] for $\mathbf{f}'_T$ versus $\mathbf{v}'$. If we have one flow pattern $\mathbf{v}'(\mathbf{r}')$, $\mathbf{f}'_T(\mathbf{r}')$ that satisfies Eq. (12.6) and the boundary condition, then $A\mathbf{v}'(\mathbf{r}')$, $A\mathbf{f}'_T(\mathbf{r}')$ is also a solution, where A is any constant. Obviously when we average the flow velocities and driving forces over macroscopic regions (as we did for θ), this microproportionality yields Darcy's law: Macroscopic $\mathbf{v}$ is proportional to macroscopic $\mathbf{f}_T$. It is a straightforward consequence of similitude applied to Eq. (12.6) and its boundary conditions. A few words about the averaging, which is often discussed at great length—but not here. First the driving force: When $\mathbf{f}'_T$ is broken down into $\mathbf{f}$ and $\nabla p'_w$, the body force $\mathbf{f}$ is of course constant and being due to gravity (unless you want to consider centrifuging), could have been written as $\rho\mathbf{g}$, where ρ is the density of water and $\mathbf{g}$ the acceleration of gravity. The water pressure p'_w can be resolved into $p + p_a + \delta p'_w$, where the air pressure p_a has already been assumed constant, even macroscopically, for this discussion, and p, as used earlier, is the tensiometer pressure. It is constant on the scale of pores. The residue $\delta p'_w$ just signifies some small, pore-scale ripple responsible for speeding up the flow through the narrower spaces; it averages to zero. Thus macroscopically we can simplify the driving force to $\mathbf{f} - \nabla p$. Second the velocity: If we average (over a plane of macroscopic size—the tight rope act again) the normal component of $\mathbf{v}'$ to get

---

[2] It is also linear if one assumes a "surface viscosity" of any stiffness from zero (free surface) to infinity (our previous assumption).

flux density of water flowing across the plane, then tilt this plane every which way, we shall always find one direction giving a maximum flux (volume rate of water passing per unit area). This defines both a direction and a size (flux density) for the macroscopic velocity **v**. For isotropic media the directions of **v** and $\mathbf{f}_T$ must correspond. (For anisotropic soils we must deal with a tensor form of Darcy's law, which is not given here.) Now we are ready to write down Darcy's law formally for this discussion,

$$\mathbf{v} = K_H(p)(\mathbf{f} - \nabla p) \tag{12.7}$$

Of course K denotes the Darcy constant of proportionality between **v** and $\mathbf{f}_T$, but where did the hysteresis function of p come from? Obviously K must depend on the size, number, and tortuosity of connected, water-filled channels through which flow occurs, and these in turn depend on the macroscopic interface shapes, which we have already seen depend hysteretically upon $p(t)$. Thus K must be a hysteresis function of p, as was indicated in Eq. (12.7).

5. Scaling Darcy's Law for Similar Media

Having used the concept of similar media for scaling of static characteristics we cannot stop there. What can we predict for the dynamic characteristics of similar media? Again we must introduce the microlength λ into the microdifferential equation [the truncated Navier–Stokes equation (12.6)] in order to "dissolve" into the action variables the visible system parameters, in this case only the viscosity η. The ∇^2 operator in Eq. (12.6) represents a second derivative (spatial); therefore it is a reciprocal length squared. Temporarily, let us reference the driving force to any arbitrary corresponding value f_0 and thereby reduce Eq. (12.6) in the form

$$\{\mathbf{f}_T'/f_0\} = \{\lambda^2 \nabla^2\}\{[\eta/(\lambda^2 f_0)]\mathbf{v}'\} \qquad (\mathbf{v}' \to 0 \quad \text{on boundaries}) \tag{12.8}$$

A reduced flow pattern $\mathbf{v}_*'$, $\mathbf{f}_{T*}$ that satisfies (12.8) is a telescoping of many physical systems with different values of λ, σ, and η (f_0 being an arbitrary choice here, not a physical parameter of the system). For macroscopic comparison we can first take f_0 to be the magnitude of the macroscopic driving force, i.e., of $\mathbf{f}_T \equiv (\mathbf{f} - \nabla p)$. Therefore patterns of special-reduced velocity $\{\eta \mathbf{v}'/(\lambda^2 \mathbf{f}_T)\}$ that satisfy (12.8) apply to all similar media in similar states which are subjected to the same reduced macroscopic driving force $\{\mathbf{f}_T/f_T\}$. Averaging the microscopic $\mathbf{v}'$ into the macroscopic **v** does not disturb this relation; i.e., the similar media above all have the same value of $\{(\eta/\lambda^2)(\mathbf{v}/f_T)\}$ when driven by the same $\{\mathbf{f}_T/f_T\}$. But the magnitude of **v** divided by the magnitude f_T is just the Darcy K. Hence we deduce that for all similar media in similar states the quantity $\{(\eta/\lambda^2)K\}$ has the same numerical value. Further, since similar media will also be in similar states if their histories $p_*(t)$ are the same (within the time-scale invariant loophole), we can reduce K as a hysteresis function of p_*:

$$K_{H_*}(p_*) \equiv \{(\eta/\lambda^2)K_H(\{p\lambda/\sigma\})\} \tag{12.9}$$

The hysteresis function is exactly the same for all similar media (for liquids giving the same γ).

D. Telescoping of Flow System Solutions

Darcy's law (12.7) is a macroscopic relation containing a gradient operator. In order to achieve a differential equation with which we can solve general boundary-value problems for whole flow systems—soil columns or profiles, for example—we must first bring in additional simple knowledge that we have—that matter is conserved—in the well-known differential form of the continuity equation,

$$\partial\theta/\delta t = -\operatorname{div} \mathbf{v} \tag{12.10}$$

Plugging in for $\mathbf{v}$ from Darcy's law (12.7) gives the Richards equation,

$$\partial K_H(p)/\partial t = -\operatorname{div}[\theta_H(p)(\mathbf{f} - \nabla p)] \tag{12.11}$$

Here we have *macro*scopic differential relations (12.7), (12.10), and (12.11), which we again want to reduce in order to telescope solutions for many different physical systems into one corresponding reduced solution.

We see divergence and gradient operators, which are of course reciprocal lengths, appearing in these equations. But—and here is an absolutely crucial point—we are now back to the tightrope act in which our concern is with rates of change, on a *macro*scopic scale, for quantities such as θ, $\mathbf{v}$, and p, quantities which already represent *averages* over a large number of individual pores. Therefore the microscopic length λ is no longer a relevant reference for these operators. In one sense, λ has been used up in the similitude and averaging used to obtain $\theta_H(p_*)$ and $K_{H_*}(p_*)$. Since we would obviously like to scale-model whole flow systems which might differ in size—scale the soil profile depths, for example—we shall now introduce a new length, a *macro*scopic characteristic length L, for comparing such flow system solutions. At once we see that ∇_* and div_* must become $\{L\, \nabla\}$ and $\{L\operatorname{div}\}$.

Applying this first to Darcy's law (12.7), which incorporates our previous conclusions for reducing p and $K_H(p)$ in order to cover systems of similar media, we see that the macroscopic $\mathbf{v}$ must now be reduced somewhat more elegantly than earlier when, in considering the reduction of K, the brute force, temporary reduction in terms of f_0 was used. There is no choice left. When we multiply (12.7) through by the necessary parametric factors for reducing K, $\mathbf{V}$, and p where they appear, we have

$$\{(\eta/\lambda^2)(L)(\lambda/\sigma)\mathbf{v}\} = [(\eta/\lambda^2)(K_\mathrm{H}\{(\lambda/\sigma)p\})][\{(L\lambda/\sigma)\mathbf{f}\} - \{L\,\mathbf{V}\}\{(\lambda/\sigma)p\}] \quad (12.12)$$

or

$$\mathbf{v}_* \equiv \{(\eta/\sigma)(L/\lambda)\mathbf{v}\} \qquad \text{and} \qquad \mathbf{f}_* \equiv \{(L\lambda/\sigma)\mathbf{f}\}$$

Finally, using this reduction of $\mathbf{v}_*$ along with that for div_* in the continuity equation (12.10) yields the expression for reduced time, unambiguously in this context:

$$\partial\theta_\mathrm{H}\{(\lambda/\sigma)p\}/\partial\{(\sigma\lambda/\eta L^2)t\} = \{L\,\mathrm{div}\}\{(\eta L/\sigma\lambda)\mathbf{v}\} \quad (12.13)$$

or

$$t_* \equiv \{(\sigma/\eta)(\lambda/L^2)t\}$$

Now let us assemble all our deductions for convenient reference:

$$\theta_* \equiv \{\theta\} \quad \text{or} \quad \theta_* = \{\phi\}\{s\} \qquad (\phi,\ \text{porosity};\quad s,\ \text{saturation})$$
$$p_* \equiv \{(\lambda/\sigma)p\} \qquad\qquad \mathbf{f}_* \equiv \{(L\lambda/\sigma)\mathbf{f}\}$$
$$K_* \equiv \{(\eta/\lambda^2)K\} \qquad\qquad \mathbf{r}_* \equiv \{\mathbf{r}/L\}$$
$$\mathbf{v}_* \equiv \{(\eta/\sigma)(L/\lambda)\mathbf{v}\} \qquad \mathbf{V}_* \equiv \{L\,\mathbf{V}\}$$
$$t_* \equiv \{(\sigma/\eta)(\lambda/L^2)t\} \qquad \mathrm{div}_* \equiv \{L\,\mathrm{div}\}$$

Darcy: $\qquad\qquad\qquad \mathbf{v}_* = K_{\mathrm{H}*}(p_*)(\mathbf{f}_* - \mathbf{V}_* p_*)$

Continuity: $\qquad \partial\theta_{\mathrm{H}*}(p_*)/\partial t_* = \mathrm{div}_*\,\mathbf{v}_*$

Richards: $\qquad \partial\theta_{\mathrm{H}*}(p_*)/\partial t_* = \mathrm{div}_*[K_{\mathrm{H}*}(p_*)(\mathbf{f}_* - \mathbf{V}_*(p_*)]$

E. Consequence: What Does It Mean?

Introducing the macroscopic L for similitude of whole flow systems has brought scaling relations bursting out like popcorn. It is time to step back and stare at these results for awhile to see what they mean.

The first obvious remark is that the two characteristic lengths—microscopic λ and macroscopic L—appear in a variety of combinations: λ, λ^{-2}, L, L^{-1}, $L\lambda$, $L\lambda^{-1}$, and λL^{-2}. These were deduced from similitude analysis; obviously they could not have been developed by dimensional analysis.

Let us next sample a few of the quite diverse kinds of comparisons that are permitted by this analysis.

1. SINGLE-SOIL COMPARISONS

For all similitude cases, θ is the same for different systems, and (unless of course σ of the liquid is changed) this means that p is also the same for all same-soil comparisons.

a. *Variation of Liquid Parameters η,σ*

Historically, one of the early tests of these scaling relations (tests, not really of the analysis but of the extent to which the STVF assumptions are applicable to various actual soil categories) was one in which the liquid was changed from water to butyl alcohol (Elrick *et al.*, 1959). It was found, as expected, that the greater the clay content the less the accuracy of this aspect of scaling; for low-clay soils the success was quite striking, including for the hysteresis loops. (Of course the chemistry of clay and water differs very basically from that of clay and organic liquids.)

For practical soil problems we can normally regard η and σ as fixed properties of soil water. This leaves the body force $\mathbf{f}_g$, the system size L, and (for similar media) the particle size λ available for playing our games.

b. *Variation of Body Force $\mathbf{f}_g$*

There are some practical situations in which the body force $\mathbf{f}_g$ changes to an adjustable parameter $\mathbf{f}$. For example, one might be interested in a flat layer of soil, having its normal vector tilted at some angle β with the vertical. If the layer is thin enough, we may consider the body force to be acting down slope, parallel to the soil layer, with strength $f = f_g \sin \beta$. As a very different example, one might wish to scale tiny flow systems carried on a centrifuge, provided that r is roughly constant over the system depth Δr. In such a system, the magnitude of g in $\mathbf{f}_g = \rho \mathbf{g}$ would be replaced by $\omega^2 r$. In either example we would have to hold $\mathbf{f}_*$ constant in scaled comparisons. For single-soil systems (λ fixed), $\mathbf{f}_*$ constant requires that fL be held constant. For example, weakening $\mathbf{f}$ by the $\sin \beta$ factor means a corresponding increase of the length of the system downgrade. Alternatively, on a centrifuge with $\omega^2 r \gg g$ we obviously scale to a correspondingly smaller system. In either case, turning to the time scale, we find that t_* contains t/L^2, which means that tf^2 is the same for scaled systems in this category. If β is $30°$ for example, $\mathbf{f}$ is half of $\mathbf{f}_g$ along the slope, which itself is twice as long; thus the time scale is stretched four times longer. Whatever would happen in a vertical column in 1 hr will take 4 hr along the $30°$ slope.

c. *Special Case; Body Forces Negligible*

An important special case is reached at the limit $\mathbf{f}_g \to 0$, which arises in thin horizontal plane layers or is approximated during the early stages of any sudden wetting transient for as long a time as the gravity forces can be considered negligible compared to the steep initial gradients of capillary potential. Without the body force, the reduced Richards equation becomes

$$\partial\theta_{\mathrm{H}}(p_*)/\partial t_* = \mathrm{div}_*(K_{\mathrm{H}*}(p_*)\nabla_* p_*) \tag{12.14}$$

so for same-soil comparisons we can obviously vary the system size—vary L—at will, provided we soak up the L variation in the t_* term, within which the t/L^2 factor is the relevant portion. Thus two force-free systems of the same soil but differing in physical size can be compared using a t/L^2 time scale. If one system is twice as big, we merely have to wait four times as long for the same things to happen; if the boundary conditions are changed with time they too must be tailored to fit this plan, of course. (For example if we cut off water after 10 min in one system we would have to cut it off after 40 min in a system twice as big, and so on.)

d. *Internal Similitude of Solutions with No Space Time Markers*

A special force-free case is that of unlimited system size, for which obviously no scale for L is imposed by the size of the boundaries. A timed boundary condition scenario (like turning off the water after 10 min) would of course still impose macroscopic scale L's on the comparison system through the necessity of matching t/L^2 for the two. (If we decided to turn off the water after 40 min in the second system we would thereby force the macroscopic scale to be twice as large in the second system, even though there are no physical size boundaries present.) For boundary scenarios lacking any such time-scale features—let us say one just applies water at $t = 0$ and then sustains it forever unchanged—*any two stages of the process* can be regarded simply as scale models *of each other*. The most familiar example is the famous problem of horizontal infiltration in one (Cartesian) dimension—an infinite column initially uniform at wetness θ_0 and potential p_0, to which, for $t > 0$, water is abruptly applied at $x = 0$ and held thenceforth at constant pressure p_i. Suppose at some later time t_1 a water depth Q_1, in (cm^3 water)/(cm^2 of face), has been imbibed. We can regard t_1 as the final moment for a system 1, forcing L_1 to be proportional to the water depth Q_1. We can then compare this configuration $\theta(x)_1$ to the configuration $\theta(x)_2$ developed when the same system has been continued on past t_1 to a later "final moment" t_2, when Q_2 has been imbibed so that L_2 is proportional to Q_2. Since the final t_* for both "systems" must be made the same for scaling purposes, i.e., since we must take $t_2/L_2^2 = t_1/L_1^2$ and $Q_2/L_2 = Q_1/L_1$, we get $Q_2/\sqrt{t_2} = Q_1/\sqrt{t_1}$. Aha, the famous Boltzmann $\sqrt{t}$ idea has just dropped out of similitude considerations like a ripe plum. Solutions of this general class are for that reason sometimes called "similitude solutions." Two other examples of similitude solutions are mentioned in

Chapter 11 of this book: infinite force-free one-dimensional infiltration in cylindrical and spherical coordinates. An important point is that these two latter cases cannot tolerate a finite entrance hole—cylindrical or spherical, respectively—because that hole size would at once define a fixed macro-length scale and destroy the solution similitude. A theoretical hole of zero size is required, and that implies a pole of infinite pressure at the origin. Hence these solutions should be of strictly theoretical interest. Nevertheless, we can sidestep this impracticality for the cylindrical case by noting that the theoretical solution exhibits a constant water *flux* at the input so that we can approximate the similitude solution, even with a finite hole size, if we fundamentally *impose a constant flux boundary condition* in the first place, and then just wait for the initial perturbations due to the finite size of the hole to become unimportant.

We can compare force-free imbibition into holes of two different sizes— right down to $t = 0$ without approximation—but of course the sizes of the two holes just define the sizes of the two L's; with, say, a constant p_i condition, we no longer have the simple similitude solutions for Q versus t of either system alone (not even approximately). Indeed so long as two systems are force free, we could use holes of any shape whatever that are macroscopic scale models of each other and scale one solution to the other, using L's defined from the sizes of the scale-modeled holes.

2. SIMILAR MEDIA COMPARISONS

In all the foregoing comparisons we could now add the complication of comparing similar media rather than same-soil systems.

a. *No-Gravity Cases*

In the horizontal infiltration cases exhibiting a similitude solution, the sorptivity S (defined as the constant ratio of $Q/\sqrt{t}$) could be directly compared for two similar media if the value of p_i applied at $t = 0$ is zero, in which case p_{i_*} is conveniently also zero for both. Then since $Q_* \equiv \{Q/L\}$ and $t_* \equiv \{(\sigma/\eta)(\lambda/L^2)t\}$, we have

$$Q_*^2/t_* = \{(Q^2/t)(1/L^2)(\eta/\sigma)(L^2/\lambda)\} = \{(\eta/\sigma)S^2/\lambda\} = S_*^2. \quad (12.15)$$

Thus we can measure sorptivity for two similar soils—as Reichardt *et al.* (1972) have done—and deduce the ratio of their λ's to be the same as the ratio of their sorptivities squared. If we want to compare such similitude imbibition for two similar media in cases where $p_i \neq 0$, we must also impose the condition on the p_i values that $(p_{i_*})_1 = (p_{i_*})_2$ or $p_{i1}\lambda_1 = p_{i2}\lambda_2$. Space does not justify discussing further examples in these categories here.

b. *Constant-Gravity Cases with Similar Media*

Scaling solutions for similar media in cases where the gravity force must be taken into account gives a result that may seem surprising at first. For similitude of the Darcy (or Richards) equation containing the gravity-force term, in order that the reduced versions of these differential equations be alike, the reduced constant-force terms $\mathbf{f}_{g*}$ must be alike. Forgetting the unchanged constant σ, we must have the same value of $(L\lambda)\mathbf{f}_g$ for both.

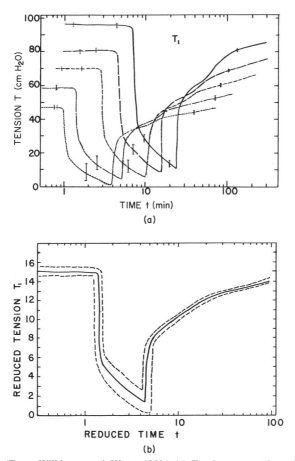

Fig. 12.3. (From Wilkinson and Klute, 1959.) (a) Tension versus time observed at the same reduced depth using five similar soils with appropriately scaled sizes and infiltration/ drainage scenarios at the boundaries: ——, 104–125μ; ——, 125–149μ; ———, 149–177μ; —·—, 177–210μ; ···, 210–250μ. (b) The same data plotted as reduced tension versus reduced time.

But $f_g = \rho g$ is already constant in practice, which means that $L\lambda$ must be held the same for two system solutions to be scalable. In other words, L and λ *scale oppositely*. A similar soil with half the particle size must be packed up into a scale-model flow system that is twice as big. Klute and Wilkinson (1959) procured five nearly similar soils with a progression of values of λ, and then scaled their system sizes L as λ^{-1} and scaled time for the water programs applied at the boundaries as λ^{-3}. As shown in Fig. 12.3, at one selected reduced depth they observed physical transients of tension that were all quite different, but when plotted in reduced terms they were all alike within experimental accuracy.

F. Applications

1. GENERAL INSIGHT

What is the practical value of all these scaling concepts? For same-soil comparisons we are restricted to thin slabs or columns either horizontal or tilted—not a run-of-the-mill situation. Ordinarily the fixed force of gravity is involved, in which case we are restricted to comparisons of similar media and must contrive somehow to alter the flow-system size in opposite ratio to particle size. Nevertheless, as the hangman said when he was himself about to be hanged, one's point of view is changed considerably. For example, one can now understand that 1 in. of rain on a coarse silt can best be compared to 2 in. on a finer silt (having half the particle size), L being set not by system boundaries, which are never reached, but by the depth of water imbibed. Further, the redistribution pattern after 10 hr for the coarser soil should be compared with that (at doubled depths) after 80 hr for the finer soil. (The factor of 2^3 comes from reduced time $(\lambda/L^2)t$, where since λL is fixed for reduced gravity $\lambda^3 t$ must be fixed.) Becoming familiar with this sort of telescoping comparison takes a little of the mystery out of these nonlinear, hysteretic soil flow systems; it contributes an element of justified intuition.

The same sort of insight led to the development recently of a new approach for a teaching lab in soil physics. To demonstrate the strange behavior of soil profiles during infiltration, redistribution, drainage, etc., one can drastically coarsen the soil—say 20-fold to some conveniently obtainable sand—whereupon the profile depths to be packed are made 20-fold shallower, the tensions needed for tensiometry become 20-fold less, and the time for things to happen is shortened by 20^3 or 8000-fold! In practice, then, one can demonstrate phenomena in one afternoon's lab that would take days or weeks and would require much more sophistication of equipment to demonstrate with typical agricultural soils.

2. DEFINING λ AS A BASIC SOIL PARAMETER EVEN FOR DISSIMILAR SOILS

It is clear that similarity of two unrelated soils cannot be taken for granted. This can be seen merely by plotting up the p, θ, and k characteristics of various soils using log–log paper. Two soils can only be scalable if their characteristics can be made to match by multiplying p, θ, and k by suitable (different) scale factors. On log–log paper any such scaling reduces to mere translation (without rotation). Even on log–log paper, the simple drainage curve for a monodisperse soil such as sand will be considerably more abrupt than that for a polydisperse soil such as loam. Hence the comparisons utilized above as a soil-physics lab strategy must be considered qualitative —or perhaps be regarded as being accurately compared only to a fairly monodisperse silt profile.

None the less, when one considers developing an approximate "equation of state" for the hysteretic (p, θ, k) characteristics of practical soils (an equation in which the specific values of a finite number of constant parameters will specify adequately the totality of (p, θ, k) for any specific soil)— and when accomplished this will be a monumental practical step forward— one of these few parameters must obviously represent the microscopic characteristic length λ that will enter p as $\lambda^1 p$, θ as $\lambda^0 \theta$, and k as $\lambda^{-2} k$.

How best to define λ in general for soils that are not similar is not yet clear. It may well be dictated by the yet to be discovered functional approximation of the (p, θ, k) equation. One useful basis is either the saturated or the "satiated" (i.e., cracks and wormholes drained) conductivity, taking $(\eta/\lambda^2)k_{\text{sat}} = 1$. Another basis that has received recent attention is the sorptivity S, i.e., the $Q/\sqrt{t}$ slope for force-free inhibition, taking $S^2/\lambda = 1$. These two choices give different values of λ; for two similar media the ratio of the two λ choices is the same; for two dissimilar media their ratio may be different.

3. HETEROGENEITY AS λ VARIATION OF SIMILAR MEDIA?

Warrick *et al.* (1977), using S to characterize λ, have found that within a given field—of what would ordinarily be considered one kind of soil— the actual value of λ varies markedly, and on a scale of a few meters. However, when the detailed moisture characteristics of their separate samples are plotted in reduced form they all turn out to be essentially similar media. This sort of painstaking investigation will require decades of experimental work in many parts of the world to achieve any sort of generality, but certainly it already provides a ray of hope for the difficult problem of encompassing the effects of short-range heterogeneity within soil physics. Heterogeneity is one of the major roadblocks still hindering the widespread application of technological soil physics to such gigascale problems as watershed modeling for management and planning of land use.

13 Spatial Variability of Soil Physical Properties in the Field

A. W. Warrick

The University of Arizona
Tucson, Arizona

D. R. Nielsen

The University of California
Davis, California

A. Introduction

Soils vary. They are a product of the factors of formation and as a consequence continuously change over the earth's surface. There are over 10,000 soil series identified in the United States alone. It should be obvious that if we wish to fully describe the heterogeneity of a field, all that is necessary is to locate an accurate soil survey. This is a logical first step and for planning purposes is invaluable. However, a point to remember is that the taxonomic separations are designed for broad applications and, consequently, based on many factors of which any specific physical property plays only a minor role. In addition, the soil is not composed of homogeneous units as the map would indicate, but rather always has some degree of variability. Thus, the survey integrates many aspects, but regardless of scale or criteria there remains missing information which may be better included using an alternative approach, especially when we have a specific application in mind.

From the 1970s to the present there has been a major upsurge of interest in the heterogeneity of soil physical parameters. Roughly speaking, the efforts are to describe what exists in nature and to explore ramifications. Future applications will be to aid in management practices—both short and long term. The grower needs sound scientific input to support his own good judgment. Even a moderately sized operation has hundreds of thousands of dollars invested and someone has to make decisions in order to optimize available water, fertilizer, pesticides, energy, and labor. The rules of operation are constantly changing. The farmer must avoid natural disasters such as drought and plant diseases, as well as adjust to the demands of modern-day society, including, simultaneously, a clean environment, an ever-increasing competition for land resources, and cheap food. Similarly, the manager of grazing and recreational areas also has to make decisions, often while most of society watches over his shoulder and offers constant and conflicting advice.

Modern-day computer technology has resulted in numerous models of natural systems. Important environmental "problems" are controlled by the soil's capacity to transmit, store, and react with salts, pesticides, and waste products. Soil physical properties play a central role in such phenomena and the quality and form of data is of ever-increasing importance. For example, the movement of salt and water is described by the hydraulic conductivity, hydraulic gradients, porosity, and reaction kinetics. Scientists have long realized that results can be no more dependable than the quality of the input. Also, we wish to assess the answers, one way being to establish confidence limits. Not only is a final value important, but also whether it is likely to be true within 1% or 10%, or even 1000%! Both data quality and confidence limits are strongly influenced by spatial variability.

Our first step is to show how variability may be expressed. Examples follow of calculated coefficients of variations and of how to determine sample numbers. Later scaling is demonstrated for data reduction and problem solving. Lastly, we shall discuss spatial structure.

B. Expressing Variability

As a first approximation, we can say something about a statistical population knowing only its mean value and standard deviation. If we collect n samples with measured values of $x_1, x_2, \ldots, x_n$, estimates for the mean and standard deviation are μ_E and σ_E given by

$$\mu_E = \sum x_i / n \tag{13.1}$$

$$\sigma_E = [\sum (x_i - \mu_E)^2 / (n - 1)]^{1/2} \tag{13.2}$$

where we assume the summation is taken over all n values. The mean value

μ_E gives us an average value of the parameter and the standard deviation tells us about the range or scatter of the parameter. Large values of σ_E correspond to samples which are dissimilar, small values to samples which are mostly close to the estimated mean. These two sample statistics do *not* in themselves say anything about the type of distribution.

A population is more completely defined by its frequency distribution— whether discussing soil physical properties or how tall people are. Given the frequency distribution, we can determine all sorts of things—including averages, dispersions, and even the probability that a randomly drawn value will be within specified limits.

Figure 13.1a shows the distribution for bulk density determined by 180 cores of Pima clay loam. The point show the number of observations found in a range of 0.1 g/cm^3. Thus, the leftmost point shows two samples to be

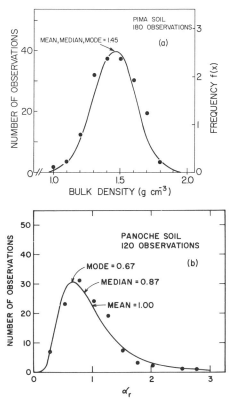

Fig. 13.1. (a) Normal frequency distribution along with experimentally determined frequency for Pima clay loam. (From Coelho, 1974.) For the right-hand ordinate, x is the bulk density. (b) Distribution of scaling coefficient α_r for 180 cm depth of Panoche soil. The smooth curve is for a log-normal distribution, used in Example 2 (pp. 333–336).

between 0.95 and 1.05. The top two points show that about 36–38 samples are in the classes 1.35–1.45 and 1.45–1.55. The estimated mean and standard deviation are 1.45 and 0.16, respectively.

The smooth curve in Fig. 13.1a is for a (the idealized) normal probability density function. The equation of $f(x)$ is

$$f(x) = (1/\sigma \sqrt{2\pi}) \exp[-(x - \mu)^2/2\sigma^2] \tag{13.3}$$

where μ is the *mean* and σ the *standard deviation*. In our example, the variable x is the bulk density value and μ and σ are taken equal to the estimates, 1.45 and 0.16. The curve is bell shaped, with mean μ at the center of the bell. Also, the *median* value, for which half of the population values are smaller and half are larger, falls at the central point. The *mode*, for which $f(x)$ is a maximum (the range for the most frequently measured values), coincides with both the mean and median. The mean value μ determines where the center of the bell is located. The σ controls the spread—a large σ gives an $f(x)$ curve which is low and wide, and small values give narrow, tall peaks. A property of the probability density function is that the area under the curve is unity, which is equivalent to stating that we are 100% certain any value is included somewhere in the distribution.

If we take a random sample X from the population, the probability of its value being between x_1 and x_2 is

$$P\{x_1 < X \le x_2\} = \int_{x_1}^{x_2} f(x)\, dx \tag{13.4}$$

From Eq. (13.4), if we used $f(x)$ from Eq. (13.3), we could compare the theoretical results to the measured classes by Figure 13.1a.

If x_1 approaches $-\infty$ and $x_2 = x$, we have the probability the random value will be less than some value x

$$P\{X \le x\} = F(x) = \int_{-\infty}^{x} f(x)\, dx \tag{13.5}$$

The function $F(x)$ is called the cumulative probability function. As x becomes large, $F(x)$ approaches 1. Values of both $f(x)$ and $F(x)$ are given in statistical and mathematical tables (cf. *Handbook of Chemistry and Physics*, 1968, especially p. A-158).

A log-normal distribution is shown in Fig. 13.2a for 20 observations of pore-water velocity v_0 (cm/day) during steady-state infiltration conditions for Panoche soil (Nielsen *et al.*, 1973). The equation for the smooth curve is

$$N = \frac{n\,\Delta v_0}{v_0 \sigma_{\ln} \sqrt{2\pi}} \exp[-(\ln v_0 - \mu_{\ln})^2/2\sigma_{\ln}^2] \tag{13.6}$$

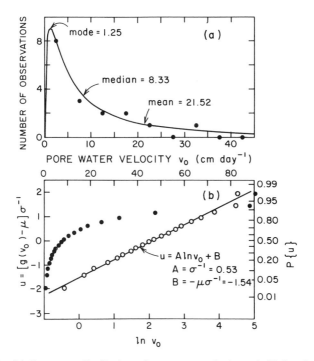

Fig. 13.2. (a) Frequency distribution of pore-water velocity and (b) fractile diagram of pore-water velocity for Panoche soil.

where N is the number of observations expected to fall within a class Δv_0 ($= 5$ cm day^{-1}), n is the total number of observations (20), and $\mu_{\ln}$ and $\sigma_{\ln}$ are estimates of the mean ($= 2.12$) and standard deviation ($= 1.38$) of the $\ln v_0$ values, respectively. The points show the number of v_0 observations whose values fall within a class of 5 cm/day. For example, eight observations of v_0 between 0 and 5 cm/day are represented by the dot plotted at $v_0 = 2.5$ cm/day. It is immediately obvious that the log-normal distribution is not bell shaped but is skewed, and hence it manifests a relatively small number of observations of v_0 that are sufficiently large to yield a mean value substantially larger than the majority of the observations. Further, unlike a normal distribution, the mode, the median, and the mean have different values. Numerically, the mode is equal to $\exp(\mu_{\ln} - \sigma_{\ln}^2)$, the median equal to $\exp(\mu_{\ln})$, and the mean equal to $\exp(\mu_{\ln} + \frac{1}{2}\sigma_{\ln}^2)$. The mean calculated from the latter formula is 21.52 cm/day, compared with the arithmetic mean of 19.19 cm/day calculated on the assumption of a normal distribution. Not recognizing that the observed values of v_0 were log-normally distributed

Table 13.1

OBSERVED VALUES ($n = 20$) OF PORE WATER VELOCITY v_0 (CM DAY^{-1}),
SOIL WATER CONTENT θ (CM^3 CM^{-3}), AND CORRESPONDING
STATISTICAL PARAMETERS[a]

i	$(v_0)_i$	$(\ln v_0)_i$	θ_i	$(i - 0.5)/n$[b]	$(x - \mu)/\sigma$[c]
1	0.66	−0.42	0.297	0.025	−1.96
2	1.16	0.15	0.310	0.075	−1.44
3	1.52	0.42	0.318	0.125	−1.15
4	2.29	0.83	0.326	0.175	−0.93
5	2.94	1.08	0.330	0.225	−0.76
6	3.46	1.24	0.333	0.275	−0.60
7	4.35	1.47	0.339	0.325	−0.45
8	4.95	1.60	0.341	0.375	−0.32
9	6.11	1.81	0.345	0.425	−0.19
10	7.39	2.00	0.350	0.475	−0.06
11	8.41	2.13	0.351	0.525	0.06
12	10.38	2.34	0.356	0.575	0.19
13	12.81	2.55	0.359	0.625	0.32
14	15.64	2.75	0.363	0.675	0.45
15	18.64	2.92	0.366	0.725	0.60
16	24.53	3.20	0.372	0.775	0.76
17	32.14	3.47	0.375	0.825	0.93
18	47.47	3.86	0.382	0.875	1.15
19	88.23	4.48	0.390	0.925	1.44
20	90.75	4.51	0.404	0.975	1.96
μ_E	19.19	2.12[d]	0.350	—	—
σ_E	26.80	1.38[e]	0.027	—	—

[a] Used to construct the frequency distribution and the fractile diagram given in Fig. 2.

[b] $(i - 0.5)/n$ approximates the value of the cumulative probability function $P\{u\}$ where

$$P\{u\} = \frac{1}{\sqrt{2\pi}} \int_{-\infty}^{u} \exp(-x^2/2)\, dx$$

[c] Values of $[g(v_0) - \mu]\sigma^{-1} = u$ are obtained from tables of $P\{u\}$ for each value of $(i - 0.5)/n$.

[d] For $g(v_0) = \ln v_0$, $\mu_E = \sum_{i=1}^{20} [(\ln v_0)_i/20] = 2.12$.

[e] For $g(v_0) = \ln v_0$

$$\sigma_E = \left\{ \sum_{i=1}^{20} [(\ln v_0)_i - \mu_E]^2/(n - 1) \right\}^{1/2} = 1.38$$

would cause the mean to be underestimated by about $11\%[(21.52 - 19.19) \cdot 100/21.52]$.

How can we determine whether or not a set of observations is normally distributed? One of the easiest methods is to use a fractile diagram, illustrated in Fig. 13.2b for the distribution shown in Fig. 13.2a. The original 20 observations of v_0 together with all calculations necessary for constructing the fractile diagram are given in Table 13.1. The observed values of v_0 arranged in increasing order are found in the second column. The cumulative probability for each value of v_0 is approximated by $(i - 0.5)/n$, given in the fifth column. For example, for the least v_0 value of 0.66 cm/day having a logarithm of -0.42, the cumulative probability is $[(1 - 0.5)/20]$ or 0.025. For a value of $P\{u\}$ equal to 0.025, u has a value of -1.96, obtained from tables of cumulative probability. Notice that values given in column 6 are not calculated from observations of v_0 (or $\ln v_0$) but stem directly from the values in column 5. The fractile diagrams given in Fig. 13.2b are obtained by plotting values given in column 6 on the left-hand ordinate versus values of v_0 (or $\ln v_0$) on the top abscissa (or bottom abscissa). If cumulative probability graph paper is used, as suggested by the right-hand ordinate, values given in column 5 can be plotted against v_0 or $\ln v_0$.

When values of v_0 are used in the fractile diagram [i.e., $g(v_0) = v_0$], notice that the v_0 values do not fall on a straight line, and hence the 20 observations are not normally distributed. On the other hand, when values of $\ln v_0$ are used [$g(v_0) = \ln v_0$] a straight line results, indicating that the logarithmic values are normally distributed, and hence the 20 observations of v_0 stem from a log-normal population. Similarly, it can be shown that the 20 oservations of θ are normally distributed.

Normal and log-normal distributions are commonly assumed for soil physical properties. Of course, the idealized distributions do not fit the measured values exactly but are an approximation. Other theoretical distributions $g(x)$ can be considered, or alternatively, empirical distributions which are fitted more or less exactly to measured values.

C. Measured Values

The *coefficient of variation* (CV) is useful for expressing variability on a relative basis:

$$CV = (\sigma/\mu)100\%$$

The CV is dimensionless and will be the same regardless of what units are used for the measurements. Furthermore, the values from one parameter to another can be compared. For some soil parameters, the CV is consist-

Table 13.2
VALUES OF ESTIMATED MEANS, STANDARD DEVIATIONS, AND COEFFICIENTS OF VARIATION FOR TEN SOIL PROPERTIES

Parameter	Mean	Standard deviation	Coefficient of variation (%)	Source
Low variation				
1. Bulk density (g/cm^3)	1.3	0.09	6.9	Gumaa (1978), 5 combined depths on 64 cores.
	1.4	0.095	6.8	Arizona, with 15 hectares, contained 5 series, each a "typic torrifluvent." Nielsen *et al.* (1973), combined data for 6 depths. California, Panoche, 20 sites within 150 ha (had additional replicates).
	1.5	0.11	7.3	Cassel and Bauer (1975) 30–60 cm depth, North Dakota, Maddock sandy loam within 1.3 ha.
2. Water content (%) at zero tension (cm^3/cm^3)	40/45	4.5/4.8	11/11	Nielsen *et al.*, 30 cm depth and combined depths
	47	4.8	10	Cameron (1978), 15–30 cm depth Saskatchewan, Bainsville, 225 m^2
Medium variation				
3. Sand/silt/clay (%)	53/28/19	15/9.1/6.8	28/32/36	Gumaa, 30 cm, 64 sites
	59/29/12	22/18/6.4	37/62/53	Gumaa, 5 depths, at 64 sites
	26/27/47	11/6/8	42/22/17	Nielsen *et al.*, 30–45 cm depths
	24/30/45	14/8/10	58/27/22	Nielsen *et al.*, 12 combined depths
4. 0.1/15 bar (% water content, g/g)	27/9.5	5.4/3.1	20/33	Gumaa, 30 cm depth
	23/7.5	9.2/3.8	40/51	Gumaa, 5 combined depths
0.2 bar (cm^3/cm^3)	32	5.4	17	Nielsen *et al.*, 30 cm depth
	32	7.7	24	Nielsen *et al.*, combined depths
2.2 bar (cm^3/cm^3)	34	4.1	12	Cameron, 15–30 cm depth
15 bar (g/g)	4.5	1.4	31	Cassel and Bauer, 30–60 cm depth

High variation

5. Saturated hydraulic conductivity (cm/hr)	14	26	190	Gumaa, 5 combined depths on 64 cores
Saturated hydraulic conductivity (cm/day)	20 35	22 30	110 86	Nielsen et al., 30 cm depth Nielsen et al., combined depths
Saturated hydraulic conductivity as (in./hr)	0.62	0.64	100	Willardson and Hurst (1965), Calif., 0–90 cm depth, Gila-Vinton, 330 values, scattered area.
6. Unsaturated hydraulic conductivity (cm/day) (90% and 60% of saturation)	0.63/ 0.0026	1.75/ 0.011	280/420	Nielsen et al., 30 cm depth
7. Apparent diffusion coefficient (cm²/day)	4.9/0.12 370	8.1/0.47 $(2.4)(10)^6$	170/400 $(6.5)(10)^6$	Nielsen et al., combined depths Biggar and Nielsen (1976), all depths, 150 hectare Panoche site.
8. Pore water velocity (cm/day) (from water and solute)	44/40	7300/4400	$(1.7)(10)^4/$ $(1.1)(10)^4$	Biggar and Nielsen, all depths.
9. Electrical conductivity in μmho/cm for 1:1 extract and saturation extract	3200/ 3100	8100/3900	250/130	Wagenet and Jurinak (1978), 7.5–15 cm depth, Utah, Mancos Shale, within 777 km².
10. Log of scaling coefficient (no units)	−0.136	0.512	380	Warrick et al. (1977), Panoche combined depths. From Nielsen et al., above.

ently less than 10%, for others greater than 1000%! In order to obtain a reliable "answer," many more samples are generally needed for the parameter with the large coefficient of variability. Perhaps so many samples are needed that most methods of determination take too much time to be practical.

In Table 13.2 are estimated means, standard deviations, and coefficients of variation for ten different parameters. The parameters are grouped according to relative amounts of variability. In the lowest class are bulk density and percent porosity, with a CV of about 7–10%. For porosity, the values are given in pairs, say, 40 and 45, corresponding to the 30 cm and combined depths, respectively. Note the examples given are by four different investigators at four widely scattered locations. As we shall see shortly, low CV values imply that relatively few samples give a good estimate of the mean value. For other sites, we could safely assume the CV values for the low variation group to be less likely than for the other listed parameters. The measuring processes are also relatively precise.

On the other end of the spectrum, we have six parameters which are much more highly variable, each with a CV greater than 100%. With only one exception, the saturated hydraulic conductivities are in the 100–200% range. The CV for the unsaturated hydraulic conductivity, apparent diffusion coefficient, pore water velocity, and scaling coefficient (which we shall discuss later) are even larger. The measured scatter is a combination of the inherent variation and the measuring process. The measurement is difficult for these parameters and much harder than for some other factors which can be easily replicated.

Variations for the particle size fractions and water retained at suctions 0.1–15 bar are intermediate. They have CV values between 10% and 100%. Results for sand, silt, and clay are given together. Thus, for the first set, 53, 28, and 19, we have mean values of 53%, 28%, and 19% sand, silt, and clay, respectively. The Nielsen *et al.* CV values are considerably lower than Gumaa's, probably because the latter's soils were more stratified. Water-content values for the same soil tend to show larger CVs for larger suctions; for example, Gumaa showed 20% and 40% for 0.1 bar, compared to 33% and 51% for 15 bar. Also, larger variations are shown here for combined depths than for single depths, a consequence of horizonation, at least to a first approximation.

D. Sample Numbers

We now address the problem of matching the number of samples to the accuracy of estimating the mean. Two assumptions simplify the problem:

(1) The samples are idependent and (2) the number of samples are large enough that the *central limit theorem* applies. The number of samples necessary to be within d units of the mean $(1 - \alpha')100\%$ of the time is

$$N = x_{\alpha'}^2 \sigma^2 / d^2 \qquad (13.7)$$

where σ is the standard deviation. The normalized deviate $x_{\alpha'}$ can be found tabulated alone or as the two-tailed, student's t value with infinite degrees of freedom. It is defined by $x_{\alpha'} = (x - \mu)/\sigma$ with x such that Eq. (13.5) gives $F(x) = 1 - \frac{1}{2}\alpha'$. Let us illustrate with an example. Values for $\alpha' = 0.05$, 0.1, and 0.5 are $x_{\alpha'} = 1.960$, 1.645, and 0.842, respectively.

EXAMPLE 1. SAMPLE NUMBER DETERMINATION

Suppose we wish to estimate the mean value of 15 bar water at the 30 cm depth in Gumaa's field. We wish to be within 15% of the correct value at a 0.05 confidence level. We first find his estimated mean $\mu_E = 9.5$ and standard deviation $\sigma_E = 3.1$. We set d equal to 15% of our best estimate of the population mean or $d = (0.15)(9.5) = 1.4$. Similarly, we do not know the true population standard deviation, so we use our estimate $\sigma_E = 3.1$. The value of $x_{0.05}$ from above is 1.96. Thus, the estimated sample size is

$$N = (1.96)^2(3.1)^2/(1.4)^2 = 19$$

What is the interpretation of the sample number? If the assumptions are valid, the number calculated will give an estimate of the mean which will be within d of the true answer $(1 - \alpha')100\%$ of the time. σ^2 reflects the dispersion within the basic population, $x_{\alpha'}$ the degree of confidence desired for the estimate, and d how close we wish to estimate the mean. The assumptions are that the samples are independent, the central limit applies, and σ can be represented by σ_E. The independence is satisfied provided the sampling locations are random. We would not, for example, take all of our samples from one small corner and say they represent an entire field. The sampling number given by Eq. (13.7) assumes that the distributions of estimated mean values based on repeated random samples of N each will be normally distributed. This is always true when the underlying population is normal. Such is also the case for any other population distribution if N is large enough, perhaps 20 or more, according to central limit theorem.

In Table 13.3 are the approximate sample numbers calculated from Gumaa's 1978 results. For the bulk density, Eq. (13.7) gives an N value of only 2. Obviously, the central limit theorem does not apply for a sample size of 2. Nevertheless, many fewer samples are needed to estimate bulk density than the other parameters. The 0.1 bar water, 15 bar water, and textural groups require about 50–100 samples and the hydraulic conduc-

Table 13.3
SUMMARY OF THE APPROXIMATE NUMBER OF SAMPLES REQUIRED
TO ESTIMATE MEAN VALUES

Property	N	Property	N
Bulk density	2	Silt (%)	150
0.1 bar water	61	Sand (%)	53
15 bar water	98	Hydraulic conductivity	1300
Clay (%)	110		

[a] Estimation within 10% at 0.05 significance level for combined depths. (After Gumaa, 1978.)

tivity a gigantic 1300 samples. Of course, these numerical values do not transfer directly to all other fields; however, we should expect somewhat the same relative values among the parameters.

E. Scaling as a Tool for Data Synthesis

In Chapter 12, the physical interpretation and basis for scaling physical properties were discussed. Although scaling is extensively used in hydraulics, its use in soils has been limited primarily because the exact rigorous assumptions are never met in nature. We shall now show, however, that heterogeneity from location to location within a field may be approximated using a scaling coefficient for each site. The values of the scaling coefficient will not be chosen by comparing the samples under a microscope, such as might be used following the basic definitions, but rather by *best fitting* the relationships, considered as a total group.

In terms of the pressure head h_r at a given site r, the scaling relationship is

$$h_r = (\bar{\lambda}/\lambda_r)h_m = h_m/\alpha_r \tag{13.8}$$

where $\bar{\lambda}$ is a mean characteristic length and λ_r is the specific length corresponding to site r. We define the scaling coefficient α_r by

$$\alpha_r = \lambda_r/\bar{\lambda} \tag{13.9}$$

The value of h_m is a mean pressure head corresponding to the water content for which h_r is also measured. Values of α_r at a given location are constant for all values of water content for the idealized case.

A second relationship in terms of the unsaturated hydraulic conductivity is

$$K_r = \alpha_r^2 K_m \tag{13.10}$$

where α_r is constant for each site and is the same as that for Eq. (13.8).

Again, the K_r and K_m correspond to the same water content. Thus, for different values of water content, the K_r and K_m would change but the α_r would not.

Typically, for soil sampling, a soil characteristic curve is determined experimentally as a first step towards describing the hydraulic properties. This was done by Nielsen *et al.* (1973) on the 150 ha Panoche locations described in Table 13.2. Core samples were taken from six soil depths at each of 20 sites in the field for 120 depth–site combinations. In the laboratory, the cores were saturated and the water contents corresponding to pressure heads of -10, -30, -60, -90, -120, -150, and -200 bar were found. Figure 13.3a shows the results where degree of saturation s

$$s = \theta/\theta_s \qquad (13.11)$$

was used with θ_s the saturated water content. The use of s normalizes the

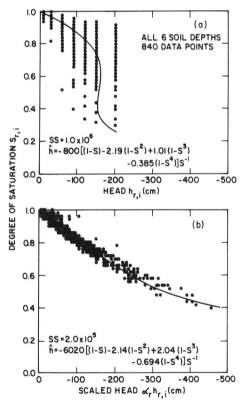

Fig. 13.3. Soil-water characteristic data for six depths of Panoche soil: (a) unscaled and (b) scaled. (After Warrick *et al.*, 1977.)

water contents to a common value of 1 at the zero-pressure head. Our motivation is convenience and does not depend on whether the scaling approach should be used. If only the depth or horizon had been used, there would be a smaller, but still appreciable, scatter. The 840 data points (some of which represent more than one data point) are widely dispersed. An attempt to "average" the results by fitting a function $\hat{h}(s)$, results in the solid line, where $\hat{h}$ is of the form

$$\hat{h} = (1/s) \sum_{p=1}^{4} a_p (1 - s^P) \tag{13.12}$$

If we choose simultaneously the four values of a_p and 120 values of α_r by minimizing the sum of squares (SS) given by

$$SS = \sum_{r,i} (\hat{h}_{r,i} - \alpha_r h_{r,i})^2 \tag{13.13}$$

the points are coalesced into the relatively narrow band of Fig. 13.3b, after Warrick *et al.* (1977). The sum of Eq. (13.13) is taken over all combinations of depth, sites, and pressure heads, which here is for $6 \times 20 \times 7 = 840$ points. The value $\hat{h}_{r,i}$ is $\hat{h}$ evaluated at $s_{r,i}$, the degree of saturation for the r,ith measurement. The α_r and a_p were solved using an iterative process on the computer. A great improvement in the way the data falls in line is obvious; this is confirmed quantitatively by comparing an SS in Fig. 13.3a to the SS in Fig. 13.3b. That in Fig. 13.3a is as if the α_r are all taken as 1, i.e., without scaling, and is five times as large as when the α_r are fitted. Of course there had to be reduction, but the effect is dramatic.

The cumulative frequency distribution of α_r and $\ln \alpha_r$ corresponding to Figure 13.3b is given as Figure 13.4. Each point is for one (or more) α_r value. The ordinate may be read as the percentage of population less than each plotted point. If the population of α_r values is normal, the plot should show nearly a straight curve; if log-normal, the $\ln \alpha_r$ should plot on a straight curve. Neither is clearly the case, but the tendency is for a highly skewed population and a log-normal seems to be a better approximation of the two. Incidentally, graph paper for both a normal and log-normal plot is readily available.

Let us list some of the good features of scaling, followed by some words of caution. First, large amounts of data can be dealt with systematically. If some points are missing, it really does not matter in terms of the applications. It allows a meaningful average curve to be found, and if we wish to characterize the heterogeneity quantitatively, we can discuss the α_r distribution. This latter feature will be exploited in the next section. Scaling, of course, does not magically make the system perfect. The exact scaling hypothesis will never be met. Results will depend on how samples are chosen,

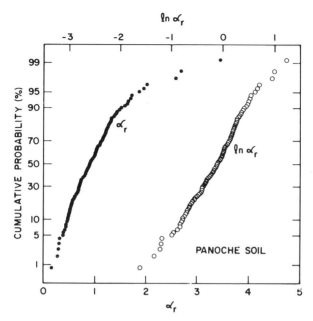

Fig. 13.4. Cumulative probability plots of α_r and $\ln \alpha_r$ calculated from soil water characteristic data of Panoche soil. (After Warrick *et al.*, 1977.)

what variable is used for scaling, and even on the form of the averaging function (e.g., $\hat{h}$).

F. Ramifications of Variability—Numerical Examples

Let us now look at the consequences of variability for infiltration (Example 2) and salt movement (Example 3).

EXAMPLE 2. INFILTRATION USING PHILIP'S SEMIANALYTICAL SOLUTION

We consider the soil-water flow equation with gravity to be

$$\frac{\partial \theta}{\partial t} = \frac{\partial}{\partial x}\left(D \frac{\partial \theta}{\partial x} \right) - \frac{\partial K}{\partial x} \tag{13.14}$$

By substituting degree of saturation $s = \theta/\theta_s$ from Eq. (13.12) and the scaled depth X and time T given by

$$X = \alpha_r x, \qquad T = (\alpha_r^3/\theta_s)t \tag{13.15}$$

the flow equation becomes

$$\frac{\partial s}{\partial T} = \frac{\partial}{\partial X}\left(\hat{D}\,\frac{\partial s}{\partial X}\right) - \frac{\partial \hat{K}}{\partial X} \qquad (13.16)$$

where $K = \alpha_r^2 \hat{K}$, $h = \hat{h}/\alpha_r$, and the scaled soil water diffusivity is defined by

$$\hat{D} = \hat{K}(d\hat{h}/ds) \qquad (13.17)$$

We assume that the soil moisture is initially constant, giving s as a constant s_i,

$$s(X, 0) = s_i \qquad (13.18)$$

and assume that the surface is maintained at a constant water content, or that s is a constant s_f

$$s(0, T) = s_f \qquad (13.19)$$

For Eqs. (13.18) and (13.19) we have implicitly assumed that θ_s is everywhere constant, a debatable point, but one which does not really affect the results.

Calculations were performed utilizing the best-fitted reduced functions for pressure head $\hat{h}$ for the Panoche soil (Fig. 13.3b),

$$\hat{h} = -6020[0.206 - s + 2.14s^2 - 2.04s^3 + 0.694s^4]/s \qquad (13.20)$$

and the corresponding best-fitting conductivity function $\hat{K}$,

$$\ln \hat{K} = -23.3 + 75s - 103s^2 + 55.7s^3 \qquad (13.21)$$

The solution is of the form (Philip, 1969)

$$X = \lambda(s)T^{1/2} + \chi(s)T + \psi(s)T^{3/2} + \cdots \qquad (13.22)$$

Values for λ, χ, and ψ are in Table 13.4. The maximum value $s_f = 0.939$ is the exact degree of saturation for which the expected value of K_r, given by Eqs. (13.10) and (13.21), is the average intake rate of the field, 14.6 cm/day.

The single set of values in Table 13.4 is valid for all α values. Thus, infiltration for any specific conditions can be found once its specific α_r value is known. Figure 13.5 shows moisture profiles for choices $\alpha = 1.23$, 0.87, and 0.62. These are for the 75, 50, and 25 percentile values, respectively, when $\ln \alpha$ is assumed normal with mean -0.136 and standard deviation 0.512. Therefore, 25% of the α values are larger than 1.23, 25% are between 1.23 and 0.87, 25% are between 0.87 and 0.62, and 25% are smaller than 0.62.

The large α corresponds to a more permeable region, and we note in Figure 13.5 that for any given t the profiles are deeper than for the other

Table 13.4
VALUES FOR λ, χ, AND ψ^a

s	λ	χ	ψ
0.939	0.0	0.0	0.0
0.867	18.4	4.92	0.778
0.798	22.8	4.15	0.607
0.728	24.2	3.79	0.592
0.657	24.8	3.63	0.591
0.587	25.1	3.55	0.591
0.517	25.3	3.50	0.592
0.446	25.5	3.47	0.592
0.376	25.5	3.45	0.592
0.305	25.6	3.44	0.592
0.235	25.6	3.43	0.592
0.164	25.6	3.43	0.592
0.0939	25.6	3.43	0.592
0.0235	25.6	3.43	0.592
0.0	∞	3.43	0.592

^{a} After Warrick and Amoozegar-Fard, 1979.

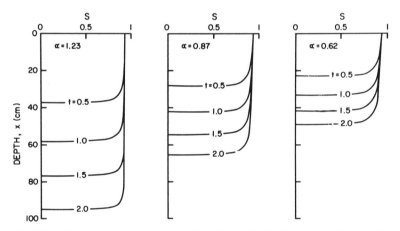

Fig. 13.5. Moisture profiles for $\alpha = 1.23$, 0.873, and 0.618 (75, 50, and 25 percentile values, respectively). The distribution of α is shown as Fig. 13.1b. (After Warrick and Amoozegar-Fard, 1979.)

two plots. For example, when $t = 1.5$ days, the left plot shows a wetting front at approximately 77 cm whereas the wetting front is only at about 42 cm for $\alpha = 0.62$. Thus, the heterogeneous field is a composite of such profiles for infiltration under ponded conditions. The scaling coefficients α quantitatively reflect the distribution of regions where intake is slow, medium, and fast. The distribution of the intake velocity (as a function of time and space) is also discussed by Warrick and Amoozegar-Fard (1979).

EXAMPLE 3. SOLUTE MOVEMENT

Movement of salts is strongly dependent on water velocities. Highly soluble ions such as chlorides, nitrates, and sodium move with water readily. Even the strongly adsorbed ions, organic molecules, and reactive materials are transported with the water.

The appropriate differential equation describing noninteracting solute flow is

$$\frac{\partial c}{\partial t} = D \frac{\partial^2 c}{\partial x^2} - v \frac{\partial c}{\partial x} \tag{13.23}$$

with c the soil solute concentration, D the apparent diffusion coefficient (units L^2/T), v the pore water velocity (units L/T and equal to the Darcian velocity divided by water-filled porosity), x the vertical depth of the soil, and t time.

The boundary conditions for a solute introduced continuously at the surface is

$$c(0, t) = 0, \qquad t < 0$$
$$c(0, t) = c_0, \qquad t > 0 \tag{13.24}$$

The initial condition is taken to be

$$c(x, 0) = 0 \tag{13.25}$$

The solution of Eq. (13.23) subject to Eqs. (13.24) and (13.25) is

$$c(x, t) = \tfrac{1}{2} c_0 \{ \text{erfc}[(x - vt)/2\sqrt{Dt}] + \exp(vx/D) \text{erfc}[(x + vt)/2\sqrt{Dt}] \} \tag{13.26}$$

where erfc is the complementary error function [cf. Eq. (7.1.2) of Abramowitz and Stegun, 1964].

A plot of the normalized concentration c/c_0 is given in Figure 13.6 and labeled deterministic. The values $D = 370$ cm^2/day and $v = 44$ cm/day are the mean values measured for the Panoche soil in Table 13.2 by Biggar and Nielsen (1976). The salt front is sigmoidal with $c/c_0 = 0.5$ at approximately 220 cm. For piston displacement and no dispersion, the front would have been a square wave exactly, with the front at $vt = 44 \times 5$ cm $= 220$ cm. The effect of the dispersion is to spread the salt front out.

A second average concentration was calculated assuming v and D were each log-normally distributed using a mean and standard deviation for

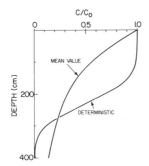

Fig. 13.6. Normalized average solute concentration c/c_0 as a function of depth after 5 days (see Example 3).

ln v of 3.01 and 1.25, respectively, and a mean and standard deviation for ln D of 4.42 and 1.74, respectively. These are also based on experimental results from the Panoche and correspond to the same population means as for the deterministic case. The calculations were carried out on the computer using a Monte Carlo simulation. The steps were

(1) Draw two random values from a population which is normally distributed with a variance of 1 and mean of zero.

(2) Calculate the corresponding random v and D by setting the random values equal to (ln v − 3.01)/1.25 and (ln D − 4.42)/1.74, respectively.

(3) Calculate a random c/c_0 from Eq. (13.26) for each t and x value of interest.

(4) Repeat (1)–(3) 1000 times and find the mean of these results for each space and time combination. (If necessary, we could go higher than 1000.)

The result is the mean value curve of Figure 13.6. Note c/c_0 is no longer sigmoidal but almost hyperbolic. At large depths, the salt is more concentrated than before; at shallow depths it is less concentrated. The depth for which $c/c_0 = 0.5$ is at about 130 cm, compared to about 220 for the deterministic case. This is primarily due to the large variation in v. The sites with large values of v carry the salt deep in the profile, and hence the c/c_0 is larger at deep depths. On the other hand, the very slow velocities contribute salt only at very shallow depths, and as a consequence the medium region has a lower concentration than the deterministic average. In fact, no single velocity can be chosen to give the shape of the true mean. The effect of D on the true mean is not as great; in fact, $D = 0$ results in almost the same answer for this example.

G. Autocorrelation and Spatial Analysis

The previous Monte Carlo examples made no mention of where the samples were located or how close together they might be. If the value of the parameters (such as convective velocity in the last case) were independent this would seem appropriate. However, intuitively we would anticipate measurements close together in the field to give values of approximately the same magnitude, while measurements farther apart would yield values differing by a greater order of magnitude. The framework for expressing such relationships has been developed in time series and spatial analysis. In particular, the field of geostatistics as developed by the South African and French mining engineers (Journel and Huijbregts, 1978) includes elaborate procedures for expressing such phenomena as well as including the effects in sampling schemes, resource estimates, and management practices of ore bodies.

An illustration for soils follows. In Table 13.5 are values of water retained at 0.1 bar on a clay loam in Arizona (Gajem and Warrick, unpublished data). One hundred samples were all collected on a line with a 7.5 cm diameter bucket auger from the 40–60 cm depth. Each sample was 20 cm away from the next, giving a total sampling length of 2000 cm. The tabulated values are arranged in order of collection and show a maximum value of 43 and a minimum of 29 percent based on the gravimetric water content. The experimental mean and standard deviations are 35 and 2.8, respectively, giving a coefficient of variation of (2.8/35)(100) or 8%. Are the values which are close together more similar than those from farther apart? To answer

Table 13.5

SEQUENCE OF 100 VALUES FOR PERCENT WATER (G/G) RETAINED AT 0.1 BAR[a]

37.5	37.6	36.7	39.0	35.0	36.0	36.8	36.5
37.1	36.6	37.9	38.8	40.0	40.6	37.6	38.7
37.9	42.1	35.9	38.0	38.5	38.3	37.4	37.1
41.8	37.2	35.4	39.1	35.3	34.0	39.0	35.5
35.0	35.7	35.1	34.1	34.0	34.7	34.2	35.2
31.7	30.7	32.7	33.7	35.3	32.6	33.5	31.4
30.9	31.4	35.5	34.2	32.5	30.5	28.9	33.0
30.9	31.7	32.0	30.9	42.8	31.0	34.3	31.2
29.1	31.7	32.9	32.1	32.8	32.2	34.4	32.2
32.2	32.4	33.3	34.3	31.8	36.9	33.9	35.2
36.4	34.7	38.9	35.0	35.6	36.9	35.3	36.4
33.7	34.2	30.9	34.3	33.2	35.3	35.6	35.1
34.4	34.0	33.6	36.6				

[a] Samples were at 50 cm depth, 20 cm apart on Pima clay loam. Values are ordered left to right.

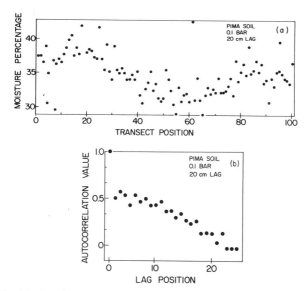

Fig. 13.7. (a) Plot of 0.1 bar water of Pima clay loam. (b) Autocorrelation for bar water of Pima clay loam.

this question, we first look at the numbers in the table or better yet plot them as a function of space. The 100 values are plotted as Fig. 13.7. Obviously, a lot of scatter is present, but there are more large values towards the left, lesser values in the middle section, and somewhat larger values towards the end.

An analytical procedure of examining the effects quantitatively is by the autocorrelation function r_k

$$r_k = c_k/\sigma_E^2 \tag{13.27}$$

with

$$c_k = \left(\frac{1}{n-k-1}\right) \sum_{i=1}^{n-k} (x_i - \sigma_E)(x_{i+k} - \sigma_E) \tag{13.28}$$

In our example, the series $x_1, x_2, \ldots, x_n$ is the 100 values of water content of Table 13.5. The index k is for the separation of k intervals (lag k). The function c_k is an autocovariance and σ_E the standard deviation [Eq. (13.2)]. The resulting values of r_k are shown as Table 13.6 for $k = 1$–25. When $k = 0$, c_k reduces simply to σ_E^2 and the autocorrelation is $r_0 = 1$, which is a trivial but limiting case. If we use $k = 1$, then we correlate x_1 with x_2, x_2 with $x_3, \ldots, x_{99}$ with x_{100} and find $r_1 = 0.52$. For $k = 2$, we correlate x_1 with x_3, x_2 with $x_4, \ldots, x_{98}$ with x_{100} and find $r_2 = 0.58$. We continue

Table 13.6
AUTOCORRELATION COEFFICIENTS[a]

Lag position	Autocorrelation coefficient	Lag position	Autocorrelation coefficient
0	1.00	14	0.35
1	0.52	15	0.28
2	0.58	16	0.25
3	0.54	17	0.30
4	0.45	18	0.15
5	0.55	19	0.14
6	0.47	20	0.13
7	0.53	21	0.06
8	0.47	22	0.14
9	0.44	23	0.00
10	0.49	24	−0.01
11	0.38	25	−0.03
12	0.40		
13	0.31		

[a] For 0.1 bar water-table data of Table 13.4.

to r_{25}, which correlates x_1 with x_{26}, x_2 with $x_{27}, \ldots, x_{75}$ with x_{100}, resulting in $r_{25} = -0.03$.

A plot of the autocorrelation is given as Fig. 13.7b. The maximum is 1 at zero lag ($k = 0$) and the values tend to decrease for larger lags. For our purposes Fig. 13.7b indicates that the values are indeed correlated over space. Had the samples been independent, we should expect the autocorrelation values to all be near zero, and clearly this is not the case.

For a well-behaved system (i.e., second-order stationary), an equivalent parameter is the semivariogram function γ_k

$$\gamma_k = c_0 - c_k \tag{13.29}$$

For zero lag ($k = 0$), the semivariogram function is 0 or shows no variance; as k increases γ_k approaches c_0 or the population variance, or *sill*. The semivariogram offers an advantage of existing for more general situation than the correlogram.

The shape of the semivariogram gives an indication of the spatial dependence of the soil physical property. If for all values of lag greater than zero γ_k is approximately constant ($\gamma_k = c_0$), the semivariogram indicates that the observations are spatially independent. If, on the other hand, values of γ_k approach c_0 gradually or in some consistent manner as the value of the lag increases, the observations are spatially dependent. Figure 13.8 shows a semivariogram constructed from 76 data points stemming from

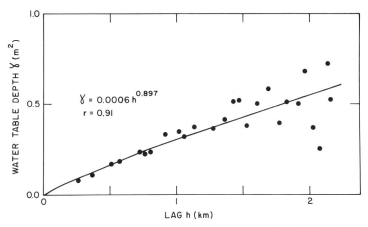

Fig. 13.8. Semivariogram of 76 water table depth observations within a 675 ha field.

water-table depth observations taken on a grid of 250 m in both directions in a 675 ha field in northern India (Dahiya, 1979). The variogram manifests a well-behaved variance structure with pairs of water-table depths differing on the average by less than 1 m over a horizontal distance of 2 km. Delhomme (1976) has provided an excellent discussion on the interpretation of semivariograms.

EXAMPLE 4. USING THE SEMIVARIOGRAM TO INTERPOLATE BETWEEN
            MEASURED VALUES

If the semivariogram indicates that the observations of the soil physical property have a spatial structure, it is advantageous to consider that spatial dependence in future sampling programs and to describe the distribution of observations over the area and to interpolate between observed values. The geostatistic extrapolation process developed by Matheron (1971) known as kriging (after D. G. Krige) is only briefly introduced here.

Let us assume that we have measured a soil-water property at n locations in a regular grid that covers the entire field under investigation. In order to interpolate between measured values, we seek an equation linking a given interpolated point with its neighbors. Owing to the fact that the semivariogram gives the expected relation between pairs of neighborhood points, it is obvious that different weights should be given to the surrounding points, depending on their distance from the one to be interpolated. Hence,

$$Z^*(x_0) = \sum_{j=1}^{n} \lambda_j Z(x_j) \tag{13.30}$$

where $Z^*(x_0)$ is the interpolated soil-water property value at point x_0, $Z(x_j)$ are the measured values at locations x_j, and λ_j are the weights calculated from the semivariogram under the condition

$$\sum_{j=1}^{n} \lambda_j = 1. \qquad (13.31)$$

The weights λ_j are calculated such that the expected value of $Z^*(x_0) - Z(x_0) = 0$ and the variance of $Z^*(x_0) - Z(x_0)$ is a minimum, where $Z(x_0)$ is the true value of the soil water property at point x_0. Details of the calculations for this simple kriging as well as for universal kriging are available (Matheron, 1963). The major advantages of kriging are that an independent estimate of the error of interpolation can be calculated and that the kriged values will be equal to the observations made at experimental locations. Moreover, the advantage of kriging over polynomial techniques is obvious inasmuch as the degree of the polynomial assumed is not related to the variance structure.

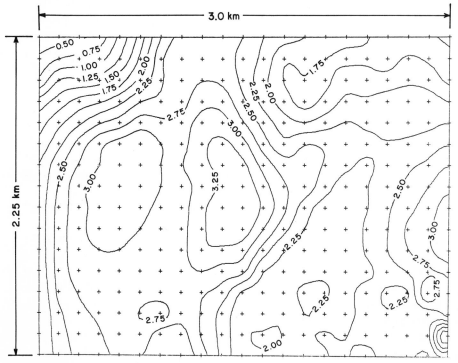

Fig. 13.9. Kriged water-table depth contours stemming from the semivariogram in Fig. 7. Water-table depths are given in meters above an arbitrary datum.

An example of a kriged set of data is given in Figure 13.9 for the water-table depths analyzed as a semivariogram in Figure 13.8. The figure shows water-table contours drawn through the 76 measured and 336 kriged points using a universal kriging technique. The plus marks indicate the 336 locations at which water tables were estimated by kriging. Not shown in the figure, but easily calculated, are the errors of estimate of each kriged value. From such estimates, and their spatial distribution over the entire field, it is relatively simple to determine whether or not additional measurements (and their locations) would be required to improve the accuracy and precision of the water-table contours.

H. Discussion

We have examined soil variability, presented numbers describing variation, estimated samples necessary for a given accuracy, demonstrated ramifications, and shown that a spatial structure exists at least for two parameters in two fields. Where does all of this lead to? Luckily the farmer has not had to wait for the perfect descriptive model to exist before producing food. Similarly, the soil mapper and surveyor has diligently pursued his course in an effective, productive manner.

Roughly, past and future efforts can be lumped into the interdependent categories of (1) survey and evaluation and (2) integration and problem solving. The use of geostatistics would seem to offer a logical quantitative framework for both categories of studies. If nothing else, it at least allows us to evaluate difficult and nebulous concepts which have been previously bypassed for expediency and lack of practical alternatives. For example, taking a sample seems simple at first, but allocation of resources and expected accuracy require the correct method, size, and location, each of which is a challenging problem in itself.

Regarding survey and evaluation, our efforts include patterns of frequency distributions, amounts of gross variability, and spatial structure (i.e., correlograms and semivariograms). Apparently, we need documentation for contrasting soils, spacings, and in some cases even methods of analysis—each for a long list of parameters. The subject offers areas of mutual interest between soil surveyors and other soil scientists. The surveyor has a rich store of data as well as a comprehensive taxonomic system. The methods of spectral analysis and spatial analysis have been used, incidentally, by a small group of researchers as a tool for classifying soils. Closely related problems include comparisons of several parameters as well as establishing "ground truths" for remote sensing by satellites and aircrafts.

The variation in soil physical parameters can be integrated to give the grower answers to his specific set of problems in order to help decide the best choices of management. On a larger scale, geographical areas can be considered in a comprehensive fashion to examine such things as runoff, flooding, erosion, forage production, or movements of salts and hazardous compounds.

The desired result is a solution to contemporary problems in harmony with economic and social constraints. The scientific components can be more clearly stated and more accurately analyzed and the results can be utilized with greater confidence.

14   *Solute Transport during Infiltration into Homogeneous Soil*

David E. Elrick

University of Guelph
Ontario, Canada

A. Introduction

Our objective in this chapter is to look a little deeper and to pursue a little further the analysis of water and solute movement during both horizontal and vertical infiltration. In order to keep the mathematics simple and the physics visible the analysis will be restricted to homogeneous soil under isothermal conditions. Hysteresis will also not be a factor because the water-content changes to be discussed are monotonic (the water content is always increasing during infiltration).

B. Horizontal Infiltration

The movement of solutes during horizontal infiltration can be analyzed in a similar way to that developed for the flow of water in Chapter 2. There the Boltzmann transformation was used to reduce the partial differential equation describing water movement [Eq. (2.22)] to the ordinary differential form of Eq. (2.32). We shall use the symbol λ (note that B is used in Part I) to represent the Boltzmann transformation [Eq. (2.27)]:

$$\lambda = xt^{-1/2} \tag{14.1}$$

Equation (10.19) in Hillel (1980) describes the mixing of solutes during one-dimensional water flow. If we now make the assumption that D_{sh} is a function of θ only and that any dependence on $\bar{v}$ can be neglected, Eq. (10.19) becomes

$$\frac{\partial(\theta c)}{\partial t} = \frac{\partial}{\partial x}\left[D_{sh}(\theta) \frac{\partial c}{\partial x} \right] - \frac{\partial(qc)}{\partial x} \tag{14.2}$$

The next step is to simplify Eq. (14.2) by making use of the one-dimensional form of the continuity equation [Eq. (9.4) in Hillel, 1980]:

$$\partial\theta/\partial t = -\partial q/\partial x \tag{14.3}$$

Use of Eq. (14.3) in Eq. (14.2) gives

$$\theta \frac{\partial c}{\partial t} = \frac{\partial}{\partial x}\left[D_{sh}(\theta) \frac{\partial c}{\partial x} \right] + D(\theta) \frac{\partial\theta}{\partial x} \frac{\partial c}{\partial x} \tag{14.4}$$

where use has been made of Eq. (9.21) in Hillel (1980):

$$q = -D(\theta) \, \partial\theta/\partial x$$

The substitution $\lambda = xt^{-1/2}$ into Eq. (14.4) removes both x and t and gives

$$-\frac{g(\theta)}{2} \frac{dc}{d\lambda} = \frac{d}{d\lambda}\left[D_{sh}(\theta) \frac{dc}{d\lambda} \right] \tag{14.5}$$

where

$$g(\theta) = \theta\lambda + 2D(\theta) \frac{d\theta}{d\lambda} = \theta\lambda - \int_{\theta_i}^{\theta} \lambda \, d\theta \tag{14.6}$$

In Eq. (14.6) use was made of the relationship

$$D(\theta) = -\frac{1}{2} \frac{d\lambda}{d\theta} \int_{\theta_i}^{\theta} \lambda \, d\theta \tag{14.7}$$

If $D(\theta)$ values are given, $g(\theta)$ can be obtained from $\theta\lambda + 2D(\theta) \, d\theta/d\lambda$. If on the other hand experimental $\theta(\lambda)$ values are given, $g(\theta)$ can be obtained from

$$\theta\lambda - \int_{\theta_i}^{\theta} \lambda \, d\theta.$$

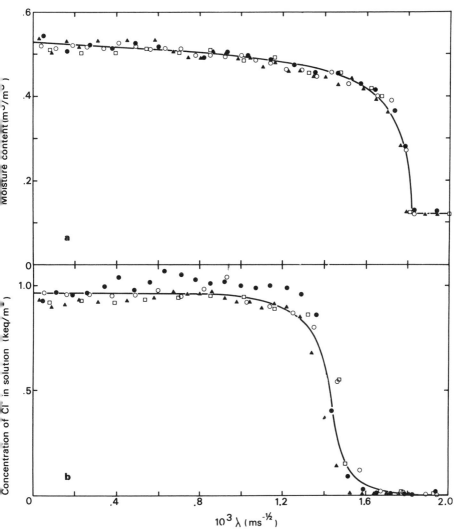

Fig. 14.1. Experimental (a) $\theta(\lambda)$ and (b) $c(\lambda)$ profiles during horizontal infiltration in a clay loam soil for four different time periods: (□) 3600, (●) 7200, (○) 14,400, and (▲) 21,600 sec.

Note that Eq. (14.5) observes similarity in terms of $\lambda = xt^{-1/2}$. Both $\theta(\lambda)$ and $c(\lambda)$ plots should therefore scale; i.e., $\theta(x)$ and $c(x)$ plots for various values of t should reduce to single $\theta(\lambda)$ and $c(\lambda)$ curves. The results of plotting such data are shown in Fig. 14.1 for a clay loam soil (Elrick *et al.*, 1979).

The appropriate boundary conditions for a semi-infinite soil column of constant initial concentration c_i where the end at $x = 0$ is instantly changed to concentration c_0 are given by

$$
\begin{aligned}
c &= c_i, & x \geq 0, \quad t = 0 \\
c &= c_0, & x = 0, \quad t \geq 0.
\end{aligned} \tag{14.8}
$$

Use of $\lambda = xt^{-1/2}$ reduces Eq. (14.8) to

$$
\begin{aligned}
c &= c_i, & \lambda \to \infty \\
c &= c_0, & \lambda = 0.
\end{aligned} \tag{14.9}
$$

The converse problem of solving for $c(\lambda)$ depends first on the calculation of $\theta(\lambda)$. This is required in order to obtain $g(\theta)$ in Equation (14.6). Because D_{sh} is a function of θ only and independent of the dependent variable c, Eq. (14.5) is linear and can be solved directly subject to Eq. (14.9) to give (Smiles et al., 1978)

$$
(c - c_0)/(c_i - c_0) = M(\lambda)/M(\infty) \tag{14.10}
$$

where

$$
M(\lambda) = \int_0^\lambda \left\{ D_{sh}^{-1}(\lambda) \exp\left[-\frac{1}{2} \int_0^\lambda D_{sh}^{-1}(\lambda) g(\lambda) \, d\lambda \right] \right\} d\lambda \tag{14.11}
$$

One can solve for $D_{sh}(\theta)$ in Eq. (14.5) by multiplying Eq. (14.5) by $d\lambda$ and integrating from $\lambda = \infty$ $(c = c_i)$ to $\lambda = \lambda$ $(c = c)$. Making use of the condition that $D_{sh}(\theta) \, dc/d\lambda = 0$ at $c = c_i$ gives

$$
D_{sh}(\theta) = -\frac{1}{2} \frac{d\lambda}{dc} \int_{c_i}^c g(c) \, dc \tag{14.12}
$$

Note that $g(c)$ can be obtained from the $g(\theta)$, $\theta(\lambda)$, and $c(\lambda)$ relationships. Therefore $D_{sh}(\theta)$ can be obtained from the appropriate slopes and areas indicated in Eq. (14.12).

Use of the experimental data in Fig. 14.1 and the application of Eq. (14.7) and (14.12) gave the D and D_{sh} data of Fig. 14.2. Note that $\Theta \equiv (\theta - \theta_i)/(\theta_0 - \theta_i)$. In this figure the high values of D_{sh} do not have a physical basis and are likely an anomaly of the analysis technique. The most valid values of $D_{sh}(\theta)$ are found in the region where the slope and area calculations of Eq. (14.12) are measurable. This restricts the Θ range to approximately $0.66 < \Theta < 0.88$. In this situation one can think of D_{sh} remaining constant at about 2×10^{-8} m²/sec.

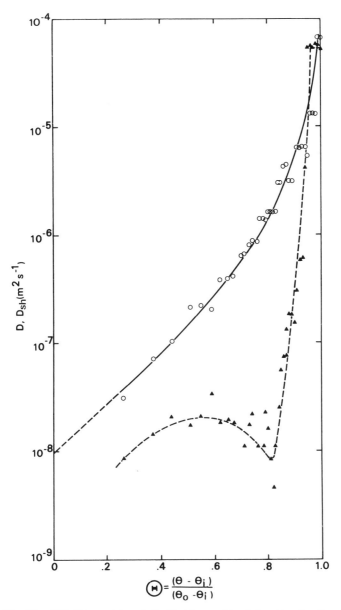

Fig. 14.2. The $D(\theta)$ ($\bigcirc$——$\bigcirc$) and $D_{sh}(\theta)$ ($\blacktriangle$– – –$\blacktriangle$) relationships for the clay loam soil of Fig. 14.1.

C. Vertical Infiltration

The vertical infiltration of water and solutes differs from horizontal infiltration only by the additional effect of gravity. The physics remains the same but the mathematics becomes considerably more complex. Specifically, the equations describing water and solute transport become

$$\frac{\partial \theta}{\partial t} = \frac{\partial}{\partial z}\left[D(\theta)\frac{\partial \theta}{\partial z}\right] - \frac{\partial K(\theta)}{\partial z} \tag{14.13}$$

and

$$\theta\frac{\partial c}{\partial t} = \frac{\partial}{\partial z}\left[\theta D_{sh}(\theta)\frac{\partial c}{\partial z}\right] + D(\theta)\frac{\partial \theta}{\partial z}\frac{\partial c}{\partial z} - K(\theta)\frac{\partial c}{\partial z} \tag{14.14}$$

Note that Eq. (14.13) is just Eq. (2.29).

The boundary conditions for a semi-infinite, vertical soil column are similar to those described previously for horizontal infiltration with x replaced by z. Note that the vertical space coordinate z is taken as positive downward.

Table 14.1
ORDINARY DIFFERENTIAL EQUATIONS FOR WATER AND SOLUTE TRANSPORT

Variable	Water	Solute
λ	$-\dfrac{\lambda_w}{2} = \dfrac{d}{d\theta}\left[D(\theta)\dfrac{d\theta}{d\lambda}\right]$	$\dfrac{1}{2}\theta\lambda_s + D(\theta)\dfrac{d\theta}{d\lambda} = -\dfrac{d}{dc}\left[D_s(\theta)\dfrac{dc}{d\lambda}\right]$
χ	$\chi_w = \dfrac{d}{d\theta}\left[P(\theta)\dfrac{d\chi_w}{d\theta}\right] + \dfrac{dK(\theta)}{d\theta}$	$\theta\chi_s - \left[P(\theta)\dfrac{d\chi_w}{d\theta} + K(\theta)\right] = \dfrac{d}{dc}\left[P_s(\theta)\dfrac{d\chi_s}{dc}\right]$
ψ	$\dfrac{3}{2}\psi_w = \dfrac{d}{d\theta}\left[P(\theta)\dfrac{d\psi_w}{d\theta} - Q(\theta)\right]$	$\dfrac{3}{2}\theta\psi_s - \left[P(\theta)\dfrac{d\psi_w}{d\theta} - Q(\theta)\right] = \dfrac{d}{dc}\left[P_s(\theta)\dfrac{d\psi_s}{dc} - Q_s(\theta)\right]$
ω	$2\omega_w = \dfrac{d}{d\theta}\left[P(\theta)\dfrac{d\omega_w}{d\theta} - R(\theta)\right]$	$2\theta\omega_s - \left[P(\theta)\dfrac{d\omega_w}{d\theta} - R(\theta)\right] = \dfrac{d}{dc}\left[P_s(\theta)\dfrac{d\omega_s}{dc} - R_s(\theta)\right]$
	$P(\theta) = D(\theta)\left(\dfrac{d\theta}{d\lambda}\right)^2$	$P_s(\theta) = \theta D_s(\theta)\left(\dfrac{dc}{d\lambda}\right)^2$
	$Q(\theta) = D(\theta)\dfrac{d\theta}{d\lambda}\left(\dfrac{d\chi_w}{d\lambda}\right)^2$	$Q_s(\theta) = \theta D_s(\theta)\dfrac{dc}{d\lambda}\left(\dfrac{d\chi_s}{d\lambda}\right)^2$
	$R(\theta) = D(\theta)\dfrac{d\theta}{d\lambda}\dfrac{d\chi_w}{d\lambda}\left[2\dfrac{d\psi_w}{d\lambda} - \left(\dfrac{d\chi_w}{d\lambda}\right)^2\right]$	$R_s(\theta) = \theta D_s(\theta)\dfrac{dc}{d\lambda}\dfrac{d\chi_s}{d\lambda}\left[2\dfrac{d\psi_s}{d\lambda} - \left(\dfrac{d\chi_s}{d\lambda}\right)^2\right]$

With respect to water the following conditions apply:

$$\theta = \theta_i, \quad z > 0, \quad t = 0$$
$$\theta = \theta_0, \quad z = 0, \quad t \geq 0 \tag{14.15}$$

With respect to the solute, similar conditions apply:

$$c = c_i, \quad z > 0, \quad t = 0$$
$$c = c_0, \quad z = 0, \quad t \geq 0 \tag{14.16}$$

In solving Eq. (14.13), Philip (1957) rewrote the equation with z as the

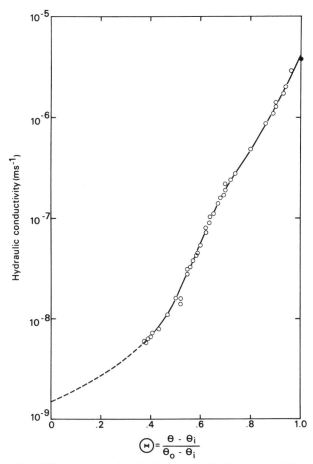

Fig. 14.3. The $K(\theta)$ relationship for the clay loam soil. The saturated hydraulic conductivity is shown as a filled circle at $\theta = 1.0$.

dependent variable and then developed a power series form for the solution [Eq. (2.31)]:

$$z = \lambda_w(\theta)t^{1/2} + \chi_w(\theta)t + \psi_w(\theta)t^{3/2} + \omega_w(\theta)t^2 + \cdots \qquad (14.17)$$

The subscript w refers to water. By a procedure similar to that developed by Philip (1957), a power series in $t^{1/2}$ may also be used to obtain a solution to Eq. (14.14) (Elrick *et al.*, 1979):

$$z = \lambda_s(c)t^{1/2} + \chi_s(c)t + \psi_s(c)t^{3/2} + \omega_s(c)t^2 + \cdots \qquad (14.18)$$

Here the subscript s refers to solute.

The functions $\lambda_w(\theta)$, $\lambda_s(c)$, $\chi_w(\theta)$, $\chi_s(c)$, etc. are each governed by separate ordinary differential equations. These equations can be obtained by substituting Eq. (14.17) and (14.18) into the respective differential equations written in the form where z is the dependent variable or by following the procedure outlined by Philip (1957). The equations are listed in Table 14.1.

Philip (1955, 1957) developed specific numerical procedures for solving the water-associated equations. Elrick *et al.* (1979) used simulation techniques based on the computer program System/360 Continuous System Modeling Program (CSMP) (IBM, 1972; Speckhart and Green, 1976) for solving both the water- and solute-associated equations.

The boundary conditions on λ are as follows:

$$\begin{aligned} \theta = \theta_i, \qquad c = c_i, \qquad \lambda \to \infty \\ \theta = \theta_0, \qquad c = c_0, \qquad \lambda = 0 \end{aligned} \qquad (14.19)$$

For χ, the conditions become

$$\begin{aligned} P\frac{d\chi_w}{d\theta} \to 0, \qquad P_s\frac{d\chi_s}{dc} \to 0, \qquad \lambda \to \infty \\ \chi_w = 0, \qquad \chi_s = 0, \qquad \lambda = 0 \end{aligned} \qquad (14.20)$$

The conditions on ψ and ω are given by Elrick *et al.* (1979).

The differential equations for χ_w and χ_s both explicity contain $K(\theta)$. The $K(\theta)$ relationship for the clay loam soil is shown in Fig. 14.3.

We now have all the inputs required to solve the associated water and solute differential equations of Table 14.1. The solutions are shown in Fig. 14.4. The $\Theta(\lambda)$ and $C(\lambda)$ [note that $C = (c - c_i)/(c_0 - c_i)$] curves are the same as the smoothed experimental data of Fig. 14.1. The $\chi_w, \chi_s, \psi_w, \psi_s, \omega_w,$ and ω_s data were obtained by use of the CSMP program referred to earlier. Note that the water-associated functions all extend to larger values of $\chi, \psi,$ and ω than the solute-associated functions. This is to be expected since the salt front lags behind that of the water for both horizontal and vertical infiltration. Also note that the orders of magnitude of $\lambda, \chi, \psi,$ and ω for both

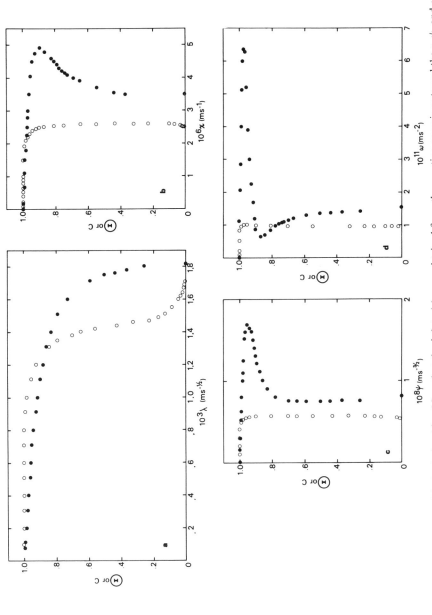

Fig. 14.4. The Θ (●) and C (○) relationships. Note that the λ data (a) were obtained from the sorption experiments and the χ, ψ, and ω data [(b)–(d) respectively] were calculated using the CSMP program.

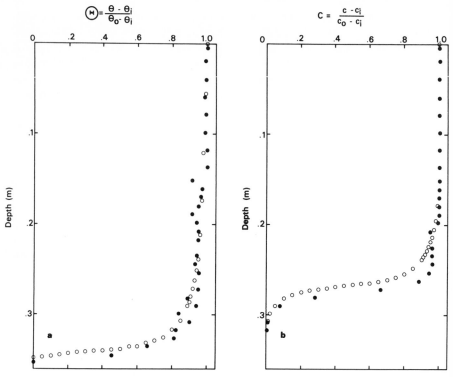

Fig. 14.5. Experimental (●) and predicted (○) infiltration profiles of $\Theta(z)$ and $C(z)$; $t = 19{,}260$ sec.

water and chloride are 10^{-3}, 10^{-6}, 10^{-8}, and 10^{-11} respectively. The domination of the leading terms of the expansions in Eq. (14.17) and (14.18) obviously depends upon the choice of time t and decreases as t increases in value.

An experiment was then carried out on vertical infiltration in the same clay loam soil. The experimental and predicted profiles of $\Theta(z)$ and $C(z)$ at 19,260 sec are shown in Fig. 14.5. The contribution of the ωt^2 term to the depth z for both the Θ and C profiles averaged about 1.5%. The agreement between the predicted and the experimental curves is as good as one might expect due to experimental differences in such things as packing the soil columns and in small initial water content changes from one experiment to another.

D. Summary

Both water and solute movement can be successfully predicted by the analysis presented here. The applicability is limited, however, because of the restrictions of homogeneous soil, of isothermal conditions, of constant initial water content, and of the constant water and salt contents of the soil surface. The thrust of these studies, however, was to develop appropriate models for situations that can be tested rigorously in the laboratory. If confirmed by laboratory experiments, the model can then be applied to the more complex situations found in the field. Numerical methods involving the method of lines (the compartmentalization approach), finite differences, and finite elements have been successfully used in these situations (Hillel, 1977; van Genuchten and Wierenga, 1974; Bresler and Hanks, 1969; van Genuchten, 1978).

Bibliography

Aase, J. K., and Kemper, W. D. (1968). Effect of ground color and microwatersheds on corn growth. *J. Soil Water Conserv.* **23,** 60–62.

Abramowitz, M., and Stegun, I. A. (1964). Handbook of mathematical functions. Nat. Bur. Stand. Appl. Math. Ser., Vol. 55, U.S. Government Printing Office, Washington, D.C.

Acevedo, E., Hsiao, T. C., and Henderson, D. W. (1971). Immediate and subsequent growth response of maize leaves to changes in water status. *Plant Physiol.* **48,** 631–636.

Adams, J. E. (1970). Effect of mulches and bed configuration. *Agron. J.* **62,** 785–790.

Adams, J. E., and Hanks, R. J. (1964). Evaporation from soil shrinkage cracks. *Soil Sci. Soc. Am. Proc.* **28,** 281–284.

Adrian, D. D., and Franzini, J. B. (1966). Impedance to infiltration by pressure build-up ahead of the wetting front. *J. Geophys. Res.* **71,** 5857–5862.

Alway, F. J., and McDole, G. R. (1917). Relation of the water-retaining capacity of a soil to its hygroscopic coefficient. *J. Agr. Res.* **9,** 27–71.

Anat, A., Duke, H. R., and Corey, A. T. (1965). Steady upward flow from water tables. Colorado State Univ. Hydrol. Paper No. 7, June.

Anderson, D. M., and Tice, A. R. (1971). Low-temperature phases of interfacial water in clay-water systems. *Soil Sci. Soc. Am. Proc.* **35,** 47–54.

Bachmat, Y., and Elrick, D. E. (1970). Hydrodynamic instability of miscible fluids in a vertical porous column. *Water Resources Res.* **6,** 156–171.

Barrs, H. D. (1968). Determination of water deficits in plant tissues. *In* "Water Deficits and Plant Growth" (T. T. Kozlowski, ed.), pp. 235–368. Academic Press, New York.

Bear, J. 1972. "Dynamics of Fluids in Porous Media." Elsevier, New York.

Belmans, C., Feyen, J., and Hillel, D. (1979). An attempt at experimental validation of macroscopic-scale models of soil moisture extraction by roots. *Soil Sci.* **127,** 174–186.

Bernacki, H., Haman, J., and Kanafojski, Cz. (1967). "Agricultural Machines: Theory and Construction." PWRIL, Warsaw, Poland. (Translated from Polish by U.S. Department of Agriculture and National Science Foundation, TT-69-50019).

Bertrand, A. R. (1965). Rate of water intake in the field. *In* "Methods of Soil Analysis" (C. A. Black, ed.). Monograph No. 9, Am. Soc. Agron., Madison, Wisconsin.

Bertrand, A. R. (1967). Water conservation through improved practices. *In* "Plant Environment and Efficient Water Use." Am. Soc. Agron., Madison, Wisconsin.

Beskow, Gunnar. (1935). Soil freezing and frost heaving with special applications to roads and railroads. *Swed. Geol. Soc.* **26,** C, 375. Translation (with special supplement of progress 1935–1946) by J. O. Osterberg. Tech. Inst., Northwestern Univ., Evanston, Illinois.

Biggar, J. W., and Nielsen, D. R. (1976). Spatial variability of the leaching characteristics of a field soil. *Water Resour. Res.* **12,** 78–84.

Bishop, A. W., and Blight, G. E. (1963). Some aspects of effective stress in saturated and partly saturated soils. *Geotechnique* **13,** 177–197.

Black, T. A., Gardner, W. R., and Thurtell, G. W. (1969). The prediction of evaporation, drainage and soil water storage for a bare soil. *Soil Sci. Soc. Am. Proc.* **33,** 655–660.

Black, T. A., Thurtell, G. W., and Tanner, C. B. (1968). Hydraulic load-cell lysimeter, construction, calibration, and tests. *Soil Sci. Soc. Am. Proc.* **32,** 632–639.

Black, T. A., Tanner, C. B., and Gardner, W. R. (1970). Evaporation from a snap bean crop. *Agron. J.* **62,** 66–69.

Blaney, H. F., and Criddle, W. D. (1950). Determining water requirements in irrigated areas from climatological and irrigation data. U.S. Soil Conservat. Serv. Tech. Publ. 96.

Bodman, G. B., and Colman, E. A. (1944). Moisture and energy conditions during downward entry of water into soils. *Soil Sci. Soc. Am. Proc.* **8,** 116–122.

Bolt, G. H., and Peech, M. (1953). The application of the Gouy theory to soil-water systems. *Soil Sci. Soc. Am. Proc.* **17,** 210–213.

Boltzmann, L. (1894). Zur integration des diffusiongleichung bei variabeln diffusions coefficienten. *Ann. Phys.* **53,** 959–964.

Bond, J. J., and Willis, W. O. (1969). Soil water evaporation: Surface residue rate and placement effects. *Soil Sci. Soc. Am. Proc.* **33,** 445–448.

Bouwer, H. (1959). Theoretical aspects of flow above the water table in tile drainage of shallow homogeneous soil. *Soil Sci. Soc. Am. Proc.* **23,** 200–203.

Bouwer, H. (1962). Analyzing groundwater mounds by resistance network analog. *J. Irrig. Drain. Div. Proc. Am. Soc. Civ. Eng.* **88,** IR3, 15–36.

Bouwer, H. (1964). Resistance network analogs for solving ground-water problems. *Ground-Water* **2** (3), 1–7.

Bouwer, H. (1978). "Groundwater Hydrology." McGraw-Hill, New York.

Bower, H., and van Schilfgaarde, J. (1963). Simplified method of predicting fall of water table in drained land. *Trans. Am. Soc. Agr. Eng.* **6,** 196, 288–291.

Braester, C. (1973). Linearized solution of infiltration at constant rate. *In* "Physical Aspects of Soil, Water and Salts in Ecosystems" (A. Hadas *et al.*, eds.). Springer-Verlag, Berlin and New York.

Bresler, E. (1972). Interacting diffuse layers in mixed mono-divalent ionic systems. *Soil Sci. Soc. Am. Proc.* **36,** 891–896.

Bresler, E., and Hanks, R. J. (1969). Numerical method for estimating simultaneous flow of water and salt in unsaturated soils. *Soil Sci. Soc. Am. Proc.* **33,** 827–832.

Bresler, E., and Miller, R. D. (1975). Estimation of pore blockage induced by freezing of unsaturated soil. *Proc. Conf. on Soil-Water Prob. in Cold Reg. Calgary.* 162–175.

Bresler, E., Kemper, W. D., and Hanks, R. J. (1969). Infiltration, redistribution, and subsequent evaporation of water from soil as affected by wetting rate and hysteresis. *Soil Sci. Soc. Am. Proc.* **33,** 832–840.

Briggs, L. J., and Shantz, H. L. (1921). The relative wilting coefficient for different plants. *Bot. Gaz.* (*Chicago*) **53,** 229–235.

Bronowski, J. (1977). "A Sense of the Future." MIT Press, Cambridge, Massachusetts.

Browning, G. M. (1950). Principles of soil physics in relation to tillage. *Agr. Eng.* **31,** 341–344.

Bruce, R. R., and Klute, A. (1956). The measurement of soil-water diffusivity. *Soil Sci. Soc. Am. Proc.* **20,** 458–562.

Bruce, R. R., and Whisler, F. D. (1973). Infiltration of water into layered field soils. *In* "Physical Aspects of Soil, Water and Salts in Ecosystems" (A. Hadas *et al.*, eds.). Springer-Verlag, Berlin and New York.

Brustkern, R. L., and Morel-Seytoux, H. J. (1970). Analytical treatment of two-phase infiltration. *J. Hydraul. Div. ASCE* **96,** 2535–2548.

Burwell, R. E., and Larson, W. E. (1969). Infiltration as influenced by tillage-induced random roughness and pore space. *Soil Sci. Soc. Am. Proc.* **33,** 449–452.

Busscher, W. J. (1979). Simulation of infiltration from a continuous and intermittent subsurface source. *Soil Sci.* **128** (in press).

Byers, G. L., and Webber, L. R. (1957). Tillage practices in relation to crop yields, power requirements, and soil properties. *Can. J. Soil Sci.* **37,** 71–75.

Campbell, G. S. (1977). "An Introduction to Environmental Biophysics." Springer-Verlag, New York.

Carslaw, H. S., and Jaeger, J. C. (1959). "Conduction of Heat in Solids." Oxford Univ. Press, London and New York.

Cary, J. W. (1963). Onsager's relations and the non-isothermal diffusion of water vapor. *J. Phys. Chem.* **67,** 126–129.

Cary, J. W. (1964). An evaporation experiment and its irreversible thermodynamics. *Int. J. Heat Mass Transfer* **7,** 531–538.

Cary, J. W. (1968). An instrument for *in situ* measurement of soil moisture flow and suction. *Soil Sci. Soc. Am. Proc.* **32,** 3–5.

Cameron, D. R. (1978). Variability of soil water retention curves and predicted hydraulics conductivities on a small plot. *Soil Sci.* **126,** 364–371.

Carvallo, H. O., Cassell, D. K., Hammond, J., and Bauer, A. (1976). Spatial variability of *in situ* unsaturated hydraulic conductivity of Maddock sandy loam. *Soil Sci.* **121,** 1–7.

Casagrande, A. (1937). Seepage through dams. *J. New England Water Works Assoc.* **51,** 131–172.

Cassel, D. K., and Bauer, A. (1975). Spatial variability in soils below depth of tillage. Bulk density and fifteen atmosphere percentage. *Soil Sci. Soc. Am. Proc.* **39,** 247–250.

Cedergren, H. R. (1967). "Seepage Drainage and Flow Nets." Wiley, New York.

Chahal, R. L. S. (1961). Supercooling of water in glass capillaries. Ph.D. Thesis, Cornell Univ., Univ. Microfilms.

Chahal, R. S. and Miller, R. D. (1965). Supercooling of water in glass capillaries. *Brit. J. Appl. Phys.* **16,** 231–239.

Chen, Y., and Banin, A. (1975). Scanning electron microscope (SEM) observations of soil structure changes induced by sodium-calcium exchange in relation to hydraulic conductivity. *Soil Sci.* **120,** 428–436.

Childs, E. C. (1947). The water table equipotentials and streamlines in drained land. *Soil Sci.* **63,** 361–376.

Childs, E. C. (1969). "An Introduction to the Physical Basis of Soil Water Phenomena." Wiley, New York.

Childs, E. C., and Poulovassilis, A. (1962). The moisture profile above a moving water table. *J. Soil Sci.* **13,** 272–285.

Chow, V. T. (ed.) (1964). "Handbook of Applied Hydrology." McGraw-Hill, New York.

Clarke, R. T. and Newson, M. D. (1978). Some detailed water balance studies of research catchments. *Proc. R. Soc. London Series A* **363,** 21–42.

Coelho, M. A. (1974). Spatial variability of water related soil physical properties. Ph.D. Dissertation, Univ. of Ariz., Tucson, Arizona (available as 75-11,061 from Xerox Univ. Microfilms, Ann Arbor, Michigan).

Coleman, E. A., and Bodman, G. B. (1945). Moisture and energy conditions during downward entry of water into moist and layered soils. *Soil Sci. Soc. Am. Proc.* **9,** 3–11.

Collis-George, N., and Youngs, E. G. (1958). Some factors determining water table heights in drained homogeneous soils. *J. Soil Sci.* **9**, 332–338.

Cooper, A. W., and Trouse, A. C., and Dumas, W. T. (1969). Controlled traffic in row crop production. *Proc. Int. Cong. Agric. Eng. (CIGR), 7th*, Baden-Baden, Germany. Section III, 1–6.

Corey, A. T. (1977). "Mechanics of Heterogeneous Fluids in Porous Media." Water Resources Publications, Fort Collins, Colorado.

Covey, W. (1963). Mathematical study of the first stage of drying of a moist soil. *Soil Sci. Soc. Am. Proc.* **27**, 130–134.

Cowan, I. R. (1965). Transport of water in the soil-plant-atmosphere system. *J. Appl. Ecol.* **2**, 221–229.

Cowan, I. R. and Milthorpe, F. L. (1968). Plant factors influencing the water status of plant tissues. *In* "Water Deficits and Plant Growth" (T. T. Kozlowski, ed.), pp. 137–193. Academic Press, New York.

Crank, J. (1956). "The Mathematics of Diffusion." Oxford Univ. Press, London and New York.

Dahiya, I. S. (1979). Personal communication, Haryana University, India.

Davies, J. A., and Allen, C. D. (1973). Equilibrium, potential and actual evaporation from cropped surfaces in S. Ontario. *J. Appl. Meteorol.* **12**, 649–657.

Day, P. R., and Luthin, J. (1956). A numerical solution of the differential equation of flow for a vertical drainage problem. *Soil Sci. Soc. Am. Proc.* **20**, 443–447.

Deacon, E. L., Priestley, C. H. B., and Swinbank, W. C. (1958). Evaporation and the water balance. Climatol.-Rev. Res. Arid Zone Res. 9–34 (UNESCO).

Decker, W. L. (1959). Variations in the net exchange of radiation from vegetation of different heights. *J. Geophys. Res.* **64**, 1617–1619.

Delhomme, J. P. (1976). Applications de la theorie des variables regionalises dans les sciences de l'eau. Thesis de Docteur-Ingenieur, Université Pierre et Marie Curie, Paris. 160 pp.

Denmead, O. T. and Shaw, R. H. (1962). Availability of soil water to plants as affected by soil moisture content and meteorological conditions. *Agron. J.* **54**, 385–390.

de Vries, D. A. (1963). Thermal properties of soils. *In* "Physics of the Plant Environment" (van Wijk, W. R., ed.). North–Holland, Amsterdam.

De Wiest, R. J. M. (ed.) (1969). "Flow Through Porous Media," Academic Press, New York.

de Wit, C. T. (1958). Transpiration and plant yields. *Versl. Landbouwk. Onderz.* **646**, 59–84.

de Wit, C. T., and van Keulen, H. (1972). "Simulation of Transport Processes in Soils." PUDOC, Wageningen, Netherlands.

d'Hollander, E., and Impens, I. (1975). Hybrid simulation of a dynamic model for water movement in a soil-plant-atmosphere continuum. *In* "Computer Simulation of Water Resources Systems" (G. C. Vansteenkiste, ed.), pp. 349–360. North-Holland Publ., Amsterdam.

Dirksen, C., and Miller, R. D. (1966). Closed-system freezing of unsaturated soil. *Soil Sci. Soc. Am. Proc.* **30**, 168–173.

Dixon, R., and Linden, M. (1972). Soil air pressure and water infiltration under border irrigation. *Soil Sci. Soc. Am. Proc.* **36**, 948–953.

Domenico, P. A. (1972). "Concepts and Models in Groundwater Hydrology." McGraw-Hill, New York.

Donahue, R. L., Shickluna, J. C., and Robertson, L. S. (1971). "Soils—An Introduction to Soils and Plant Growth," 3rd ed. Prentice-Hall, Englewood Cliffs, New Jersey.

Donnan, W. W. (1947). Model tests of a tile-spacing formula. *Soil Sci. Soc. Am. Proc.* **11**, 131–136.

Doorenbos, J., *et al.* (1975). Irrigation. FAO, Rome.

Dumas, W. T., Trouse, A. C., Smith, L. A., Kummer, F. A., and Gill, W. R. (1975). Traffic control as a means of increasing cotton yields by reducing soil compaction. Paper No. 75-1050, 1975 presented at the Annual Meeting of Am. Soc. Agric. Eng., Davis, California.

Eagleson, P. S. (1970). "Dynamic Hydrology." McGraw-Hill, New York.

Edlefsen, N. E., and Anderson, A. B. C. (1943). Thermodynamics of soil moisture. *Hilgardia* **15,** 31–298.

Edwards, W. M., and Larson, W. E. (1970). Infiltration of water into soils as influenced by surface seal development. *Soil Sci. Soc. Am. Proc.* **34,** 101.

Ekern, P. C. (1950). Raindrop impact as the force initiating soil erosion. *Soil Sci. Soc. Am. Proc.* **15,** 7–10.

Elrick, D. E., and Laryea, K. B. (1979). Sorption of water in soils: a comparison of techniques for solving the diffusion equation. *Soil Sci.* **128,** 210.

Elrick, D. E., Scandrett, J. H., and Miller, E. E. (1959). Tests of capillary flow scaling. *Soil Sci. Soc. Am. Proc.* **23,** 329–332.

Elrick, D. E., Laryea, K. B., and Groenevelt, P. H. (1979). Hydrodynamic dispersion during infiltration of water into soil. *Soil Sci. Soc. Am. J.* **43,** 856–865.

Emerson, W. W. (1959). The structure of soil crumbs. *J. Soil Sci.* **10,** 235.

Epstein, E. (1973). Roots. *Sci. Am.* **228,** 48–58.

Epstein, E. (1977). The role of roots in the chemical economy of life on earth. *Bioscience* **27,** 783–787.

Erickson, A. E. (1972). Improving the water properties of sand soil. *In* "Optimizing the Soil Physical Environment Toward Greater Crop Yields" (D. Hillel, ed.), pp. 35–42. Academic Press, New York.

Ernst, L. F. (1962). Groundwater flow in the saturated zone and its calculation when parallel horizontal open conduits are present. *Versl. Landb. Ond.* **67.15,** 189 p.

Evenari, M., and Koller, D. (1956). Masters of the desert. *Sci. Am.* **194,** 39–45.

Evenari, M., Aharone, Y., Shanan, L., and Tadmor, N. H. (1958). The ancient agriculture of the Negev. III. Early beginnings. *Israel Explor. J.* **8,** 231–268.

Everett, D. H. (1961). The thermodynamics of frost damage to porous solids. *Trans. Faraday Soc.* **57,** 1541–1551.

Fairbourn, M. L., and Cluff, C. B. (1974). Use gravel mulch to save water for crops. *Crops and Soil Magazine* April–May.

Fairbourn, M. L., and Kemper, W. D. (1971). Microwatersheds and ground color for sugarbeet production. *Agron. J.* **63,** 101–104.

Feddes, R. A., Bresler, E., and Neuman, S. P. (1974). Field test of a modified numerical model for water uptake by root systems. *Water Resources Res.* **10,** 1199–1206.

Feddes, R. A., Neuman, S. P., and Bresler, E. (1976). Finite element analysis of two-dimensional flow in soils considering water uptake by roots: II. Field applications. *Soil Sci. Soc. Am. Proc.* **39,** 231–237.

Feyen, J., Belmans, C., and Hillel, D. (1980). Comparison between measured and simulated plant water potential during soil water extraction by potted ryegrass. *Soil Sci.* **129,** 180.

Fireman, M. (1957). Salinity and alkali problems in relation to high water tables in soils. *In* "Drainage of Agricultural Lands," pp. 505–513. Monograph 7, Am. Soc. Agron., Madison, Wisconsin.

Fisher, R. A. (1923). Some factors affecting the evaporation of water from soil. *J. Agr. Sci.* **13,** 121–143.

Fisher, R. A. (1928). Further note on the capillary forces in an ideal soil. *J. Agric. Sci.* **18,** 406–410.

Fleming, G. (1972). "Computer Simulation Techniques in Hydrology." American Elsevier, New York.

Fok, Y. S. (1970). One-dimensional infiltration into layered soils. *J. Irrig. Drainage Div. ASCE* **96**, 121–129.

Forchheimer, P. (1930). "Hydraulik," 3rd ed. Teubner, Leipzig and Berlin.

Forsgate, J. A., Hosegood, P. H., and McCullock, J. S. G. (1965). Design and installation of semienclosed hydraulic lysimeters. *Agr. Meteorol.* **2**, 43–52.

Fountaine, E. R., and Payne, P. C. J. (1952). The effect of tractors on volume weight and other soil properties. Rept. 17, Nat'l Inst. Agr. Eng., Silsoe, England.

Free, J. R., and Palmer, V. J. (1940). Relationship of infiltration, air movement, and pore size in graded silica sand. *Soil Sci. Soc. Am. Proc.* **5**, 390–398.

Freeze, R. A. (1969). The mechanism of natural groundwater recharge and discharge. *Water Resour. Res.* **5**, 153–171.

Freeze, R. A., and Cherry, J. A. (1979). "Groundwater." Prentice–Hall, Englewood Cliffs, New Jersey.

Frenkel, H., Goertzen, J. O., and Rhoades, J. D. (1978). Effects of clay type and content, exchangeable sodium percentage, and electrolyte concentration on clay dispersion and soil hydraulic conductivity. *Soil Sci. Soc. Am. J.* **42**, 32–39

Fried, J. J. (1976). "Ground Water Pollution." Elsevier, Amsterdam.

Fritton, D. D., Kirkham, D., and Shaw, R. H. (1970). Soil water evaporation, isothermal diffusion, and heat and water transfer. *Soil Sci. Soc. Am. Proc.* **34**, 183–189.

Fuchs, M., and Tanner, C. B. (1967). Evaporation from a drying soil. *J. Appl. Meteorol* **6**, 852–857.

Fuchs, M., Tanner, C. B., Thurtell, G. W., and Black, T. A. (1969). Evaporation from drying surfaces by the combination method. *Agron. J.* **61**, 22–26.

Gardner, H. R., and Hanks, R. J. (1966). Evaluation of the evaporation in soil by measurement of heat flux. *Soil Sci. Soc. Am. Proc.* **30**, 425–428.

Gardner, W. R. (1956). Calculation of capillary conductivity from pressure plate outflow data. *Soil Sci. Soc. Am. Proc.* **20**, 317–320.

Gardner, W. R. (1950). Some steady state solutions of the unsaturated moisture flow equation with application to evaporation from a water table. *Soil Sci.* **85**, 228–232.

Gardner, W. R. (1959). Solutions of the flow equation for the drying of soils and other porous media. *Soil Sci. Soc. Am. Proc.* **23**, 183–187.

Gardner, W. R. (1960). Dynamic aspects of water availability to plants. *Soil Sci.* **89**, 63–73.

Gardner, W. R. (1962a). Approximate solution of a non-steady state drainage problem. *Soil Sci. Soc. Am. Proc.* **26**, 129–132.

Gardner, W. R. (1962b). Note on the separation and solution of diffusion type equations. *Soil Sci. Soc. Am. Proc.* **26**, 404.

Gardner, W. R. (1964). Relation of root distribution to water uptake and availability. *Agron. J.* **56**, 35–41.

Gardner, W. R., and Fireman, M. (1958). Laboratory studies of evaporation from soil columns in the presence of a water table. *Soil Sci.* **85**, 244–249.

Gardner, W. R., and Hillel, D. (1962). The relation of external evaporative conditions to the drying of soils. *J. Geophys. Res.* **67**, 4319–4325.

Gardner, W. R., and Mayhugh, M. S. (1958). Solutions and tests of the diffusion equation for the movement of water in soil. *Soil Sci. Soc. Am. Proc.* **22**, 197–201.

Gardner, W. R., Hillel, D., and Benyamini, Y. (1970a). Post irrigation movement of soil water: I. Redistribution. *Water Resources Res.* **6**, 851–861.

Gardner, W. R., Hillel, D., and Benyamini, Y. (1970b). Post irrigation movement of soil water: II. Simultaneous redistribution and evaporation. *Water Resources Res.* **6**, 1148–1153.

Gash, J. H. C., and Stewart, J. B. (1975). The average surface resistance of a pine forest. *Boundary-Layer Meteorol.* **8**, 453–464.

Gill, W. R. (1974). Tillage and soybean root growth. *Proc. Conf. on Soybean Production, Marketing and Use* (sponsored by TVA, Am. Soybean Assoc. and Natl. Soybean Crop Improvement Council), pp. 66–72. Chattanooga, Tennessee.

Gill, W. R. (1979). Tillage. *In* "The Encyclopedia of Soil Science," Part 1 (Fairbridge, R. W., and Finkl, C. W., Jr., eds.). pp. 566–571. Dowden, Hutchinson & Ross, Stroudsburg, Pennsylvania.

Gill, W. R., and Cooper, A. (1972). Soil compaction—its causes and remedies. *Proc. Western Cotton Production Conf.*, Bakersfield, California, pp. 11–13.

Gill, W. R., and VandenBerg, G. E. (1967). "Soil Dynamics in Tillage and Traction," U.S. Department of Agriculture Handbook No. 316.

Gold, L. W. (1957). A possible force mechanism associated with the freezing of water in porous materials. *High. Res. Board, Bull.* **168,** 65–72.

Goldberg, D., and Shmueli, M. (1970). Drip irrigation—a method used under arid and desert conditions of high water and soil salinity. *Trans. ASAE* **13,** 38–41.

Goryachkin, W. P. (1968). "Collected Works," Vols. 1 and 2. Kolos, Moscow. (Translated from Russian by U.S. Department of Agriculture and National Science Foundation, TT-72-50023 and TT-71-50087).

Gouy, M. (1910). Sur la constitution de la charge electrique a la surface d'un electrolyte. *Ann. Phys.* **9,** 457–468.

Grace, J. (1978). "Plant Response to Wind." Academic Press, New York.

Graham, W. G., and King, K. M. (1961). Fraction of net radiation utilized in evapotranspiration from a corn crop. *Soil Sci. Soc. Am. Proc.* **25,** 158–160.

Greacen, E. L., Ponsana, P., and Barley, K. P. (1976). Resistance to water flow in the roots of cereals. *In* "Water and Plant Life, Ecological Studies 19," pp. 86–100. Springer-Verlag, Berlin and New York.

Greb, B. W. (1966). Effect of surface-applied wheat straw on soil water losses by solar distillation. *Soil Sci. Soc. Am. Proc.* **30,** 786–788.

Green, W. H., and Ampt, G. A. (1911). Studies on soil physics: I. Flow of air and water through soils. *J. Agr. Sci.* **4,** 1–24.

Gumaa, G. S. (1978). Spatial variability of *in situ* available water. Ph.D. Dissertation. The Univ. of Ariz., Tucson, Arizona. (available as 78-24365 from Xerox, University Microfilms, Ann Arbor, Michigan).

Hadas, A., and Hillel, D. (1968). An experimental study of evaporation from uniform columns in the presence of a water table. *Trans. Int. Congr. Soil Sci.*, *9th, Adelaide* **I,** 67–74.

Haines, W. B. (1930). Studies in the physical properties of soils. V. The hysteresis effect in capillary properties and the modes of moisture distribution associated therewith. *J. Agr. Sci.* **20,** 97–116.

Halstead, M. H., and Covey, W. (1957). Some meteorological aspects of evapotranspiration. *Soil Sci. Soc. Am. Proc.* **21,** 461–464.

"Handbook of Chemistry and Physics." (1968). 49th ed. The Chemical Rubber Co., Cleveland, Ohio.

Hanks, R. J. (1980). Yield and water use relationships. *In* "Efficient Water Use in Crop Production." Am. Soc. Agron., Madison, Wisconsin (in press).

Hanks, R. J., and Woodruff, N. P. (1958). Influence of wind on water vapor transfer through soil, gravel and straw mulches. *Soil Sci.* **86,** 160–164.

Hanks, R. J., and Bowers, S. A. (1960). Nonsteady-state moisture, temperature, and soil air pressure approximation with an electric simulator. *Soil Sci. Soc. Am. Proc.* **24,** 247–252.

Hanks, R. J., and Bowers, S. A. (1962). Numerical solution of the moisture flow equation for infiltration into layered soils. *Soil Sci. Soc. Am. Proc.* **26,** 530–534.

Hanks, R. J., and Gardner, H. R. (1965). Influence of different diffusivity-water content relations on evaporation of water from soils. *Soil Sci. Soc. Am. Proc.* **29**, 495–498.

Hanks, R. J., Bowers, S. B., and Boyd, L. D. (1961). Influence of soil surface conditions on the net radiation, soil temperature, and evaporation. *Soil Sci.* **91**, 233–239.

Hanks, R. J., Gardner, H. R., and Fairbourn, M. L. (1967). Evaporation of water from soils as influenced by drying with wind or radiation. *Soil Sci. Soc. Am. Proc.* **31**, 593–598.

Hansen, G. K. (1975). A dynamic continuous simultation model of water state and transportation in the soil-plant-atmosphere system. *Acta Agr. Scand.* **25**, 129–149.

Hanway, J. J. (1963). Growth stages of corn (Zea mays, L.). *Agron. J.* **55**, 487–492.

Harlan, R. L. (1973). An analysis of coupled heat-fluid transport in partially frozen soil. *Water Resour. Res.* **9**, 1314–1325.

Harr, M. E. (1962). "Groundwater and Seepage." McGraw-Hill, New York.

Harrold, L. L. (1966). Measuring evapotranspiration by lysimetry. *In* "Evapotranspiration and its Role in Water Resources Management." Am. Soc. Agr. Eng., St. Joseph, Michigan.

Hawkins, J. C. (1959). Cultivations. *Natl. Inst. Agric. Eng. Ann. Rep.*, 1–7.

Herkelrath, W. N. (1975). Water uptake by plant roots. Ph.D. dissertation, Univ. of Wisconsin, Madison, Wisconsin.

Hesstvedt, E. (1964). The interfacial energy ice/water. Pub. #56, Norwegian Geotech. Inst.

Hide, J. C. (1954). Observation on factors influencing the evaporation of soil moisture. *Soil Sci. Soc. Am. Proc.* **18**, 234–239.

Hide, J. C. (1958). Soil moisture conservation in the Great Plains. *Adv. Agron.* **10**, 23–26.

Hill, E. D., and Parlange, J.-Y. Wetting front instability in layered soils. *Soil Sci. Soc. Am. Proc.* **36**, 697–702.

Hill, R. W., Hanks, R. J., Keller, J., and Rasmussen, V. P. (1974). Predicting corn growth as affected by water management: An example. CUSUSWASH 211(d)-6. Utah State Univ., Logan, Utah. p. 18.

Hillel, D. (1960). Crust formation in loessial soils. *Trans. Int. Soil Sci. Congr., 7th, Madison, Wisconsin* **I**, 330–339.

Hillel, D. (1964). Infiltration and rainfall runoff as affected by surface crusts. *Trans. Int. Soil Sci. Congr., 8th, Bucharest* **2**, 53–62.

Hillel, D. (1967). "Runoff Inducement in Arid Lands." Hebrew Univ. of Jerusalem, Israel.

Hillel, D. (1968). Soil Water Evaporation and Means of Minimizing It. Rep. to U.S. Dept. Agr., Hebrew Univ. of Jerusalem, Israel.

Hillel, D. (1971). "Soil and Water: Physical Principles and Processes." Academic Press, New York.

Hillel, D. (1975). Evaporation from bare soil under steady and diurnally fluctuating evaporativity. *Soil Sci.* **120**, 230–237.

Hillel, D. (1976a). On the role of soil moisture hysteresis in the suppression of evaporation from bare soil. *Soil Sci.* **122**, 309–314.

Hillel, D. (1976b). Soil management. *In* "McGraw-Hill Yearbook of Science and Technology." McGraw-Hill, New York.

Hillel, D. (1977). "Computer Simulation of Soil-Water Dynamics." Int. Dev. Res. Centre, Ottawa, Canada.

Hillel, D. (1980). "Fundamentals of Soil Physics." Academic Press, New York.

Hillel, D., and Gardner, W. R. (1969). Steady infiltration into crust-topped profiles. *Soil. Sci.* **108**, 137–142.

Hillel, D., and Gardner, W. R. (1970). Transient infiltration into crust-topped profiles. *Soil Sci.* **109**, 410–416.

Hillel, D., and Guron, Y. (1973). Relation between evapotranspiration rate and maize yield. *Water Resources Res.* **9**, 743–748.

Hillel, D., and Hornberger, G. M. (1979). Physical model of the hydrology of sloping heterogeneous fields. *Soil Sci. Soc. Am. J.* **43,** 434–439.

Hillel, D., and Rawitz, E. (1972). Soil moisture conservation. *In* "Water Deficits and Plant Growth" (T. T. Kozlowski, ed.), pp. 307–337. Academic Press, New York.

Hillel, D., and Talpaz, H. (1976). Simulation of root growth and its effect on the pattern of soil water uptake by a nonuniform root system. *Soil Sci.* **121,** 307–312.

Hillel, D., and Talpaz, H. (1977). Simulation of soil water dynamics in layered soils. *Soil Sci.* **123,** 54–62.

Hillel, D., and van Bavel, C. H. M. (1976). Dependence of profile water storage on soil hydraulic properties: a simulation model. *Soil Sci. Soc. Am. J.* **40,** 807–815.

Hillel, D., Ariel, D., Orlowski, S., Stibbe, E., Wolf, D., and Yavnai, A. (1969). Soil-crop-tillage interactions in dryland and irrigated farming. Research report submitted to the U.S. Dept. Agric. by the Hebrew University of Jerusalem, Jerusalem, Israel.

Hillel, D., Gairon, S., Falkenflug, V., and Rawitz, E. (1969). New design of a low-cost hydraulic lysimeter for field measurement of evapo-transpiration. *Israel J. Agr. Res.* **19,** 57–63.

Hillel, D., Krentos, V., and Stylianou, Y. (1972). Procedure and test of an internal drainage method for measuring soil hydraulic characteristics in situ. *Soil Sci.* **114,** 395–400.

Hillel, D., van Beek, C., and Talpaz, H. (1975). A microscopic-scale model of soil water uptake and salt movement to plant roots. *Soil Sci.* **120,** 385–399.

Hillel, D., Talpaz, H., and van Keulen, H. (1976). A macroscopic-scale model of water uptake by a nonuniform root system and of water and salt movement in the soil profile. *Soil Sci.* **121,** 242–255.

Hoekstra, P., and Miller, R. D. (1967). On the mobility of water molecules in the transition layer between ice and a solid surface. *J. Colloid Interface Sci.* **25,** 166–173.

Hoekstra, P., Osterkamp, T. E., and Weeks, W. F. (1965). The migration of liquid inclusions in single ice crystals. *J. Geophys. Res.* **70,** 5035–5041.

Holmes, J. W., Greacen, E. L., and Gurr, C. G. (1960). The evaporation of water from bare soils with different tilths. *Trans. Int. Congr. Soil Sci. 7th, Madison, Wisconsin* **1,** 188–194.

Holtan, H. N. (1961). A concept for infiltration estimates in watershed engineering. U.S. Dept. Agr., Agr. Res. Service Publ. 41–51.

Hooghoudt, S. B. (1937). Bijdregen tot de kennis van eenige natuurkundige grootheden van de grond, 6. *Versl. Landb. Ond.* **43,** 461–676.

Horton, R. E. (1940). An approach toward a physical interpretation of infiltration-capacity. *Soil Sci. Soc. Am. Proc.* **5,** 399–417.

Hubbert, M. K. (1940). The theory of groundwater motion. *J. Geol.* **48,** 785–944.

Huck, M. G., Klepper, B., and Taylor, H. M. (1970). Diurnal variation in root diameter. *Plant Physiol.* **45,** 529–530.

IBM. (1972). System/360 continuous system modeling program user's manual program number 360A-CS-16X, 5th ed. IBM Corp., Tech. Pub. Dept. White Plains, New York.

Isherwood, J. D. (1959). Water-table recession in tile-drained land. *J. Geophys. Res.* **64,** 795–804.

Israelson, O. W., and West, F. L. (1922). Water holding capacity of irrigated soils. Utah State Univ. Agr. Exp. Sta. Bull. 183.

Jackson, R. D. (1964a). Water vapor diffusion in relatively dry soil: I. Theoretical considerations and sorption experiments. *Soil Sci. Soc. Am. Proc.* **28,** 172–176.

Jackson, R. D. (1964b). Water vapor diffusion in relatively dry soil: II. Desorption experiment. *Soil Sci. Soc. Am. Proc.* **28,** 464–466.

Jackson, R. D. (1964c). Water vapor diffusion in relatively dry soil: III. Steady state experiments. *Soil Sci. Soc. Am. Proc.* **28,** 466–470.

Jackson, R. D. (1973). Diurnal changes in soil water content during drying. *In* "Field Soil Water Regime," pp. 37–55. Soil Sci. Soc. Am., Madison, Wisconsin.

Jackson, R. D., and Whisler, F. D. (1970). Approximate equations for vertical nonsteady-state drainage: I. Theoretical approach. *Soil Sci. Soc. Am. Proc.* **34,** 715–718.

Jackson, R. D., Kimball, B. A., Reginato, R. J., and Nakayama, S. F. (1973). Diurnal soil water evaporation: Time-depth-flux patterns. *Soil Sci. Soc. Am. Proc.* **37,** 505–509.

Jackson, R. D., Reginato, R. J., Kimball, B. A., and Nakayama, F. S. (1974). Diurnal soil-water evaporation: comparison of measured and calculated soil-water fluxes. *Soil Sci. Soc. Am. Proc.* **38,** 861–866.

Jensen, M. E. (1973). (Editor) "Consumptive Use of Water and Irrigation Water Requirements." Amer. Soc. Civ. Eng. New York. 215 pp.

Jensen, M. E., and Hanks, R. J. (1967). Nonsteadystate drainage from porous media. *J. Irrig. Drain. Div. Am. Soc. Civil Eng.* **93,** IR3, 209–231.

Johnston, J. R., and Hill, H. O. (1945). A study of the shrinkage and swelling of rendzina soils. *Soil Sci. Soc. Am. Proc.* **9,** 24–29.

Journel, A. G., and Huijbregts, Ch. J. (1978). "Mining Geostatistics." Academic Press, New York.

Jury, W. A., and Bellantuoni, B. (1976). Heat and water movement under surface rocks in a field soil. *Soil Sci. Soc. Am. J.* **40,** 505–513.

Jury, W. A., and Tanner, C. B. (1975). Advective modifications of the Priestly and Taylor evapotranspiration formula. *Agron. J.* **67,** 840–842.

Kanemasu, E. T., Stone, L. R., and Powers, W. L. (1976). Evapotranspiration model tested for soybean and sorghum. *Agron. J.* **68,** 569–572.

Kanemasu, E. T., Rasmussen, V. P., and Bagley, J. (1978). Estimating water requirements for corn with a "pocket calculator." *Agron. J.* **70,** 610.

Kemper, W. D., and Quirk, J. P. (1970). Graphic presentation of a mathematical solution for interacting diffuse layers. *Soil Sci. Soc. Am. Proc.* **34,** 347–351.

Kepner, R. A., Bainer, R., and Barger, E. L. (1972). "Principles of Farm Machinery," 2nd ed. AVI., Westport, Connecticut.

Kersten, Miles S. (1949). Thermal properties of soils. Univ. Minn. Bull. Eng. Exp. Stn. Ser. **28.**

Kiesselbach, T. A. (1916). Transpiration as a factor in crop production. Neb. Bull. Agric. Exp. Stn. Res. **6.**

King, K. M., Tanner, C. B., and Suomi, V. E. (1956). A floating lysimeter and its evaporation recorder. *Trans. Am. Geophys. Un.* **37,** 738–742.

Kirkham, D. (1949). Flow of ponded water into drain tubes in soil overlying an impervious layer. *Trans. Am. Geophys. Un.* **30,** 369–385.

Kirkham, D. (1957). The ponded water case. *In* "Drainage of Agricultural Lands," pp. 139–180. Monograph 7, Amer. Soc. Agron., Madison, Wisconsin.

Kirkham, D. (1958). Seepage of steady rainfall through soil into drains. *Trans. Am. Geophys. Un.* **32,** 892–908.

Kirkham, D. (1967). Explanation of paradoxes in Dupuit-Forchheimer seepage theory. *Water Resources Res.* **3,** 609–622.

Kirkham, D., and Gaskell, R. E. (1951). The falling water table in tile and ditch drainage. *Soil Sci. Soc. Am. Proc.* **15,** 37–42.

Klute, A., Whisler, F. D., and Scott, E. H. (1965). Numerical solution of the flow equation for water in a horizontal finite soil column. *Soil Sci. Soc. Am. Proc.* **29,** 353–358.

Koopmans, R. W. R., and Miller, R. D. (1966). Soil freezing and soil water characteristic curves. *Soil. Sci. Soc. Am. Proc.* **30,** 680–685.

Kostiakov, A. N. (1932). On the dynamics of the coefficient of water-percolation in soils and on the necessity of studying it from a dynamic point of view for purposes of amelioration. *Trans. Com. Int. Soc. Soil Sci., 6th, Moscow* Part A, 17–21.

Kramer, P. J., and Coile, J. S. (1940). An estimation of the volume of water made available by root extension. *Plant Physiol.* **15,** 743–747.

Lambert, J. R., and Penning de Vries, F. W. T. (1973). Dynamics of water in the soil plant atmosphere system: A model named Troika. *In* "Physical Aspects of Soil, Water and Salts in Ecosystems." Springer-Verlag, Berlin and New York.

Larson, W. E., and Gill, W. R. (1973). Soil physical parameters for designing new tillage systems. Proc. Natl. Conf. Conserv. Tillage (sponsored by Soil Conserv. Soc. Am., Am. Soc. Agron., and Am. Soc. Agr. Eng.). Des Moines, Iowa, pp. 13–21.

Law, J. P. (1964). Effect of fatty alcohol and a nonionic surfactant on soil moisture evaporation in a controlled environment. *Soil Sci. Sci. Am. Proc.* **28,** 695–699.

LeFur, B. (1962). Influences de la capillarité et de la gravité sur le déplacement nonmiscible unidimensionel dans un milieu poreux. *J. Mécan.* **1,** 59.

Lemon, E. R. (1956). The potentialities for decreasing soil moisture evaporation loss. *Soil Sci. Soc. Am. Proc.* **20,** 120–125.

Lemon, E. R. (1960). Photosynthesis under field conditions. II. An aerodynamic method for determining the turbulent carbon dioxide exchange between the atmosphere and a corn field. *Agron. J.* **52,** 697–703.

Leyser, J. P., and Loch, J. P. G. (1972). Effect of xylem resistance on the water relations of plant and soil. Dept. Theoretical Production Ecology. Agr. Univ., Wageningen, Netherlands.

Lin, C. C. (1955). "The Theory of Hydrodynamic Stability." Cambridge Univ. Press, London and New York.

Linell, K. A., and Kaplar, C. W. (1959). The factor of soil and material type in frost action. *Highw. Res. Board, Bull.* **225,** 81–128.

Lowdermilk, W. C. (1954). The use of flood water by the Nabataeans and the Byzantines. *Isr. Explora. J.* **4,** 50–51.

Luth, H. A., and Wismer, R. D. (1971). Performance of plane soil cutting blade in sand. *Trans. ASAE* **14,** 225–259, 262.

Luthin, J. N., (ed.) (1957). "Drainage of Agricultural Lands." Monograph 7; Am. Soc. Agron., Madison, Wisconsin.

Luthin, J. N. (1966). "Drainage Engineering." Wiley, New York.

Luthin, J. N. (1974). Drainage analogues. *In* "Drainage for Agriculture." Monograph 17, Am. Soc. Agron., Madison, Wisconsin.

Lyles, L., and Woodruff, U. P. (1961). Surface soil cloddiness in relation to soil density at the time of tillage. *Soil Sci.* **91,** 178–182.

Maasland, M. (1957). Soil anisotropy and land drainage. *In* "Drainage of Agricultural Lands" (J. N. Luthin, ed.). Am. Soc. Agron., Madison, Wisconsin.

McIlroy, I. C., and Angus, D. E. (1963). The Aspendale multiple weighted lysimeter installation. CSIRO Div. Meteorol. Phys. Tech. Paper No. 14. Melbourne, Australia.

McIntyre, D. S. (1958). Soil splash and the formation of surface crusts by raindrop impact. *Soil Sci.* **85,** 261–266.

Mackay, J. R. (1978). Sub-pingo water lenses, Tuktoyaktuk Peninsula, Northwest Territories. *Can. J. Earth Sci.* **15,** 1219–1227.

Manning, R. (1891). On the flow of water in open channels and pipes. *Trans. Inst. Civil Eng. Ireland* **20,** 161–207.

Marshall, T. J., and Gurr, C. G. (1966). Movement of water and chlorides in relatively dry soil. *Soil Sci.* **77,** 147–152.

Matano, C. (1933). On the relation between the diffusion coefficients and concentrations of solid materials (the nickel-copper system). *Jpn. J. Phys.* **8,** 108–113.

Matheron, G. (1963). Principles of geostatistics. *Econ. Geol.* **58,** 1246–1266.

Matheron, G. (1971). The theory of regionalized variables and its applications. Ecole des Mines, Fountainebleu, France.

Mein, R. G., and Larson, C. L. (1973). Modeling infiltration during a steady rain. *Water Resources Res.* **9,** 384–394.

Miller, E. E. (1975). Physics of swelling and cracking soils. *J. Colloid Interface Sci.* **52**, 434–443.

Miller, R. D. (1970). Ice sandwich: Functional semipermeable membrane. *Science* **169**, 584–585.

Miller, R. D. (1973a). Soil freezing in relation to pore water pressure and temperature. *Int. Conf. Permafrost, 2nd, Natl. Acad. Sci.*, 344–352.

Miller, R. D. (1973b). The porous phase barrier and crystallization. *Sep. Sci.* **8**, 521–535.

Miller, R. D. (1978). Frost heaving in non-colloidal soils. *Int. Conf. on Permafrost 3rd*, **1**, 708–720.

Miller, D. E., and Gardner, W. H. (1962). Water infiltration into stratified soil. *Soil Sci. Soc. Am. Proc.* **26**, 115–118.

Miller, E. E., and Klute, A. (1967). Dynamics of soil water. Part I—Mechanical forces. *In* "Irrigation of Agricultural Lands," pp. 209–244. Monograph 11, Am. Soc. Agron., Madison, Wisconsin.

Miller, E. E., and Miller, R. D. (1956). Physical theory for capillary flow phenomena. *J. Appl. Phys.* **27**, 324–332.

Miller, R. D., Baker, J. H., and Kolaian, J. H. (1960). Particle size, overburden pressure, pore water pressure and freezing temperature of ice lenses in soil. *Trans. Int. Congress Soil Sci.*, *7th*, **1**, 122–129.

Miller, R. D., Loch, J. P. G., and Bresler, E. (1975). Transport of water and heat in a frozen permeameter. *Soil Sci. Soc. Am. Proc.* **39**, 1029–1036.

Molz, F. J. (1975). Potential distribution in the soil-root system. *Agron. J.* **67**, 726–729.

Molz, F. J. (1976). Water transport in the soil-root system: Transient analysis. *Water Resour. Res.* **12**, 805–807.

Molz, F. J., and Ikenberry, E. (1974). Water transport through plant cells and cell walls: Theoretical development. *Soil Sci. Soc. Am. Proc.* **38**, 699–704.

Molz, F. J., and Remson, I. (1970). Extraction term models of soil moisture use by transpiring plants. *Water Resources Res.* **6**, 1346–1356.

Molz, F. J., and Remson, I. (1971). Application of an extraction term model to the study of moisture flow to plant roots. *Agron. J.* **63**, 72–77.

Monteith, J. L. (1973). "Principles of Environmental Physics." American Elsevier, New York.

Monteith, J. L. (1965). Evaporation and environment. *Symp. Soc. Exp. Biol.*, *19th*, 205–234.

Monteith, J. L. (1978). Models and measurement in crop climatology. *Proc. Int. Soil Sci. Congress*, *11th* Edmonton,

Monteith, J. L., Szeicz, G., and Waggoner, P. E. (1965). The measurement and control of stomatal resistance in the field. *J. Appl. Ecol.* **2**, 345–357.

Moore, R. E. (1939). Water conduction from shallow water tables. *Hilgardia* **12**, 383–426.

Morel-Seytoux, H. J., and Noblanc, A. (1972). Infiltration predictions by a moving strained coordinates method. *In* "Physical Aspects of Soil, Water and Salts in Ecosystems" (A. Hadas *et al.*, eds.). Springer-Verlag, Berlin and New York.

Muskat, M. (1946). "The Flow of Homogeneous Fluids Through Porous Media." Edwards, Ann Arbor, Michigan.

Myers, L. E. (1961). Water proofing soil to collect precipitation. *J. Soil Water Conserv.* **16**, 281–282.

Myers, L. E. (1963). Water harvesting. Special publication of the U.S. Water Conservation Laboratory, Tempe, Arizona.

Newman, E. I. (1969). Resistance to water flow in soil and plant: 1. Soil resistance in relation to amount of roots: Theoretical estimates. *J. Appl. Ecol.* **6**, 1–12.

Newman, E. I. (1974). Root and soil water relations. *In* "The Plant Root and Its Environment" (E. W. Carson, ed.), pp. 363–440. Univ. Press of Virginia, Charlottesville, Virginia.

Nichols, M. L. (1929). Methods of research in soil dynamics as applied to implement design. *Ala. Agric. Exp. Stn. Bull.* 229, *Agric. Eng.* (1932). **13**, 279–285.

Nichols, M. L., and Reed, I. F. (1934). Soil dynamics: VI. Physical reactions of soils to moldboard surfaces. *Agric. Eng.* **15,** 187–190.

Nielsen, D. R., Biggar, J. W., and Erh, K. T. (1973). Spatial variability of field-measured soil-water properties. *Hilgardia* **42,** 215–259.

Nimah, M. N., and Hanks, R. J. (1973). Model for estimating soil, water, plant, and atmosphere interactions: 1. Description and sensitivity. *Soil Sci. Soc. Am. Proc.* **37,** 522–527.

Nixon, P. R., and Lawless, G. P. (1960). Translocation of moisture with time in unsaturated soil profiles. *J. Geophys. Res.* **65,** 655–661.

Nobel, P. S. (1970). "Plant Cell Physiology: A Physiochemical Approach." Freeman, San Francisco, California.

Nobel, P. S. (1974). "Introduction to Biophysical Plant Physiology." Freeman, San Francisco, California.

Ogata, G., and Richards, L. A. (1957). Water content changes following irrigation of bare field soil that is protected from evaporation. *Soil Sci. Soc. Am. Proc.* **21,** 355–356.

Ogata, G., Richards, L. A., and Gardner, W. R. (1960). Transpiration of alfalfa determined from soil water content changes. *Soil Sci.* **89,** 179–182.

Oster, J. D., and Schroer, F. W. (1979). Infiltration as influenced by irrigation water quality. *Soil Sci. Soc. Am. J.* (in press).

Oster, J. D., and Willardson, L. S. (1971). Reliability of salinity sensors for the management of soil salinity. *Agron. J.* **63,** 695–698.

Parlange, J.-Y. (1971). Theory of water movement in soils. I. One dimensional absorption. *Soil Sci.* **111,** 134–137.

Parlange, J.-Y., and Hill, D. E. (1976). Theoretical analysis of wetting front instability in soils. *Soil Sci.* **122,** 236–239.

Payne, P. C. J. (1956). The relationship between the mechanical properties of soil and the performance of simple cultivation implements. *J. Agr. Eng. Res.* **1,** 23–28.

Peaceman, D. W., and Rachford, H. H. Jr. (1955). The numerical solution of parabolic and elliptic differential equations, *J. Soc. Ind. Appl. Math.* **3,** 28–41.

Pearse, J. F., Oliver, T. R., and Newitt, D. M. (1949). The mechanism of the drying of solids: Part I. The forces giving rise to movement of water in granular beds during drying. *Trans. Inst. Chem. Eng. (London)* **27,** 1–8.

Peck, A. J. (1965). Moisture profile development and air compression during water uptake by bounded porous bodies: 3. Vertical columns. *Soil Sci.* **100,** 44–51.

Peck, A. J. (1970). Redistribution of soil water after infiltration. *Aust. J. Soil Res.* **7.**

Pelton, W. L. (1961). The use of lysimetric methods to measure evapotranspiration *Proc. Hydrol. Symp.* **2,** 106–134 (Queen's Printer, Ottawa, Canada. Cat. No. R32-361/2).

Penman, H. L. (1948). Natural evaporation from open water, bare soil and grass. *Proc. R. Soc. London Ser. A* **193,** 120–146.

Penman, H. L. (1949). The dependence of transpiration on weather and soil conditions. *J. Soil Sci.* **1,** 74–89.

Penman, H. L. (1953). The physical bases of irrigation control. *Rep. Int. Hort. Congress, 13th,* **2,** 913–924.

Penman, H. L. (1955). The evaporation calculations. *In* "Lake Eyre, The Great Flooding of 1949–1950." Royal Geographical Society of Australia.

Penman, H. L. (1956). Evaporation: an introductory survey. *Neth J. Agr. Sci.* **4,** 9–29.

Penner, E. (1959). The mechanism of frost heaving in soils. *Highw. Res. Board. Bull.* **225,** 1–13.

Penner, E., and Ueda, T. (1978). A soil frost-susceptibility test and a basis for interpreting heaving rates. *Int. Conf. Permafrost, 3rd,* **1,** 721–727.

Pereira, H. C., and Jones, P. A. (1954). A tillage study in Kenya coffee: Part II. The effect of tillage practices on the structure of the soil. *Emp. J. Expt. Agric.* **22,** 323–327.

Peters, D. B. (1965). Water availability. *In* "Methods of Soil Analysis" (C. A. Black, ed.), pp. 279–285. Monograph 9, Am. Soc. Agron., Madison, Wisconsin.

Philip, J. R. (1955a). Numerical solution of equations of the diffusion type with diffusivity concentration dependent. *Trans. Faraday Soc.* **51**, 885–892.

Philip, J. R. (1955b). The concept of diffusion applied to soil water. *Proc. Nat. Acad. Sci. India* **24A**, 93–104.

Philip, J. R. (1957). Numerical solution of equations of the diffusion type with diffusivity concentration-dependent II. *Aust. J. Phys.* **10**, 29–42.

Philip, J. R. (1957a). The theory of infiltration: 2. The profile at infinity. *Soil Sci.* **83**, 435–448.

Philip, J. R. (1957b). The theory of infiltration: 3. Moisture profiles and relation to experiment. *Soil Sci.* **84**, 163–178.

Philip, J. R. (1957c). The theory of infiltration: 4. Sorptivity and algebraic infiltration equations. *Soil Sci.* **84**, 257–264.

Philip, J. R. (1957d). Evaporation, moisture and heat fields in the soil. *J. Meteorol.* **14**, 354–366.

Philip, J. R. (1966a). Absorption and infiltration in two- and three-dimensional systems. *In* "Water in the Unsaturated Zone" (R. E. Rijtema and H. Wassink, eds.). Vol. 2, pp. 503–525. IASH/UNESCO Symp., Wageningen.

Philip, J. R. (1966b). Plant water relations: some physical aspects. *Ann Rev. Plant Physiol.* **17**, 245–268.

Philip, J. R. (1969a). Theory of infiltration. *Adv. Hydrosci.* **5**, 215–290.

Philip, J. R. (1969b). Hydrostatics and hydrodynamics in swelling soils. *Water Resources Res.* **5**, 1070–1077.

Philip, J. R. (1972). Hydrology of swelling soils. *In* "Salinity and Water Use" (a national symposium on hydrology sponsored by the Australian Academy of Science). Macmillan, New York.

Philip, J. R. (1974). Water movement in soil. *In* "Heat and Mass Transfer in the Biosphere" (D. A. de Vries and N. H. Afgan, eds.), pp. 29–47. Halsted Press-Wiley, New York.

Philip, J. R. (1975). Stability analysis of infiltration. *Soil Sci. Soc. Am. Proc.* **39**, 1042–1049.

Philip, J. R., and de Vries, D. A. (1957). Moisture movement in porous materials under temperature gradients. *Trans. Am. Geophys. Un.* **38**, 222–228.

Phillips, S. H., and Young, H. M. (1973). "No-tillage farming." Reiman, Milwaukee, Wisconsin.

Poulovassilis, A. (1962). Hysteresis of pore water, an application of the concept of independent domains. *Soil Sci.* **93**, 405–412.

Pruitt, W. O., and Angus, D. E. (1960). Large weighing lysimeter for measuring evapotranspiration. *Trans. Am. Soc. Agr. Eng.* **3**, 3–15, 18.

Raats, P. A. C. (1973). Unstable wetting fronts in uniform and nonuniform soils. *Soil Sci. Soc. Am. Proc.* **37**, 681–685.

Rasmussen, V. P., and Hanks, R. J. (1978). Model for predicting spring wheat yields with limited climatological and soil data. *Agron. J.* **70**, 940–944.

Ravina, I., and Low, P. F. (1972). Relation between swelling, water properties and b-dimension in montmorillonite-water systems. *Clay Miner.* **20**, 109–123.

Rawitz, E. (1969). The dependence of growth rate and transpiration on plant and soil physical parameters under controlled conditions. *Soil Sci.* **110**, 172–182.

Rawitz, E., and Hillel, D. (1974). Progress and problems of drip irrigation in Israel. *Proc. Int. Conf. Drip Irrig., San Diego, California.*

Rawitz, E., Margolin, M., and Hillel, D. (1972). An improved variable-intensity sprinkling infiltrometer. *Soil Sci. Soc. Am. Proc.* **36**, 533–535.

Rawlins, S. L., and Raats, P. A. C. (1975). Prospects for high frequency irrigation. *Science* **188**, 604–610.

Reichardt, K., Nielsen, D. R., and Biggar, J. W. (1972). Scaling of horizontal infiltration into homogeneous soils. *Soil Sci. Soc. Am. Proc.* **36**, 241–245.

Reicosky, D. C., Cassel, D. K., Blevius, R. L., Gill, W. R., and Naderman, G. C. (1977). Conservation tillage in the Southeast. *J. Soil Water Conserv.* **32**, 13–19.

Reicosky, D. C., and Ritchie, J. T. (1976). Relative importance of soil resistance and plant resistance in root water absorption. *Soil Sci. Soc. Am. J.* **40**, 293–297.

Remson, I., Drake, R. L., McNeary, S. S., and Walls, E. M. (1965). Vertical drainage of an unsaturated soil. *Am. Soc. Civil Eng. Proc. J. Hyd. Div.* **9**, 55–74.

Remson, I., Fungaroli, A. A., and Hornberger, G. M. (1967). Numerical analysis of soil moisture systems. *Am. Soc. Civil Eng. Proc. J. Irrig. Drain. Div.* **3**, 153–166.

Remson, I., Hornberger, G. M., and Molz, F. (1971). "Numerical Methods in Subsurface Hydrology." Wiley (Interscience), New York.

Retta, A., and Hanks, R. J. (1979). Corn and alfalfa production as influenced by limited irrigation. *Irrig. Sci.* (in press).

Richards, L. A. (1952). Report of the subcommittee on permeability and infiltration, Committee on Terminology, Soil Science Society of America. *Soil Sci. Soc. Am. Proc.* **16**, 85–88.

Richards, L. A. (ed.) (1954). "Diagnosis and Improvement of Saline and Alkali Soils." U.S. Dept. Agr. Handbook 60.

Richards, L. A. (1960). Advances in soil physics. *Trans. Int. Congr. Soil Sci., 7th, Madison* **I**, 67–69.

Richards, L. A., and Moore, D. C. (1952). Influence of capillary conductivity and depth of wetting on moisture retention in soil. *Trans. Am. Geophys. Un.* **33**, 4.

Richards, L. A., and Wadleigh, C. H. (1952). Soil water and plant growth. *In* "Soil Physical Conditions and Plant Growth," p. 13. Am. Soc. Agron. Monograph 2.

Richards, L. A., and Weaver, L. R. (1944). Fifteen atmosphere percentage as related to the permanent wilting percentage. *Soil Sci.* **56**, 331–339.

Richards, L. A., Gardner, W. R., and Ogata, G. (1956). Physical processes determining water loss from soil. *Soil Sci. Soc. Am. Proc.* **20**, 310–314.

Richards, S. J. (1965). Soil suction measurements with tensiometers. *In* "Methods of Soil Analysis," pp. 153–163. *Am. Soc. Agron.,* Monograph 9.

Ripple, C. D., Rubin, J., and van Hylkama, T. E. A. (1972). Estimating steady-state evaporation rates from bare soils under conditions of high water table. U.S. Geol. Survey, Water Supp. Pap. 2019-A.

Ritchie, J. T., and Adams, J. E. (1974). Field measurement of evaporation from soil shrinkage cracks. *Soil Sci. Soc. Am. Proc.* **38**, 131–134.

Robins, J. S., Pruitt, W. O., and Gardner, W. H. (1954). Unsaturated flow of water in field soils and its effect on soil moisture investigations. *Soil Sci Soc. Am. Proc.* **18**, 344–348.

Römkens, M. J. M., and Miller, R. D. (1973). Migration of mineral particles in ice with a temperature gradient. *J. Colloid Interface Sci.* **42**, 103–111.

Rose, C. W. (1966). "Agricultural Physics." Pergamon, Oxford.

Rose, C. W. (1968). Evaporation from bare soil under high radiation conditions. *Trans. Int. Congr. Soil Sci., 9th, Adelaide* **I**, 57–66.

Rose, C. W., and Stern, W. R. (1967a). Determination of withdrawal of water from soil by crop roots as a function of depth and time. *Aust. J. Soil Res.* **5**, 11–19.

Rose, C. W., and Stern, W. R. (1967b). The drainage component of the water balance equation. *Aust. J. Soil Res.* **3**, 95–100.

Rose, C. W., Byrne, G. F. and Begg, J. E. (1966). An accurate hydraulic lysimeter with remote weight recording. CSIRO Div. Land Res. and Reg. Survey, Tech. Paper 27. Canberra, Australia.

Rose, C. W., Byrne, G. F., and Hansen, G. K. (1976). Water transport from soil through plant to atmosphere: A lumped-parameter model. *Agric. Meteorol.* **16**, 171–184.

Rosenberg, N. J. (1974). "Microclimate: The Biological Environment." Wiley, New York.

Rubin, J. (1966). Theory of rainfall uptake by soils initially drier than their field capacity and its applications. *Water Resour. Res.* **2**, 739–749.

Rubin, J. (1967). Numerical method for analyzing hysteresis-affected, post-infiltration redistribution of soil moisture. *Soil Sci. Soc. Am. Proc.* **31**, 13–20.

Rubin, J., and Steinhardt, R. (1963). Soil water relations during rain infiltration: I. Theory. *Soil Sci. Soc. Am. Proc.* **27**, 246–251.

Rubin, J., and Steinhardt, R. (1964). Soil water relations during rain infiltration: III. Water uptake at incipient ponding. *Soil Sci. Soc. Am. Proc.* **28**, 614–619.

Rubin, J., Steinhardt, R., and Reiniger, P. (1964). Soil water relations during rain infiltration: II. Moisture content profiles during rains of low intensities. *Soil Sci. Soc. Am. Proc.* **28**, 1–5.

Russell, G. (1980). Crop evaporation, surface resistance, and soil water status. *Agric. Meteorol.* **21**, 213–226.

Rutter, A. J. (1967). An analysis of evaporation from a stand of Scots pine. *In* Forest Hydrology (W. E. Sopper and H. W. Lull, eds), pp. 403–417. Pergamon Press, Oxford.

Saffman, P. G., and Taylor, G. I. (1958). The penetration of a fluid into a porous medium or Hele-Shaw cell containing a more viscous liquid. *Proc. R. Soc. London Ser. A* **245**, 312–331.

Sahin, T. (1973). Transport of water in frozen soil. M.S. Thesis, Cornell Univ., Ithaca, New York.

Schofield, R. K. (1935). The pF of water in soil. *Trans. Int. Congress Soil Sci. 3rd*, **2**, 37–48.

Schofield, R. K. (1946). Ionic forces in thick films between charged surfaces. *Trans. Faraday Soc.* **42B**, 219–225.

Seaton, K. A., Landsberg, J. J., and Sedgley, R. H. (1977). Transpiration and leaf water potentials of wheat in relation to changing soil water potential. *Aust. J. Agr. Res.* **28**, 355–367.

Selim, H. M., and Kirkham, D. (1970). Soil temperature and water content changes during drying as influenced by cracks: A laboratory experiment. *Soil Sci. Soc. Am. Proc.* **34**, 565–569.

Sellers, W. D. (1965). "Physical Climatology." Univ. of Chicago Press, Chicago, Illinois.

Sellin, R. H. J. (1969). "Flow in Channels." MacMillan, New York.

Shalhevet, J., Mantell, A., Bielorai, A., and Shimshi, D. (1976). Irrigation of field and orchard crops under semi-arid conditions. Int. Irrig. Inf. Ctr. Publ. No. 1.

Shanan, L., Tadmor, N. H., Evenari, M., and Reiniger, P. (1970). Runoff farming in the desert. III. Microcatchments for improvement of desert range. *Agron. J.* **62**, 445–449.

Shantz, H. L., and Piemeisel, L. N. (1927). The water requirement of plants at Akron, Colo. *J. Agric. Res.* **34**, 1093–1190.

Sheppard, M. I., Kay, B. D. and Loch, J. P. G. (1978). Development and testing of a computer model for heat and mass flow in freezing soils. *Soil Sci. Soc. Am. J.* **42**, 38.

Shuttleworth, W. J. (1976). A one-dimensional theoretical description of the vegetation-atmosphere interaction. *Boundary Layer Meteorol.* **10**, 273–302.

Simmons, C. S., Nielsen, D. R., and Biggar, J. W. (1979). Scaling of field-measured soil water properties. *Hilgardia* (in press).

Skaggs, R. W., Huggins, L. F., Monke, E. J., and Foster, G. R. (1969). Experimental evaluation of infiltration equations. *Trans. Am. Soc. Agr. Eng.* **12**, 822–828.

Skapski, A., Billups, R. and Rooney, A. (1957). The capillary cone method for determination of surface tension of solids. *J. Chem. Phys.* **26**, 1350.

Slater, P. J., and Williams, J. B. (1965). The influence of texture on the moisture characteristics of soils. I. A critical comparison of techniques for determining the available water capacity and moisture characteristic curve of a soil. *J. Soil Sci.* **16**, 1–12.

Slatyer, R. O. (1967). "Plant Water Relationships." Academic Press, New York.

Slatyer, R. O., and McIlroy, I. C. (1961). "Practical Microclimatology." CSIRO, Australia.

Smiles, D. E. (1974). Infiltration into swelling material. *Soil Sci.* **117**, 110–116.

Smiles, D. E. (1976). On the validity of the theory of flow in saturated swelling materials. *Aust. J. Soil Res.* **14**, 389–395.

Smiles, D. E., and Rosenthal, M. J. (1968). The movement of water in swelling materials. *Aust. J. Soil Res.* **6**, 237–248.

Smiles, D. E., Philip, J. R., Knight, J. H., and Elrick, D. E. (1978). Hydrodynamic dispersion during sorption of water by soil. *Soil Sci. Soc. Am. J.* **42**, 229–234.

Smith, G. D. (1965). "Numerical Solution of Partial Differential Equations." Oxford Univ. Press, London and New York.

Smith, R. E., and Woolhiser, D. A. (1971). Overland flow on an infiltrating surface. *Water Resources Res.* **5**, 144–152.

Smythe, W. R. (1950). "Static and Dynamic Electricity." McGraw-Hill, New York.

So, H. B., Aylmore, L. A. C., and Quirk, J. P. (1976). The resistance of intact maize roots to water flow. *Soil Sci. Soc. Am. J.* **40**, 222–225.

Söhne, W. H. (1956). Some basic considerations of soil mechanics applied to agricultural engineering. *Grundl. d. Landtech.* **7**, 11–27.

Söhne, W. H. (1966). Characterization of tillage tools. *Särtryck ur Grundförbättring* **1**, 31–48.

Sorensen, V. M., Hanks, R. J., and Cartee, R. L. (1979). Cultivation during early season and irrigation influences on corn production. *Agron. J.* (in press).

Southwell, R. V. (1946). "Relaxation Methods in Engineering Science." Oxford Univ. Press, London and New York.

Speckhart, F. H. and Green, W. L. (1976). "A Guide to Using CSMP." Prentice-Hall, Englewood Cliffs, New Jersey.

Stanhill, G. (1965). Observation on the reduction of soil temperature. *Agr. Meteorol.* **2**, 197–203.

Staple, W. J. (1969). Comparison of computed and measured moisture redistribution following infiltration. *Soil Sci. Soc. Am. Proc.* **33**, 206.

Stewart, R. B. and Rouse, W. R. (1977). Substantiation of the Priestley-Taylor parameter for potential evaporation in high latitudes. *J. Appl. Meteorol.* **16**, 649–650.

Stewart, J. I., Danielson, R. E., Hanks, R. J., Jackson, E. B., Hagan, R. M., Pruitt, W. O., Franklin, W. T., and Riley, J. P. (1977). Optimizing crop production through control of water and salinity levels in the soil. Utah Water Lab. PRWG 151-1. p. 191. Logan, Utah.

Sutcliffe, J. (1968). "Plants and Water." Edward Arnold, London.

Swartzendruber, D. (1969). The flow of water in unsaturated soils. *In* "Flow Through Porous Media" (R. J. M. DeWiest, ed.), Chapter 6, pp. 215–292. Academic Press, New York.

Swartzendruber, D., and Hillel, D. (1973). The physics of infiltration. *In* "Physics of Soil, Water and Salts in Ecosystems" (A. Hadas *et al.*, eds.). Springer-Verlag, Berlin and New York.

Swartzendruber, D., and Hillel, D. (1975). Infiltration and runoff for small field plots under constant intensity rainfall. *Water Resources Res.* **11**, 445–451.

Szeicz, G., van Bavel, C. H. M., and Takami, S. (1973). Stomatal factor in water use and dry matter production by sorghum. *Agr. Meteorol.* **12**, 361–389.

Taber, S. (1930). The mechanics of frost heaving. *J. Geol.* **38**, 303–317.

Tackett, J. L., and Pearson, R. W. (1965). Some characteristics of soil crusts formed by simulated rainfall. *Soil Sci.* **99**, 407–413.

Tadmor, N. H., Evenari, M., Shanan, L., and Hillel, D. (1957). The ancient desert agriculture of the Negev. *Isr. J. Agric. Res.* **8**, 127–151.

Takagi, S. (1960). Analysis of the vertical downward flow of water through a two-layered soil. *Soil Sci.* **90**, 98–103.

Talsma, T. (1963). The control of saline ground water. *Med. Landb. Wageningen* **63**(10), 1–68.

Tanner, C. B. (1957). Factors affecting evaporation from plants and soils. *J. Soil Water Conserv.* **12,** 221–227.

Tanner, C. B. (1960). Energy balance approach to evapotranspiration from crops. *Soil Sci. Soc. Am. Proc.* **24,** 1–9.

Tanner, C. B. (1968). Evaporation of water from plants and soil. *In* "Water Deficits and Plant Growth." Academic Press, New York.

Tanner, C. B., and Lemon, E. R. (1962). Radiant energy utilized in evaporation. *Agron. J.* **54,** 207–212.

Tanner, C. B., and Pelton, W. L. (1960). Potential evapotranspiration estimates by the approximate energy balance method of Penman. *J. Geophys. Res.* **65,** 3391–3413.

Taylor, G. I. (1950). The instability of liquid surfaces when accelerated in a direction perpendicular to their planes. *Proc. R. Soc. London Ser. A* **201,** 192–196.

Taylor, H. M., and Klepper, B. (1976). Water uptake by cotton root systems: An examination of assumptions in the single root model. *Soil Sci.* **120,** 57–67.

Taylor, S. A., and Cary, J. W. (1960). Analysis of the simultaneous flow of water and heat or electricity with the thermodynamics of irreversible processes. *Trans. Int. Congr. Soil Sci., 7th, Madison, Wisconsin* **I,** 80–90.

Thom, A. S. (1975). Momentum, mass and heat exchange of plant communities. *In* "Vegetation and the Atmosphere" (ed. J. L. Monteith). Academic Press, New York.

Thom, A. S., and Oliver, H. R. (1977). On Penman's equation. *Q. J. R. Meteorol. Soc.* **103,** 345–357.

Thorthwaite, C. W. (1948). An approach toward a rational classification of climate. *Geograph. Rev.* **38,** 55–94.

Todd, D. K. (1967). "Ground Water Hydrology," 6th printing. Wiley, New York.

Tovey, R. 1963. Consumptive use and yield of alfalfa in the presence of static water tables. Tech. Bull. 232. *Nev. Agric. Exp. Stn.*

Trouse, A. C. (1978). Tillage and traffic effects on soil. Paper presented at 33rd Annu. Meet. Soil Conserv. Soc. Am., Denver, Colorado.

Tystovich, X. (1975). "The Mechanics of Frozen Ground" (Swinzow, G. K., ed.). McGraw-Hill, New York.

Ulyanov, N. A. (1969). "Theory of Self-Propelled Wheeled Earth Working Transport Machines" (Teoriya Samokhodnykh Zemleroi otransportnykh Mashiny). Machine Construction, Moscow.

Vachaud, G., and Thony, J. L. (1971). Hysteresis during infiltration and redistribution in a soil column at different initial water contents. *Water Resources Res.* **7,** 111–127.

van Bavel, C. H. M. (1966). Potential evaporation: The combination concept and its experimental verification. *Water Resources Res.* **2,** 455–467.

van Bavel, C. H. M., and Ahmed, J. (1976). Dynamic simulation of water depletion in the root zone. *Ecol. Modelling* **2,** 189–212.

van Bavel, C. H. M., and Hillel, D. (1975). A simulation study of soil heat and moisture dynamics as affected by a dry mulch. *Proc. Summer Comput. Simulat. Conf., San Francisco, California.* Simulation Councils, LaJolla, California.

van Bavel, C. H. M., and Hillel, D. (1976). Calculating potential and actual evaporation from a bare soil surface by simulation of concurrent flow of water and heat. *Agr. Meteorol.* **17,** 453–476.

van Bavel, C. H. M., and Myers, L. E. (1962). An automatic weighing lysimeter. *Agr. Eng.* **43,** 580–583.

van Bavel, C. H. M., Stirk, G. B., and Brust, K. J. (1968a). Hydraulic properties of a clay loam soil and the field measurement of water uptake by roots: I. Interpretation of water content and pressure profiles. *Soil Sci. Soc. Proc.* **32,** 310–317.

van Bavel, C. H. M., Brust, K. J., and Stirk, G. B. (1968b). Hydraulic properties of a clay loam soil and the field measurement of water uptake by roots: II. The water balance of the root zone. *Soil Sci. Soc. Am. Proc.* **23,** 317–321.

van Genuchten, M. Th. (1978). Numerical solutions of the one-dimensional saturated-unsaturated flow equation. Res. Rep. 78-WR-9, Water Resour. Prog., Dept. Civil Eng., Princeton Univ., Princeton, New Jersey.

van Genuchten, M. Th., and Wierenga, P. J. (1974). Simulation of one-dimensional solute transfer in porous media. *Agric. Exp. Stn. Bull.* 628, New Mexico State Univ., Las Cruces, New Mexico.

van Keulen, H., and Hillel, D. (1974). A simulation study of the drying-front phenomenon. *Soil Sci.* **118,** 270–273.

van Schilfgaarde, J. (1957). Approximate solutions to drainage flow problems. *In* "Drainage of Agricultural Lands," pp. 79–112. *Am. Soc. Agron.*, Monograph 7.

van Schilfgaarde, J. (ed.) (1974). Drainage for Agriculture. Monograph 17, Am. Soc. Agron., Madison, Wisconsin.

Varga, R. S. (1962). "Matrix Iterative Analysis." Prentice Hall, Englewood Cliffs, New Jersey.

Veihmeyer, F. J., and Hendrickson, A. J. (1927). Soil moisture conditions in relation to plant growth. *Plant Physiol.* **2,** 71–78.

Veihmeyer, F. J., and Hendrickson, A. H. (1931). The moisture equivalent as a measure of the field capacity of soils. *Soil Sci.* **32,** 181–193.

Veihmeyer, F. J., and Hendrickson, A. H. (1949). Methods of measuring field capacity and wilting percentages of soils. *Soil Sci.* **68,** 75–94.

Veihmeyer, F. J., and Hendrickson, A. H. (1950). Soil moisture in relation to plant growth. *Ann. Rev. Plant Physiol.* **1,** 285–304.

Veihmeyer, F. J., and Hendrickson, A. H. (1955). Does transpiration decrease as the soil moisture decreases? *Trans. Am. Geophys. Un.* **36,** 425–448.

Vennard, J. K. (1961). "Elementary Fluid Mechanics," 4th ed. Wiley, New York.

Verwey, E. J. W., and Overbeek, J. Th. G. (1948). "Theory of the Stability of Lyophobic Colloids." Elsevier, New York.

Viets, F. G., Jr. (1962). Fertilizers and the efficient use of water. *Adv. Agron.* **14,** 228–261.

Viets, F. G., Jr. (1966). Increasing water use efficiency by soil management. *In* "Plant Environment and Efficient Water Use" (W. H. Pierre, D. Kirkham, J. Pesek, and R. Shaw, eds.), pp. 259–274. Am. Soc. Agron. and Soil Sci. Soc. of Am., Madison, Wisconsin.

Visser, W. C. (1959). Crop Growth and Availability of Moisture. Inst. of Land and Water Management, Wageningen, Netherlands, Tech. Bull. No. 6.

Voorhees, W. B., and Hendrick, J. G. (1977). Compaction—good and bad effects on energy needs. *Crops and Soils Magazine* **29,** 11–13.

Vries, D. A. de. (1963). Thermal properties of soils. In "Physics of Plant Environment." North-Holland, Amsterdam.

Wadleigh, C. H. (1946). The integrated soil moisture stress upon a root system in a large container of saline soil. *Soil Sci.* **6,** 225–238.

Wagenet, R. J., and Jurinak, J. J. (1978). Spatial variability of soluble salt content in a Mancos shale watershed. *Soil Sci.* **126,** 342–349.

Wang, F. C., and Lakshminarayana, V. (1968). Mathematical simulation of water movement through unsaturated nonhomogeneous soil. *Soil Sci. Soc. Am. Proc.* **32,** 329–334.

Warrick, A. W., and Amoozegar-Fard, A. (1979). Infiltration and drainage calculations using spatially scaled hydraulic properties. *Water Resour. Res.* **15,** 1116–1120.

Warrick, A. W., Mullen, G. J., and Nielsen, D. R. (1977). Scaling field measured soil hydraulic properties using a similar media concept. *Water Resour. Res.* **13,** 355–362.

Watson, K. K. (1966). An instantaneous profile method for determining the hydraulic conductivity of unsaturated porous materials. *Water Resour. Res.* **2**, 709–715.

Whisler, F. D., and Millington, R. J. (1968). Analysis of steady-state evapotranspiration from a soil column. *Soil Sci. Soc. Am. Proc.* **32**, 167–174.

Whisler, F. D., Klute, A., and Millington, R. J. (1968). Analysis of steady-state evapotranspiration from a soil column. *Soil Sci. Soc. Am. Proc.* **32**, 167–174.

Wiegand, C. L., and Taylor, S. A. (1961). Evaporative Drying of Porous Media. Spec. Rep. 15, Agr. Exp. Sta. Utah State Univ., Logan, Utah.

Wilkinson, G. E., and Klute, A. (1959). Some tests of the similar media concept of capillary flow: II. Flow systems data. *Soil Sci. Soc. Am. Proc.* **22**, 432–437.

Willardson, L. S., and Hurst, R. L. (1965). Sample size estimates in permeability studies. *J. Irr. Am. Soc. Civil Eng.* **91**, 1–9.

Williams, P. J. (1966). Pore pressures at a penetrating frost line and their prediction. *Geotechnique* **16**, 187–208.

Williams, P. J. (1972). Use of the ice-water surface tension concept in engineering practice. *Highw. Res. Rec.* **393**, 19–29.

Williams, P. J., and Burt, T. P. (1974). Measurement of hydraulic conductivity of frozen soils. *Can. Geotech. J.* **11**, 647–650.

Willis, W. O. (1960). Evaporation from layered soils in the presence of a water table. *Soil Sci. Soc. Am. Proc.* **24**, 239–242.

Wind, G. P. (1955). Flow of water through plant roots. *Neth. J. Agric. Sci.* **3**, 259–264.

Wind, G. P. (1959). A field experiment concerning capillary rise of moisture in a heavy clay soil. *Neth. J. Agric. Sci.* **3**, 60–69.

Wolf, J. M. (1968). The Role of Root Growth in Supplying Moisture to Plants. Unpublished Doctoral Dissertation, Univ. of Rochester, Rochester, New York.

Wooding, R. A. (1965). A hydraulic model for the catchment-stream problem. 1. Kinematic wave theory. *J. Hydrol.* **3**, 254–267.

Wooding, R. A. (1969). Growth of fingers at an unstable diffusing interface in a porous medium or Hele-Shaw cell. *J. Fluid Mech.* **39**, 477–495.

Yang, S. J., and DeJong, E. (1971). Effect of soil water potential and bulk density on water uptake patterns and resistance to flow of water in wheat plants. *Can. J. Soil Sci.* **51**, 211–220.

Yatsuk, E. P. (1971). "Rotary Soil Working Machines" (Rotatsionnye Pochvoobrabatyvayushchie Mashiny). Machine Construction, Moscow.

Youngs, E. G. (1958a). Redistribution of moisture in porous materials after infiltration. *Soil Sci.* **86**, 117–125.

Youngs, E. G. (1958b). Redistribution of moisture in porous materials after infiltration. *Soil Sci.* **86**, 202–207.

Youngs, E. G. (1960a). The drainage of liquids from porous materials. *J. Geophys. Res.* **65**, 4025–4030.

Youngs, E. G. (1960b). The hysteresis effect in soil moisture studies. *Trans. Int. Soil Sci. Congr. 7th, Madison* **1**, 107–113.

Youngs, E. G. (1964). An infiltration method of measuring the hydraulic conductivity of porous materials. *Soil Sci.* **97**, 307–311.

Youngs, E. G., and Towner, G. E. (1970). Comment on paper by Philip (1969). *Water Resources Res.* **6**, 1246.

Zelenin, A. N., Balovnev, V. I., and Kerov, I. P. (1975). "Machines for Earthmoving Work" (Mashiny dlya Zemlyanteykh Rabot). Machine Construction, Moscow. 423 p.

Van De Pol, R. M., Wierenga, P. J., and Nielsen, D. R. (1977). Solute movement in a field soil. *Soil Sci. Soc. Am. J.* **41**, 10–13.

Index